T0203348

Channel Coding in Communication Networks

Channel Coding in Communication Networks

From Theory to Turbocodes

Edited by
Alain Glavieux

First published in France 2005 by Hermès Science/Lavoisier entitled "Codage de canal: des bases théoriques aux turbocodes"
First published in Great Britain and the United States in 2007 by ISTE Ltd

ISTE Ltd
6 Fitzroy Square
London W1T 5DX
UK

ISTE USA
4308 Patrice Road
Newport Beach, CA 92663
USA

www.iste.co.uk

Library of Congress Cataloging-in-Publication Data

Codage de canal, des bases théoriques aux turbocodes. English
Channel coding in communication networks: from theory to turbocodes/edited by Alain Glavieux. -- 1st ed.
p. cm.
Includes bibliographical references and index.
ISBN-13: 978-1-905209-24-8
ISBN-10: 1-905209-24-X
1. Coding theory. 2. Error-correcting codes (Information theory) I. Glavieux, Alain. II. Title.
TK5102.92.C63 2006
003'.54--dc22

2006032632

British Library Cataloguing-in-Publication Data
A CIP record for this book is available from the British Library
ISBN 10: 1-905209-24-X
ISBN 13: 978-1-905209-24-8

Printed and bound in Great Britain by Antony Rowe Ltd, Chippenham, Wiltshire.

Table of Contents

Homage to Alain Glavieux

To accomplish the sad duty of paying homage to Alain Glavieux, I have referred to his biography as much as my own memories. Two points of this biography struck me, although I had hardly paid attention to them until now. I first noted that Alain Glavieux, born in 1949, is the exact contemporary of information theory, since it was based on the articles of Shannon in 1948 and 1949. I also noted that his first research at the Ecole Nationale Supérieure de Télécommunications de Bretagne (ENST Brittany) related to underwater acoustic communications.

To work on these communications, first of all, meant to be interested in concrete local problems linked to the maritime vocation of the town of Brest. It also meant daring to face extreme difficulties because the marine environment is one of the worst transmission channels there is. Carrying out effective underwater communications can be conceived only by associating multiple functions (coding, modulation, equalizing, synchronizing) that do not only have to be optimized separately, but must be conceived together. This experience, along with the need for general solutions, which are the only effective ones in overcoming such difficulties, has prepared him, I believe, for the masterpiece of the invention of turbocodes, born from his very fruitful collaboration with Claude Berrou. Better still, no one could understand better than him that iterative decoding, the principal innovation introduced apart from the actual structure of the turbocodes, implies a more general principle of exchange of information between elements with different functions but converging towards the same goal. Admittedly, the idea of dealing with problems of reception using values representing the reliability of symbols and thus lending themselves to such an exchange, instead of simple decisions, had already been exploited by some researchers, like Joachim Hagenauer and myself, but the invention of turbocodes brought the most beautiful illustration conceivable, paving the way for a multitude of applications.

Shannon had shown in 1948 that there exists a bound for the possible information flow in the presence of noise, the capacity of the channel, but had not clarified the means of dealing with it. If the asymptotic nature of the Shannon theorem did not leave any hope to effectively reach the capacity, the attempts to approach it had remained in vain despite the efforts of thousands of researchers. Turbocodes finally succeeded 45 years after the statement of the theorem. They improved the best performances by almost 3 decibels. What would we have read in the newspapers if an athlete had broken the 100 meters record by running it in 5 seconds! If this development remained almost unknown to the general public, it resounded like a thunder clap in the community of information and coding theoreticians.

This result and the method that led to it called into question well anchored practices and half-truths, which time had solidified into dogmas. They revealed that unimportant crude restrictions had in fact excluded the best codes from the field of research. The inventors of turbocodes looked again at the basic problem in the spirit of Shannon himself, not trying to satisfy the posed *a priori* criterion to maximize the minimal distance of the code, but to optimize its real performances. To imitate random coding, a process that is optimal, but unrealizable in practice that Shannon had employed to demonstrate the theorem, Berrou and Glavieux introduced an easily controllable share of risk into coding in the form of an interleaving, whose inversion did not present any difficulty. The turbocodes scheme is remarkably simple and their realization is easy using currently available means, but it should be noted that they would have been inconceivable without the immense progress of the technology of semi-conductors and its corollary, the availability of computers. In fact, computer simulations made it possible to choose the best options and to succeed, at the end of an unprecedented experimental study into the subject, with the first turbocode. Its announced performances were accommodated with an incredulous smile by experts, before they realized that they could easily reproduce and verify them. The shock that resulted from it obliged everyone to revise the very manner of conceiving and analyzing codes. The ways of thinking and the methods were completely renewed, as testified by the true metamorphosis of the literature in the field caused by this invention.

It was certainly not easy to invent turbocodes. From a human point of view it was perhaps more difficult still to have invented them. How, indeed, could he handle the authority conferred by the abrupt celebrity thus acquired? Alain Glavieux was absolutely faithful to himself and very respectful of others. He preferred efficiency to glamour. He was very conscious of the responsibilities arising from this authority and avoided the peremptory declarations on the orientation of research, knowing that, set into dogmas, they were also likely to become blocked. He thus used this authority with the greatest prudence and, just as at the start when he had put his engineering talent to the service of people and of regional developments, he devoted himself to employ it to the benefit of the students of the ENST Brittany and of the local economy, in particular, by managing the relations of the school with companies. He particularly devoted himself to help incipient companies, schooling them in "seedbed". He was also concerned with making science and the technology of communication known, as testified, for example, by his role as the main editor this book. Some of these tasks entailed not very exciting administrative aspects. Others would have used their prestige to avoid them, but he fully accepted his responsibilities. In spite of the serious disease which was going to overpower him, he devoted himself to them until the very last effort.

The untimely death of Alain Glavieux leaves an enormous vacuum. Fruits of an exemplary friendship with Claude Berrou, turbocodes definitively marked the theory and practice of communications, with all the scientific, economic, social and human consequences that it implies. Among those, the experimental sanction brought to information theory opens the way for its application to natural sciences. The name of Alain Glavieux will remain attached to a work with extraordinary implications in the future, which, alas, offers his close relations only meager consolation.

Gérard Battail

Chapter 1

Information Theory

1.1. Introduction: the Shannon paradigm

The very title of this book is borrowed from the information theory vocabulary, and, quite naturally, it is an outline of this theory that will serve as an introduction. The subject of information theory is the scientific study of communications. To this end it defines a quantitative measurement of the communicated content, i.e. information, and deals with two operations essential for communication techniques: source coding and channel encoding. Its main results are two fundamental theorems related to each of these operations. The possibility of channel encoding itself has been essentially revealed by information theory. That shows, to which point a brief summary of this theory is essential for its introduction. Apart from some capital knowledge of its possibilities and limits, the theory has, however, hardly contributed to the invention of means of implementation: whereas it is the necessary basis for the understanding of channel encoding, it by no means suffices for its description. The reader interested in information theory, but requiring more information than is provided in this brief introduction, may refer to [1], which also contains broader bibliographical references.

To start with, we will comment on the model of a communication, known as the *Shannon paradigm* after the American engineer and mathematician Claude E. Shannon, born in 1916, who set down the foundations for information theory and established the principal results [2], in particular, two fundamental theorems. This model is represented in Figure 1.1. A source generates a message directed to a recipient. The source and the recipient are two separated, and therefore distant, entities, but between them there exists a channel, which, on the one hand, is the medium of the

---

Chapter written by Gérard BATTAIL.

propagation phenomena, in the sense that an excitation of its receptor by the source leads to a response observable by the recipient at the exit, and, on the other hand, of the disturbance phenomena. Due to the latter, the excitation applied is not enough to determine with certainty the response of the channel. The recipient cannot perceive the message transmitted other than by observing the response of the channel.

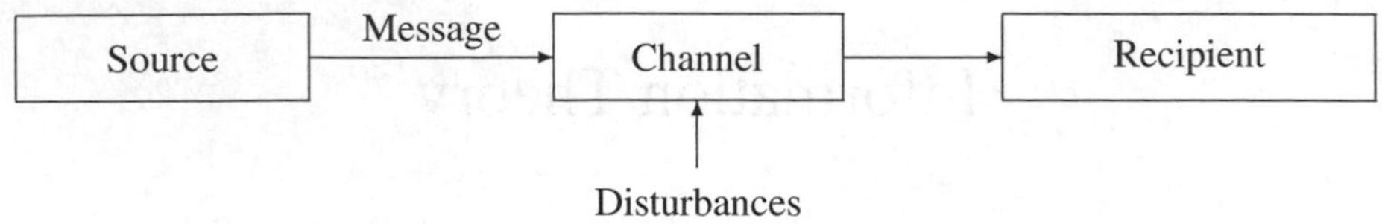

Figure 1.1. *Fundamental communication diagram: Shannon paradigm*

The source, is, for example, a person who speaks and the recipient is a person who listens, the channel being the surrounding air, or two telephone sets connected by a line; or the source may well be a person who writes, with the recipient being a reader and the channel being a sheet of paper[1], unless the script writer and the reader are connected via a conducting circuit using telegraphic equipment. The diagram in Figure 1.1 applies to a large variety of sources, channels and recipients. The slightly unusual word "paradigm" indicates the general model of a certain structure, independently of the interchangeable objects, whose relations it describes (for example in grammar). This diagram was introduced by Shannon in 1948, in a slightly different form, at the beginning of his fundamental article [2]. As banal as it may appear to us now, this simple identification of partners was a prerequisite for the development of the theory.

The principal property of the channel considered in information theory is the presence of disturbances that degrade the transmitted message. If we are surprised by the importance given to phenomena, which often pass unnoticed in everyday life, it should not be forgotten that the observation of the communication channel response, necessary to perceive the message, is a physical measurement which can only be made with limited precision. The reasons limiting the precision of measurements are numerous and certain precautions make it possible to improve these. However, the omnipresence of thermal noise is enough to justify the central role given to disturbances. One of the essential conclusions of information theory, as we will see, identifies disturbances as the factor which in the final analysis limits the possibilities of communication. Neglecting disturbances would also lead to paradoxes.

1. The Shannon paradigm in fact applies to the recording of a message as well as to its transmission, that is, in the case where the source and the recipient are separated in time and not only in space, as we have supposed up until now.

We will note that the distinction between a useful message and a disturbance is entirely governed by the finality of the recipient. For example, the sun is a source of parasitic radiation for a satellite communication system. However, for a radio-astronomer who studies the electromagnetic radiation of the sun, it is the signal of the satellite which disturbs his observation. In fact, it is convenient to locate in the "source" block of Shannon's scheme the events concerning the recipient, whereas the disturbance events are located in the "channel" block.

Hereafter we will consider only a restricted category of sources, where each event consists of the emission of a physical signal expressing the choice of one element, known as a *symbol*, in a certain finite abstract set known as an *alphabet*. It could be a set of decimal or binary digits, as well as an alphabet in the usual sense: Latin, Greek or Arabic, etc. The message generated by the source consists of a sequence of symbols and is then known as "digital". In the simplest case the successive choices of a symbol are independent and the source is said to be "without memory". In information theory we are not interested in the actual signals that represent symbols. Instead we consider mathematical operations with symbols whose results also belong to a finite alphabet physically represented in the same way. The operation, which assigns physical signals to abstract symbols, stems from *modulation* techniques.

The restriction to numerical sources is chiefly interesting because it makes it possible to build a simple information theory, whereas considering sources known as "analog" where the occurring events are represented by continuous values involves fundamental mathematical difficulties that at the same time complicate and weaken the theoretical postulates. Moreover, this restriction is much weaker than it appears, since *digitalization* techniques based on *sampling* and *quantification* operations allow an approximate digital representation, which may be tuned as finely as we wish, of signals generated by an analog source. All modern sound (word, music) and image processing in fact resorts to an analog/digital conversion, whether it is a question of communication or recording. The part of the information theory dealing with analog sources and their approximate conversion into digital sources is called *distortion or rate theory*. The reader interested in this subject may refer to references [3-5].

To clarify the subject of information theory and to introduce its fundamental concepts, before even considering quantitative measurement of information, a few observations on the Shannon paradigm will be useful. Let us suppose the source, channel and recipient to be unspecified: nothing ensures *a priori* the compatibility between the source and the channel, on the one hand, and the channel and the recipient, on the other hand. For example, in radiotelephony the source and the recipient are human but the immaterial channel symbolizes the propagation of electromagnetic waves. It is therefore necessary to supplement the diagram in Figure 1.1 with blocks representing the equipment necessary for the technical functions of conversion and adaptation. We thus obtain the diagram in Figure 1.2a. It is merely a variation of Figure 1.1, since the set formed by the source and the transmitting equipment, on the one hand, and the set of the receiving equipment and the recipient, on the other hand, may be interpreted as a new source-recipient pair adapted to the initial channel (Figure 1.2b). We can also consider the set of the transmitting equipment, the channel and the receiving equipment to constitute a new channel, adapted to the source-recipient pair provided initially (Figure 1.2c); thus, in the preceding examples, we have regarded a telephonic or telegraphic circuit as the channel, consisting in an transmitter, a transmission medium and a receiver.

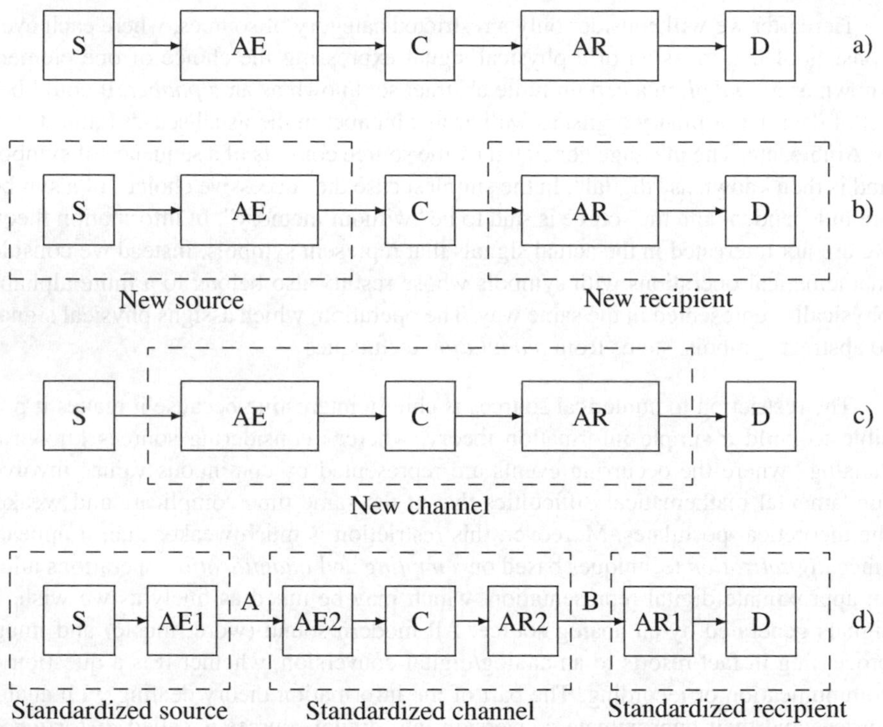

Figure 1.2. *Variants of the Shannon paradigm. S means "source", C, "channel" and D "recipient". AE means "transmitter" and AR "receiver"*

A more productive point of view, in fact, consists of dividing each transmitter and receiver into two blocks: one particular to the source (or the recipient), the other adapted to the channel input (or output). This diagram has the advantage of making it possible to *standardize* the characteristics of the blocks in Figure 1.2 thus redefined: new source aside from point A in Figure 1.2d; new channel between points A and B; new recipient beyond B. The engineering problems may then be summarized as *separately* designing the pairs of adaptation blocks noted AE1 and AR1 in the figure, on the one hand, and AE2 and AR2 on the other hand. We will not specify what the mentioned standardization consists of until after having introduced the concepts of source and channel coding.

Generally speaking we are free to redefine the borders of the blocks in Figure 1.2 for the purposes of analysis; the section of any circuit connecting a source to a recipient in two points – such that the origin of the message useful for the recipient is on the left of the figure, and all the links where disturbances are present are in its central part – defines a new source-channel-recipient triplet.

We will often have to resort to a schematization of the blocks in Figure 1.2, which sometimes may be very simplistic. However, the conclusions drawn will be general enough to be applicable to the majority of concrete situations. Indeed, these simplifications will most often be necessary only to make certain fundamental values calculable, whose existence remains guaranteed under relatively broad assumptions. Moreover, even if these assumptions are not exactly satisfied (it is often difficult, even impossible, to achieve experimental certainty that they are), the solutions of communication problems obtained in the form of device structures or algorithms generally remain usable, perhaps at the cost of losing the exact optimality afforded by the theory when the corresponding assumptions are satisfied.

1.2. Principal coding functions

The message transmitted by the source can be replaced by any other, provided that it is deduced from it in a certain and reversible manner. Then there is neither creation nor destruction of information, and information remains invariant with respect to the set of messages that can be used to communicate it. Since it is possible to assign messages with various characteristics to the same information, transformations of an initial message make it possible to equip it with desirable properties *a priori*. We will now examine what these properties are, and what these transformations, known as *coding* procedures, consist of and how, in particular, to carry out the standardization of the "source", "channel" and "recipient" blocks introduced above.

We may *a priori* envisage transforming a digital message by source coding, channel coding and cryptography.

1.2.1. *Source coding*

Source coding aims to achieve *maximum concision*. Using a channel is more expensive the longer the message is, "cost" being taken here to mean very generally the requirement of limited resources, such as time, power or bandwidth. In order to decrease this cost, coding can, thus, aim at substituting the message transmitted by the source by the shortest possible message. It is required that the coding be reversible, in the sense that the initial message can be restored exactly on the basis of its result.

Let us take an example to illustrate the actual possibility that coding makes the message more concise. Let us suppose that the message transmitted by the source is binary and that the successive symbols are selected independently of each other with very unequal probabilities, for example, $\Pr(0) = 0.99$ and $\Pr(1) = 0.01$. We can transform this message by counting the number of zeros between two successive "1" (supposing that the message is preceded by a fictional "1") and, if it is lower than $255 = 2^8 - 1$ (for example), we can represent this number by a word with 8 binary digits. We also agree on a means of representing longer sequences of zeros by several words with 8 binary digits. We thus replace on average 100 initial symbols by 8.67 coded symbols, that is, a saving factor of approximately 11.5 [1, p. 12].

1.2.2. *Channel coding*

The goal of *channel coding* is completely different: to protect the message against channel noise. We insist on the need for taking channel noise into account to the point of making their existence its specific property. If the result of this noise is a symbol error probability incompatible with the specified restitution quality, we propose to transform the initial message by such a coding that it increases transmission security in the presence of noise. The theory does not even exclude the extreme case where specified quality is the total absence of errors.

The actual possibility of protecting messages against channel noise is not obvious. This protection will be the subject of this entire book; here we will provide a very simple example of it, which is only intended to illustrate the possibility.

Let us consider a channel binary at its input and output, where the probabilities of an output symbol conditioned by an input symbol, known as "transition", are constant (this channel is stationary), and where the probability that the output symbol differs from the input symbol, i.e. of an error, is the same regardless of the input symbol (it is symmetric). It is the *binary symmetric* channel represented in Figure 1.3. The probability of error there is, for example, $p = 10^{-3}$.

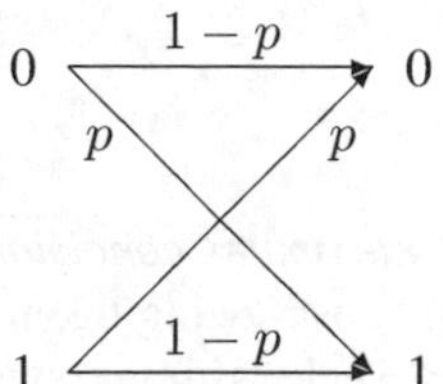

Figure 1.3. *Diagram of a binary symmetric channel with a probability of error p*

We wish to use this channel to transmit a message with a probability of error per binary symbol lower than $3 \cdot 10^{-6}$. This result can be achieved by repeating each symbol of the message 3 times, the decision taken at the receiver end being based on a majority. Indeed, the probability p_e of this decision being erroneous is equal to the probability of 2 or 3 errors out of the 3 received symbols, or:

$$p_e = 3p^2(1-p) + p^3 = 3p^2 - 2p^3 = 2,998 \cdot 10^{-6}$$

A lower probability of error would have been obtained by repeating each symbol $5, 7$, etc. times, the decision rule remaining the majority vote.

This example simultaneously shows the possibility of coding protecting the message against noise and the cost that it entails: a lengthening of the message. It is, however, a rudimentary process. Comparable results could have been obtained at a much lower redundancy cost making use of more elaborated codes called "error correcting codes", to which a large part of this book will be dedicated. However, it is generally true that protection against noise is achieved only by introducing *redundancy*, as demonstrated in section 1.5.3.

The objectives of source coding and channel coding thus appear to be incompatible. They are even contradictory, since source coding increases the vulnerability to errors while improving concision. Thus, in our example of source coding an error in one of the binary digits of the coded message would cause a shift of the entire sequence of the restored message, a much more serious error since it involves many symbols. This simple observation shows that the reduction of redundancy and the reduction of vulnerability to errors cannot be considered independently of each other in the design of a communications chain.

1.2.3. *Cryptography*

Let us note, finally, that a coding procedure can have yet another function, in theory without affecting redundancy or vulnerability to errors: *ciphering* the message, i.e. making it unintelligible to anyone but its recipient, by operating a secret transformation that only he can reverse. Deciphering, i.e. the reconstruction of the message by an indiscreet interceptor who does not know the "key" specifying the transformation making it possible to reverse it, must be difficult enough to amount to factual impossibility. Other functions also involve cryptography, for example, providing the message with properties making it possible to authenticate its origin (to identify the source without ambiguity or error), or to render any deterioration of the message by obliteration, insertion or substitution of symbols detectable. Generally speaking, it is a question of protecting the message against indiscretions or fraudulent deteriorations. Cryptography constitutes a discipline in its own right, but this will be outside the scope of this book.

1.2.4. *Standardization of the Shannon diagram blocks*

It is now possible for us to specify what the standardization of blocks in Figure 1.2 presented above consists of, still restricting ourselves to a digital source. The message coming from the source initially undergoes source coding, ideally with a message deprived of redundancy as a result, i.e. where successive symbols are independent and where, moreover, all the symbols of the alphabet appear with an equal probability. The coding operation realized in this manner constitutes an adaptation only to the characteristics of the source. The result of this coding is very susceptible to noise, since each of its symbols is essential to the integrity of information. It is therefore necessary to carry out channel coding making the message emerging from the source encoder (ideally) invulnerable to channel noise, which necessarily implies reintroducing redundancy.

We can suppose that source coding has been carried out in an ideal fashion; the only role of channel coding is then to protect a message without redundancy from channel noise. If the message being coded in this way is not completely rid of redundancy, the protection obtained can only increase.

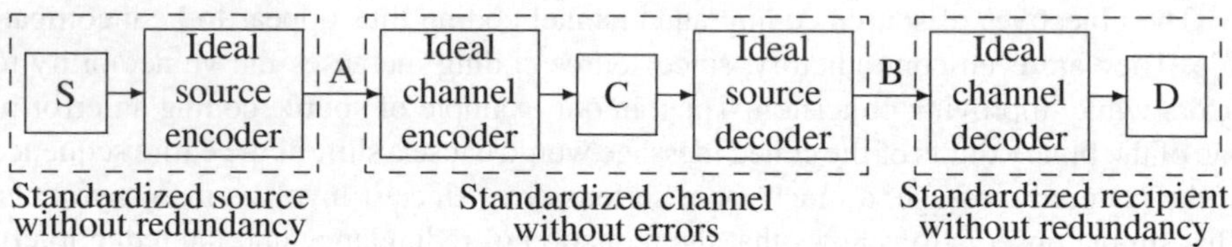

Figure 1.4. *Standardization of the "source", "channel" and "recipient" blocks of the Shannon diagram. S, C and D indicate the initial source, channel and recipient respectively*

We can then redraw Figure 1.2d as in Figure 1.4, where the standardized source generates a message without redundancy and where the standardized channel contains no errors.

The procedure consisting of removing the redundancy of the initial message by source coding, then reintroducing redundancy by channel coding can appear contradictory, but the redundant initial source is not *a priori* adapted to the properties of the channel, to which we connect it. Rather than globally conceiving coding systems to adapt a particular source to a particular channel, the standardization that we have just defined makes it possible to treat the source and the channel separately, and no longer the source-channel pair. This standardization also has a secondary advantage: the alphabet of messages at points A and B of Figure 1.4 is arbitrary. We can suppose it to be binary, for example, i.e. the simplest possible, without significantly restricting the generality.

1.2.5. *Fundamental theorems*

The examples above suggest that coding operations can yield the following results:

– for source coding, a coded message deprived of redundancy, although the initial message comes from a redundant source;

– for channel coding, a message restored without errors after decoding, although the coded message is received through a disturbed channel.

These possibilities are affirmed by the fundamental theorems of the information theory, under conditions which they specify. Their demonstration does not require clarifying the means of reaching these results. Algorithms approximating optimal source coding have been known for a long time; on the contrary, the means of approximating the fundamental limits with regards to channel coding remain unknown in general, although the recent invention of turbo-codes constitutes a very important step in this direction [6].

The fundamental theorems concern the ultimate limits of coding techniques and are expressed according to the values used for quantitative measurement of information that we now have to introduce. In particular, we will define source entropy and channel capacity. The fundamental theorems confer an operational value to these items, which confirms their adequacy for transmission problems and clarifies their significance.

1.3. Quantitative measurement of information

1.3.1. *Principle*

The description of transmitted messages, of their transformation into signals suitable for propagation, as well as of noise, belongs to signal theory. Messages and signals undergo transformations necessary for their transmission (in particular, various forms of coding and modulation), but they are merely vehicles of a more fundamental and more difficultly definable entity, invariant in these transformations: *information*. The invariance of information with respect to messages and signals used as its support implies that it is possible to choose from a set of equivalent messages representing the same information those which *a priori* have certain desirable properties. We have introduced in section 1.2 coding operations, in the various senses of this word.

We will not try now to define information, contenting ourselves to introduce its quantitative measurement that the theory proposes, a measure, which was a necessary condition of its development. We will briefly reconsider the difficult problem of its definition in the comments to section 1.3.6, in particular, to stress that, for the theory, information is dissociated from the meaning of messages. As in thermodynamics, the values considered are statistical in nature and the most important theorems establish the existence of limits.

The obvious remark that the transmission of a message would be useless if it were known by its recipient in advance leads to:

– treating a source of information as being the seat of random events whose sequence constitutes the transmitted message;

– defining the quantity of information of this message as a measure of its unpredictability, compared to its improbability.

1.3.2. *Measurement of self-information*

Let X be an event occurring with a certain probability p. We measure its uncertainty by $f(1/p)$, $f(\cdot)$ being a suitably selected increasing function. The quantity of information associated with the event x is thus

$$h(x) = f(1/p).$$

To choose the function $f(\cdot)$ it is reasonable to admit that the quantity of information brought by the joint occurrence of two independent events x_1 and x_2 is the sum of the quantities of information carried separately by each one of them. We thus wish to have:

$$h(x_1, x_2) = h(x_1) + h(x_2),$$

which implies for the function $f(\cdot)$:

$$f(1/p_1p_2) = f(1/p_1) + f(1/p_2),$$

where p_1 and p_2 are the probabilities of x_1 and x_2 occurring, respectively. Indeed, the probability of joint occurrence of x_1 and x_2 is the product p_1p_2 of their probabilities. The continuous function that associates the sum of functions having each of its terms as an argument to an argument formed by a product is the logarithm function. We are consequently led to choose:

$$h(x) = \log(1/p) = -\log(p). \qquad [1.1]$$

This choice also implies that $h(x) = 0$, if $p = 1$, so that a certain event brings a zero amount of information, which conforms to the initial observation, upon which quantitative measurement of information is based.

The logarithm function is defined only to the nearest positive factor determined by the base of the logarithms, whose choice thus specifies the unit of information. The usually selected base is 2 and the unit of information is then called the *bit*, an acronym of *binary digit*. This term is widely employed, despite the regrettable confusion that it introduces between the unit of information and a binary digit which is neither necessarily carrying information, nor, if it is, has information equal to the binary unit. Following a proposal of the *International Standards Organization (ISO)*, we prefer to indicate the binary unit of information by *shannon*, in tribute to Shannon who introduced it [2]. It will often be useless for us to specify hereafter the unit of information, and we will then also leave unspecified the logarithms base.

1.3.3. *Entropy of a source*

When the *source* is hosting repetitive and *stationary* events, i.e. if its operation is independent of the origin of time, we can define an average quantity of information produced by this source and carried by the message that it transmits: its *entropy*.

We then model the operation of a digital source by the regular, periodic emission of a random variable X, for example, subject to a certain finite number n of occurrences $x_1, x_2, \ldots, x_n$, with the corresponding probabilities $p_1, p_2, \ldots, p_n$, with $\sum_{i=1}^{n} p_i = 1$. In the simplest case, the successive occurrences of X, i.e. the choices of symbol, are independent and the source is known as "without memory". Rather than considering the quantity of information carried by a particular occurrence, we consider the average information, in the statistical sense, i.e. the value called *entropy* defined by:

$$H(X) \triangleq \sum_{i=1}^{n} p_i h(x_i) = -\sum_{i=1}^{n} p_i \log(p_i). \qquad [1.2]$$

If the successive occurrences of X are not independent, we define the entropy by symbol of a stationary source by the limit:

$$H = \lim_{s \to \infty} \frac{1}{s} H_s, \qquad [1.3]$$

with

$$H_s = -\sum_c p(c) \log p(c)$$

where c is any sequence of length s, which the source transmits with probability $p(c)$. The sum is calculated for all the possible sequences c. The stationarity of the source suffices for the existence of the limit in [1.3].

The entropy defined by [1.2] has many properties, among which:

– it is positive or zero, zero only if one of the probabilities of occurrence is equal to 1, which leads the others to zero and the random variable X reduced to a given parameter;

– its maximum is reached when all the probabilities of occurrence are equal, therefore if $p_i = 1/n$ regardless of i;

– it is convex ∩, i.e. the replacement of the initial probability distribution by a distribution where each probability is obtained by taking an average of the probabilities of the initial distribution increases entropy (it remains unchanged only if it is maximum initially).

NOTE.– A notation such as $H(X)$, is convenient but abusive, since X is not a true argument there. It is only used to identify the random variable X whose entropy is H, which, in fact, depends only on its probability distribution. This note applies throughout the remainder of the book, every time a random variable appears as a pseudo-argument of a information measure.

1.3.4. *Mutual information measure*

Up until now we have considered self-information, i.e. associated to an event or a single random variable. It is often interesting to consider pairs of events or random variables, in particular, those relating to channel input and output. In this case it is necessary to measure the average quantity of information that the data of a message received at the output of a channel brings to the message transmitted at the input. As opposed to entropy, which relates only to the source, this value depends simultaneously on the source and the channel. We will see that it is symmetric in the sense that it also measures the quantity of information which the data of the transmitted message brings to the received message. For this reason it is called *average mutual information* (often shortened to just "mutual information"), "information" already being here an acronym for "quantity of information". This value is different from the entropy of the message at the output of the channel and is smaller than it, since, far from bringing additional information, the channel can only degrade the message transmitted by the source, which suffers there from random noise.

In the simplest case, the channel is without memory like the source, in the sense that each output symbol depends only on an input symbol, itself independent of others, if the source is without memory. It can thus be fully described by the probabilities of output symbols conditioned to input symbols, referred to as transition probabilities, which are constant for a stationary channel. Let X be the random variable at the channel input, that is, represent the source symbols with possible occurrences $x_1, x_2, \ldots, x_n$, and probabilities $p_1, p_2, \ldots, p_n$, and Y be the random variable at the output of the same channel, having occurrences $y_1, y_2, \ldots, y_m$, m being an integer, perhaps, different from n, with probabilities $p_1, p_2, \ldots, p_n$ and transition probabilities $p_{ij} \triangleq \Pr(y_j \mid x_i)$, i.e.:

$$\Pr(y_j) = \sum_{i=1}^{n} \Pr(x_i, y_j) = \sum_{i=1}^{n} p_i p_{ij}. \qquad [1.4]$$

Then, the average mutual information is defined, coherently with the self-information measured by entropy, as the statistical average of the logarithmic increase in the probability of X, which stems from the given Y, i.e. by:

$$I(X;Y) \triangleq \sum_{i=1}^{n}\sum_{j=1}^{m} \Pr(x_i, y_j) \log[\Pr(x_i \mid y_j)/ \Pr(x_i)], \qquad [1.5]$$

where $\Pr(x_i, y_j)$ is the joint probability of x_i and y_j, equal to $\Pr(x_i, y_j) = \Pr(x_i)$ $\Pr(y_j \mid x_i) = p_i p_{ij}$, while $\Pr(y_j \mid x_i)$ is the conditional probability of y_j knowing x_i is realized, and $\Pr(x_i \mid y_j)$ is that of x_i, knowing y_j is realized. In the indicated form this measurement of information appears dissymmetric in X and Y, but it suffices to express in [1.5] $\Pr(x_i \mid y_j)$ according to Bayes' rule:

$$\Pr(x_i, y_j) = \Pr(x_i \mid y_j) \Pr(y_j)$$

to obtain the symmetric expression

$$I(X;Y) = \sum_{i=1}^{n}\sum_{j=1}^{m} \Pr(x_i, y_j) \log[\Pr(x_i, y_j)/ \Pr(x_i) \Pr(y_j)], \qquad [1.6]$$

where $\Pr(x_i) = p_i$ and where $\Pr(y_j)$ is given by [1.4].

The definition of average mutual information and its symmetry in X and Y means that we can write:

$$\begin{aligned} I(X;Y) &= H(X) - H(X \mid Y) = H(Y) - H(Y \mid X) \\ &= H(X) + H(Y) - H(X, Y), \end{aligned} \qquad [1.7]$$

where $H(X)$ and $H(Y)$ are the entropies of the random variables X and Y of channel input and output, respectively, and $H(X, Y)$ is the joint entropy of X and Y, defined by

$$H(X, Y) \triangleq - \sum_{i=1}^{n}\sum_{j=1}^{m} \Pr(x_i, y_j) \log \Pr(x_i, y_j), \qquad [1.8]$$

while $H(X \mid Y)$ is the entropy of X conditioned to Y, defined by:

$$H(X \mid Y) \triangleq - \sum_{i=1}^{n}\sum_{j=1}^{m} \Pr(x_i, y_j) \log \Pr(x_i \mid y_j), \qquad [1.9]$$

$H(Y \mid X)$ being obtained by a simple exchange of X and Y in this definition. We will note that in expression [1.9] the argument of the logarithm is different from its factor, whereas it is identical to it in the definition [1.8] of joint entropy.

We demonstrate that conditioning necessarily decreases entropy, i.e. $H(X \mid Y) \leq H(X)$, equality being possible only if X and Y are independent variables. It follows that average mutual information $I(X;Y)$ is positive or zero, zero only if X and Y are independent, a case where, indeed, the data of Y does not provide any information at all on X.

Valid for a source and a channel without memory, both of them discrete, these expressions are easily generalized to sources and/or channels where the successive symbols are not mutually independent.

1.3.5. *Channel capacity*

The average mutual information does not only characterize the channel, but also depends on the source. In order to measure the ability of a channel to transmit information, the theory defines its *capacity* as the maximum of the average mutual information between its input and output variables with respect to all the possible *stationary and ergodic* sources connected to its input (their existence is demonstrated under certain regularity conditions: the channel must be not only stationary, but also causal, in the sense that its output cannot depend on input symbols which have not yet been introduced, and of finite memory, in the sense that the channel output depends only on a finite number of input symbols). Ergodism is a concept distinct from stationarity and is a condition of homogeneity of the set of messages likely to be transmitted by the source. For an ergodic source, an indefinitely prolonged observation of a single message almost definitely suffices to characterize the set of the possible transmitted messages statistically.

The capacity of a channel without memory is given simply by:

$$C = \max_{\underline{p}} I(X;Y), \qquad [1.10]$$

where $I(X;Y)$ has one of the expressions [1.5] to [1.7] and where $\underline{p}$ indicates the probability distribution of the symbols at channel input, i.e., for an alphabet of size n, the set of probabilities $p_1, p_2, \ldots, p_n$ subject to the constraint $\sum_{i=1}^{n} p_i = 1$. More complicated expressions, yet without difficulty of principle, can be written in the case of a causal channel with finite memory.

If the channel is symmetric, which implies that the set of translation probabilities is independent of the input symbol considered, the calculation of the maximum of $I(X;Y)$, in [1.10] is made a lot easier, because we then know that it is obtained by assigning the same probability equal to $1/n$ to all input symbols, and the entropies $H(X)$ and $H(Y)$ are also maximal.

1.3.6. *Comments on the measurement of information*

We have only used the observation from section 1.3.1 for the quantitative measurement of information. Information defined in this manner is, thus, a very restrictive concept compared to the current meaning of the word. It should be stressed, in particular, that at no time did we consider the meaning of messages: information theory disregards semantics completely. Its point of view is that of a messenger whose function is limited to the transfer of information, about which it only needs to know a quantitative external characteristic, a point of view that is also common to engineers. Similarly, a physical subject has multiple attributes, such as its form, texture, color, internal structure, etc., but its behavior in a force field depends only on its mass. The significance of a message results from a prior agreement between the source and the recipient, ignored by the theory due to its subjective character. This agreement lies in the qualitative realm, which by hypothesis evades quantitative measurement. The transfer of a certain quantity of information is, however, a necessary condition to communicate a certain meaning, since a message is the obligatory intermediary.

Literal and alien to semantics, information appears to us as a class of equivalence of messages, such that the result of the transformation of a message pertaining to it, by any reversible coding, also belongs to it. It is thus a much more abstract concept than that of a message. The way in which we measure information involves its critical dependence on the existence of a probabilistic set of events, but other definitions avoid resorting to probabilities, in particular, Kolmogorov's theory of complexity, which, perhaps, makes it possible to base the probabilities on the concept of information by reversing the roles [7,8].

1.4. Source coding

1.4.1. *Introduction*

It appeared essential to us to provide an outline of source coding and the corresponding fundamental theorem for two main reasons: on the one hand, to avoid giving a truncated image of information theory, and on the other hand, because, as we have already observed, the two functions of source and channel coding cannot in fact be dissociated in the concrete design of a communication system, as they might be conceptually in a theoretical discourse.

A source is redundant if its entropy by symbol is lower than the possible maximum, equal to $\log q_s$ for an alphabet of size q_s. This alphabet is then misused, from the point of view of being economic with symbols. We can also say that the probabilistic set of sequences transmitted by the source with a given arbitrary length does not correspond to that of all the possible sequences formed by the symbols of the alphabet, with equal probabilities assigned. The result is similar if successive symbols are selected either with unequal probabilities in the alphabet, or not independently from each other (both cases may occur simultaneously). The fundamental theorem of source coding affirms that it is possible to eliminate all redundancy from a message transmitted by a stationary source. Coding must use the kth extension of this source for a sufficiently large k, the announced result being only reachable asymptotically. The kth extension of a source S whose alphabet has q_s elements (we will call this source q_s-ary) is the source deduced from the initial source considering the symbols which it transmits by blocks of k, each block interpreted as a symbol of the alphabet with q_s^k symbols (known as q_s^k-ary). Noted S^k, this extension is simply another way of describing S, not a different source. If coding tends towards optimality, the message obtained has an average length per symbol of the initial source which tends towards the entropy of this source expressed by taking the size q of the alphabet employed for coding as the logarithms base.

1.4.2. *Decodability, Kraft-McMillan inequality*

The principal properties required of source coding are decodability, i.e. the possibility of exploiting the coded message without ambiguity, allowing a unique way to split it into significant entities, which will be referred to as codewords and the regularity that prohibits the same codeword to represent two different symbols (or groups of symbols) transmitted by the source. Among the means of ensuring decodability let us mention, without aiming to be exhaustive:

– coding in blocks where all codewords resulting from coding have the same length,

– addition of an additional symbol to the alphabet with the exclusive function of separating the codewords,

– the constraint that no codeword is the prefix of another, that is to say, identical to its beginning. Coding using this last means is referred to as *irreducible*.

Any decodable code regardless of the means used to render it such verifies the Kraft-McMillan inequality, which is a necessary and sufficient condition for the existence of this property:

$$\sum_{i=1}^{N} q^{-n_i} \leq 1, \qquad [1.11]$$

where q denotes the size of the code alphabet, N is the number of codewords with lengths $n_1, n_2, \ldots, n_N$ symbols respectively. The demonstration of this inequality is very easy for an irreducible code. Let n_N be the largest length of codewords. It is sufficient to note that the set of all the codewords of length n_N written with an alphabet of q symbols can be represented by all the paths of a tree where q branches diverge from a single root, q branches then diverge from each end, and so on until the length of the paths in the tree reaches n_N branches. There are q^{n_N} paths of different lengths

n_N. When the ith codeword is selected as belonging to the code, the condition that no codeword is the prefix of any other codeword interdicts the $q^{n_N - n_i}$ paths whose first n_i branches represent the ith codeword. Overall, the choice of all codewords (which all must be different to satisfy the regularity condition) prohibits $\sum_{i=1}^{N} q^{n_N - n_i}$ paths, a number at most equal to their total number q^{n_N}. We obtain [1.11] by dividing the two members of this inequality by q^{n_N}.

1.4.3. *Demonstration of the fundamental theorem*

Let S be a source without memory with an alphabet of size N and entropy H; let $\overline{n}$ be the average length of the codewords necessary for decipherable coding of the symbols which it transmits, expressed in a number of q-ary code symbols. Then the double inequality

$$H/\log(q) \le \overline{n} < H/\log(q) + 1 \qquad [1.12]$$

is verified. We demonstrate it on the basis of Gibbs' inequality, a simple consequence of the convexity $\cap$ of the function $y = -x \log x$ for $1 < x \le 1$:

$$\sum_{i=1}^{N} p_i \log(q_i/p_i) \le 0, \sum_{i=1}^{N} p_i = \sum_{i=1}^{N} q_i = 1, \qquad [1.13]$$

the equality taking place if, and only if, $p_i = q_i$ for all i. We apply this inequality to the set of N codewords used to code the N source symbols, defining p_i as the probability of the ith symbol and posing

$$q_i = q^{-n_i}/\mathcal{Q}, \qquad [1.14]$$

with

$$\mathcal{Q} = \sum_{i=1}^{N} q^{-n_i}. \qquad [1.15]$$

Applying [1.13] it follows:

$$-\sum_{i=1}^{N} p_i \log(p_i) \le (\sum_{i=1}^{N} p_i n_i) \log(q) + \log(\mathcal{Q})$$

or, taking into account [1.2] and posing $\overline{n} = \sum_{i=1}^{N} p_i n_i$:

$$H \le \overline{n} \log(q) + \log(\mathcal{Q}).$$

Having to be decipherable, the code satisfies the Kraft-McMillan inequality [1.11]; the definition of $\mathcal{Q}$ by [1.15] and [1.14] thus involves $\log(\mathcal{Q}) \le 0$.

Let us first examine the conditions under which the equality in [1.12] is verified. Firstly, that implies $\mathcal{Q} = 1$, i.e. equality in [1.11], which expresses that we use all possible codewords compatible with decodability, which we will suppose; and also $p_i = q_i$ for all i, that is

$$n_i = -\log(p_i)/\log(q), 1 \leq i \leq N;$$

if there are N integers verifying this condition, the coding is referred to as *absolutely optimal*.

In general that is not so, but we can always find N integers satisfying [1.11] with equality and such that:

$$-\log(p_i)/\log(q) \leq n_i < -\log(p_i)/\log(q) + 1, \ 1 \leq i \leq N,$$

To obtain [1.12] it is enough to multiply by p_i and to sum up for i from 1 to N.

THEOREM 1.1 (THE FUNDAMENTAL THEOREM OF SOURCE CODING). *For any stationary source there is a decodable coding process where the average length $\overline{n}$ of codewords per source symbol is as close to its limit lower $H/\log(q)$ as we wish.*

If the source considered is without memory, we can write [1.12] for its kth extension. Then H is replaced by kH; dividing by k we obtain:

$$H/\log(q) \leq \overline{n}_k/k < H/\log(q) + 1/k, \qquad [1.16]$$

where $\overline{n}_k$ is the average length of codewords coding the blocks of k symbols of the initial source, from where $\overline{n}_k/k = \overline{n}$. The order k of the extension can be chosen to be arbitrarily large, proving the assertion of the theorem for a source without memory.

This result is generalized directly to any stationary source, since we defined its entropy by [1.3], as the limit for infinite s of H_s/s, H_s being the entropy of its sth extension.

1.4.4. *Outline of optimal algorithms of source coding*

Optimal algorithms, i.e. those making it possible to reach this result, are available, in particular the Huffman algorithm. Very roughly, it involves constructing the tree representing the codewords of an irreducible code, which ensures its decodability, so that shorter codewords are used for more probable symbols, and longer codewords are used for less probable symbols [9]. If optimal coding can be achieved for a finite k, this length is proportional to the inverse of the logarithm of the occurrence probability of the corresponding symbol. Otherwise the increase in k makes it possible to improve the relative precision of the approximation of real numbers by obviously integer codeword lengths. Moreover, the increase in the number of symbols of the alphabet with

k involving an increase in the number of codewords, the distribution of codeword lengths can be adapted all the better to the probability distribution of these symbols.

Another family of source coding algorithms called "arithmetic coding" subtly avoids taking recourse in an extension of the source to approximate the theoretical limit of the average length after coding, i.e. the source entropy [10,11]. We make the average length of the message after coding tend towards its limit $H/\log(q)$ by indefinitely reducing the tolerated variation between the probabilities of the symbols and their approximation by a fraction with a coding parameter for denominator, which must therefore grow indefinitely.

1.5. Channel coding

1.5.1. *Introduction and statement of the fundamental theorem*

The fundamental channel coding theorem is undoubtedly the most important result of information theory, and is definitely so for this book. We will first state it and then provide the Gallager demonstration simplified in the sense that it uses the usual assumptions with respect to coding, and in particular that of coding by blocks. Like the original Shannon demonstrations, it exploits the extraordinary idea of random coding and in addition to the proof of the fundamental theorem achieves useful exponential terminals showing how the probability of error varies after decoding according to the length of codewords. But this demonstration hardly satisfies intuition, which is why we will precede its explanation by less formal comments on the need for redundancy and random coding. Based on a simple example, they are intended to reveal the fundamental theorem as a consequence of the law of large numbers. From there we will gain an intuitive comprehension of the theorem, of random coding and also, hopefully, of channel coding in general.

The fundamental theorem of channel coding can be stated as follows:

THEOREM 1.2 (THE FUNDAMENTAL THEOREM OF CHANNEL CODING). *Using an appropriate coding process involving sufficiently long codewords, it is possible to satisfy a quality standard of message reconstruction, if it is severe, provided that the entropy H of the source is either lower than the capacity C of the channel, or:*

$$H < C \qquad [1.17]$$

In its most usual formulation, the reconstruction quality standard used is an upper limit of the word error probability.

A converse theorem states that if the inequality [1.17] is not verified, it is impossible to obtain an arbitrarily small probability of error under the same conditions (in fact, the word error probability tends towards 1 when the length of the codewords increases indefinitely).

1.5.2. *General comments*

The fundamental theorem of channel coding is undoubtedly the most original and the most important result of information theory: original in that it implies the paradoxical possibility of a transmission without error via a disturbed channel, so contrary to apparent common sense that engineers had not even imagined it before Shannon; important in theory, but also in practice, because a transmission without error is a highly desirable result. The absence of explicit means to carry it out efficiently, just as the importance of the stake, were powerful incentives to perform research in the field. Starting with Shannon's publications, they have remained active since then. Stimulated by the invention of turbo-codes, they are now more important than ever.

The mere possibility of transmitting a quantity of information through a channel, which is at most equal to its capacity C, does not suffice at all to solve the problem of communication through this channel: a message coming from a source with entropy lower or equal to C. Indeed, let us consider the first of the expressions [1.7] of mutual information for a channel without memory, rewritten here:

$$I(X;Y) = H(X) - H(X \mid Y). \qquad [1.18]$$

It appears as the difference between two terms: the average quantity of information $H(X)$ at the channel input minus the residual uncertainty with respect to X that remains when its output Y is observed, measured by $H(X \mid Y)$, in this context often referred to as "ambiguity" or "equivocation". It is clear that the effective communication of a message imposes that this term be rendered zero or negligible when $H(X)$ measures the information stemming from the source that must be received by the recipient. The messages provided to the recipient must indeed satisfy a reconstitution quality standard, for example, a sufficiently low probability of error. However, $H(X \mid Y)$ depends solely on the channel once the distribution of X has been chosen to yield the maximum $I(X;Y)$ and, if the channel is noisy, generally does not satisfy the specified criterion. The source thus cannot be directly connected to the channel input: intermediaries in the shape of an encoder and a decoder must be interposed between the source and channel input, on the one hand, and the output and the recipient, on the other hand, according to the diagram in Figure 1.4. The source message must be transformed by a certain coding, called *channel coding*, in order to distinguish it from source coding, and channel output must undergo the opposite operation of decoding intended to restore the message for the recipient.

1.5.3. *Need for redundancy*

Channel coding is necessarily redundant. Let us consider, on the one hand, the channel, with its input and output variables X and Y, and, on the other hand, the channel preceded by an encoder and followed by a decoder. We suppose that the alphabet used is the same everywhere: at the channel input and output as well as at the encoder input and the decoder output. The random variables at the encoder input and the decoder output are respectively noted U and V. The average mutual information $I(X;Y)$ is expressed by [1.18] with positive $H(X \mid Y)$ dependent on the channel. For U and V we have the homologous relation:

$$I(U;V) = H(U) - H(U \mid V)$$

but the reconstitution quality criterion now imposes $H(U \mid V) < \varepsilon$, where ε is a given positive smaller than $H(X \mid Y)$. Now the inequality:

$$I(U;V) \leq I(X;Y)$$

is true. Indeed, the encoder and the decoder do not create information and the best they can do is not to destroy it. The equality is obtained for a well conceived coding system, i.e. without information loss. It follows that:

$$H(X) - H(U) \geq H(X \mid Y) - H(U \mid V) > 0.$$

The entropy $H(U)$ is therefore smaller than $H(X)$. Let X' be the variable at the encoder output. The inequality $H(U) \geq H(X')$ where the equality is true if information is preserved in the encoder involves $H(X') < H(X)$, which expresses the need for the redundancy.

1.5.4. *Example of the binary symmetric channel*

We will now develop certain consequences of the necessarily redundant nature of channel coding in the simple, but important, case of a binary symmetric channel. Furthermore, the main conclusions reached for this channel can be generalized to almost any stationary channel. To deal with channel coding independently of the probabilities of symbols transmitted by the source we will suppose that the necessary redundancy is obtained by selecting admissible binary sequences at channel input. Moreover, we will restrict ourselves to binary codewords of constant length n, the redundancy of the code being expressed by its belonging to a subset of only 2^k codewords among the 2^n codewords of length n, with $k < n$.

1.5.4.1. *Hamming's metric*

Let E_n be the set of all binary codewords of length n. We define the Hamming weight $\mathrm{w}(\underline{a})$ of a codeword $\underline{a}$ belonging to E_n by the number of its non-zero symbols. We define the Hamming distance $d_H(\underline{a}, \underline{b})$ between two codewords $\underline{a}$ and $\underline{b}$ of the same length as the number of positions where the symbols of the two codewords differ. For example, for $n = 7, \underline{a} = [1110010]$ and $\underline{b} = [0111001]$, we have $d_H(\underline{a}, \underline{b}) = 4$. Let us define the sum of two codewords by the modulo 2 sum of symbols occupying the same positions of the codewords. Its result is still a codeword with n binary symbols. We then have:

$$d_H(\underline{a}, \underline{b}) = \mathrm{w}(\underline{a} - \underline{b}) = \mathrm{w}(\underline{a} + \underline{b}), \qquad [1.19]$$

the second equality is due to subtraction and addition carried out in modulo 2 yielding identical results.

We verify without difficulty on the basis of the definition of Hamming distance that it satisfies the axioms of a metric, that is:

$$d_H(\underline{a}, \underline{b}) \geq 0,$$
$$d_H(\underline{a}, \underline{a}) = 0,$$
$$d_H(\underline{a}, \underline{b}) = d_H(\underline{b}, \underline{a}),$$
$$d_H(\underline{a}, \underline{c}) \leq d_H(\underline{a}, \underline{b}) + d_H(\underline{b}, \underline{c}) \, \forall \underline{a}, \underline{b}, \underline{c} \in E_n.$$

Let there be a code conforming to the specifications given at the beginning of this section, employed on the binary symmetric channel with probability of error p illustrated in Figure 1.3. We can assume without loss of generality that we have $p < 1/2$. Indeed, the way in which we make the numbers 0 and 1 correspond to the received symbols is arbitrary. If a given channel has a probability of error $p > 1/2$, it suffices to swap the numbers 0 and 1 indicating the output channel symbol to get to the channel with a probability of error $1 - p < 1/2$. The case $p = 1/2$ does not present interest, because the received symbol does not provide any information on the transmitted symbol and the observation of this channel output does not serve to make a decision (it is verified that its capacity is zero).

1.5.4.2. *Decoding with minimal Hamming distance*

Applying of the operation of vectorial addition modulo 2 defined by [1.19] comes naturally for the comparison of input and output codewords in a binary symmetric channel. We can interpret their modulo 2 difference (or sum) as the "configuration of errors" generated by the channel. Then the probability $\Pr(\underline{e})$ of a particular configuration of errors $\underline{e}$ occurring is simply:

$$\Pr(\underline{e}) = p^{\mathrm{w}(\underline{e})}(1 - p)^{n - \mathrm{w}(\underline{e})}.$$

For $p < 1/2, \Pr(\underline{e})$ is a decreasing function of the weight $\mathrm{w}(\underline{e})$ of the configuration of errors, since:

$$\log[\Pr(\underline{e})] = n \log(1 - p) - \mathrm{w}(\underline{e}) \log \frac{1 - p}{p}$$

where the factor of $-w(\underline{e})$ is positive for $p < 1/2$. This weight is by definition the Hamming distance between the transmitted codeword $\underline{x}$ and the received codeword $\underline{y}$. The optimal reception at the output of a binary symmetric channel (in the sense of maximum probability, which guarantees a minimal probability of error, if the input symbols are equally probable, as we have assumed) thus consists of seeking the codeword $\underline{\hat{x}}$ belonging to the code, which is the closest to the received codeword $\underline{y}$, in the sense of the Hamming metric. This rule results, in particular, in accepting the received codeword if it belongs to the code, since the assumption of an error of zero weight is then the most probable.

1.5.4.3. *Random coding*

We will first consider the means of implementing random coding without questioning for the moment the motivations of its use. It will be enough for us to state here that the use of random coding was justified for Shannon due to the lack of an optimal or near-optimal channel coding technique (which, besides, is still the case after 50 years of research). By demonstrating the fundamental theorem for the average of a probabilistic set of codes we guarantee the existence of a code in this set which satisfies it without having to clarify its construction. Besides, we will see this idea at work in Gallager's demonstration (see section 1.5.6). An *a posteriori* reflection will enable us to better understand its significance.

As a simplifying hypothesis we will admit that the set of distances between the codewords and a given word is the same regardless of what this word is, so that, if the word formed by n zeros belongs to the code, as we will suppose, the distribution of weights is identified with that of all the distances between its codewords. This property is perfectly verified for the important class of linear codes, which will be defined in Chapter 2.

In addition we only consider the average properties of random coding, admitting that a code with an average distribution of distances of random coding is good. Let us suppose initially that we randomly draw 2^n binary codewords with a length n with the same probability and independent of each other. We thus obtain on average all the n-tuples, i.e. the average number of the codewords of weight w is equal to the number of combinations of n objects w for w, that is $C_n^w = n!/(w!(n-w)!)$. Since we have demonstrated the need for redundancy obtained by selection of codewords, we only need to draw 2^k codewords, with $k < n$, from the 2^n binary codewords of length n. The average number a_w of codewords of weight w then becomes $a_w = 2^{-(n-k)} C_n^w$, a smaller number than C_n^w, which is generally not an integer. Thus, there does not exist a redundant code with a distribution of distances exactly equal to the average distribution of weights obtained by random coding, but we can seek a code with a distribution of weight close to it, where for example the number of codewords of weight w would be the best integer approximation. Such a code that imitates random coding in that it roughly preserves the average distribution of weight and which we will refer to hereafter as *quasi-random*, can be interpreted as stemming from a decimation of the set of n-tuples only allowing 2^k of its 2^n elements to remain.

1.5.4.4. *Gilbert-Varshamov bound*

This method of construction of a quasi-random code, by decimation of the set of n-tuples, supposes that its minimal weight $w_{\min}$ is at least equal to a limit, which we will calculate as follows. For the smallest values of weight w, the number a_w is smaller than 1/2 so that its best integer approximation is 0. We are going to presume, in fact, that the integer approximation of a_w taken is equal to 0 if $a_w < \lambda$, where λ is an arbitrary positive constant, perhaps different from 1/2. Let us suppose large n and $n-k$. Then, in the inequality:

$$2^{-(n-k)} C_n^{w_{\min}} \geq \lambda,$$

which expresses that no codeword of the quasi-random code has a weight lower than $w_{\min}$, we can replace $C_n^{w_{\min}}$ by its approximation:

$$C_n^{w_{\min}} \approx \frac{1}{\sqrt{2\pi}} \frac{n^{n+1/2}}{w_{\min}^{w_{\min}+1/2}(n-w_{\min})^{n-w_{\min}+1/2}},$$

deduced from the Stirling's formula $n! \approx (n/e)^n/\sqrt{2\pi n}$, where e is the base of Napierian logarithms, from where:

$$2^{-(n-k)} \frac{n^{n+1/2}}{w_{\min}^{w_{\min}+1/2}(n-w_{\min})^{n-w_{\min}+1/2}} \geq \lambda\sqrt{2\pi}.$$

Taking base 2 logarithms dividing by n while having n tend towards infinity, neglecting the terms in $1/n$ and in $\log_2(n)/n$ and, finally, supposing that k/n tends towards a non-zero limit, we obtain the important Gilbert-Varshamov inequality:

$$\mathcal{H}_2(w_{\min}/n) \geq 1 - k/n, \qquad [1.20]$$

with

$$\mathcal{H}_2(x) \triangleq -x\log_2 x - (1-x)\log_2(1-x), \; x < 1/2. \qquad [1.21]$$

This inequality is more usually demonstrated (as it was initially) on the basis of the construction of random linear code [12]. We note that the obtained limit [1.20] is independent of the constant λ, which serves to specify the approximation of the number of codewords of a certain weight by an integer rendering it robust with respect to this approximation. The weight $w_{\min}$ tends towards infinity with n in such a way that the limit of $w_{\min}/n$ is strictly positive. The function $\mathcal{H}_2\ (x)$ is increasing for $0 \leq x < 1/2$, and the right-hand side term of [1.20] is the proportion of the redundancy symbols in the codeword. The $w_{\min}/n$ ratio is thus lower bounded by an increasing function of the code redundancy rate, which tends towards 1/2 if k/n tends towards 0.

1.5.5. *A geometrical interpretation*

For a binary symmetric channel with a probability of error p and coding by blocks where the codewords have the length n, the number of erroneous binary symbols per codeword is a random variable F with Bernoulli distribution, that is:

$$\Pr(F = i) = C_n^i p^i (1-p)^{n-i},\ 0 \le i \le n.$$

Its average is $\mu(F) = np$ and its variance $\sigma^2(F) = np(1-p)$ (these results are obtained by deriving $(x+y)^n$, equalized with its development by the binomial formula, once and twice with respect to x, and then making $x = p$ and $y = 1-p$). The standard deviation $\sigma(F)$ is then only slightly less than $\sqrt{np}$. The probability $Pr(F > np + \lambda\sqrt{np(1-p)})$, with $\lambda > 0$, is lower than $1/\lambda^2$ according to the Bienaymé-Tchebychev inequality. As small as p may be, it is possible to choose n so that $\mu(F) = np$ is large. Then, the probability that the number f of errors actually occurring exceeds $\mu(F)(1+\varepsilon)$, where ε is a positive constant, is small, which is a manifestation of the weak law of large numbers.

The set E_n of binary n-tuples can be considered, from a geometrical point of view, as a space with n dimensions. Every n-tuple is a point of this space whose coordinates are binary. The distances between these points are measured using the Hamming metric introduced at the beginning of section 1.5.4. We will suppose that the decimation carried out to pass from the set of 2^n points of space to a subset comprising only 2^k points is such that its distribution of weight is close to the average weight distribution of random coding, without further exploring the means of carrying it out.

To guarantee a small probability of error it is enough to choose a sufficiently redundant code to achieve a minimum Hamming distance $d_{\min}$ between its codewords slightly larger than $2\mu(F)$, i.e. $d_{\min} = 2(1+\delta)np$ where δ is a positive constant independent of n. Indeed, as long as $f < d_{\min}/2$, the actually transmitted codeword can be identified without ambiguity as being closer to the received codeword than any other. The probability of having $f \ge d_{\min}/2$ can be increased according to the Bienaymé-Tchebychev inequality:

$$\Pr(F \ge d_{\min}/2) \le \frac{1-p}{np\delta^2} = \frac{p(1-p)}{n(d_{\min}/2n - p)^2}, \qquad [1.22]$$

upper limit which we make as small as we want by choosing a sufficiently large n. According to section 1.5.4, quasi-random coding guarantees that $d_{\min}/n$ satisfies the Gilbert-Varshamov bound [1.20]. According to [1.22] we thus obtain that $\Pr(F \ge d_{\min}/2)$ decreases as $1/n$ following the length of the codewords. This probability is greater than that of a decoding error, but in a coarsely exaggerated fashion because, if $f \ge d_{\min}/2$, such an error occurs only if in a space with n dimensions provided by the Hamming metric, the configuration of occurring errors goes exactly in the direction of another codeword, which is at the minimum distance $d_{\min}$ from the transmitted codeword. However, the redundancy of code means that its codewords are rare in E_n, and much more so are the codewords at a minimum distance from a given codeword, which makes this occurrence highly improbable. We will obtain much better bounds of the probability of error decreasing as a function of the exponential of n in the following section.

In this geometrical interpretation, random coding appears as a means of distributing the points in E_n as regular as possible, whereas in general we know of no deterministic means of obtaining this result. The improvement of the performance of the code through an increase of the length n of the codewords can be surprising, since it tends to render certain the presence of many errors in the codeword. The important fact is that the law of large numbers also makes the received codeword almost certainly localized on the surface of a Hamming sphere centered on the transmitted codeword, with a known radius np. If the spheres with np radius centered on all the transmitted codewords are not connected, it is enough to take an n large enough to render the probability of error as small as we wish. We will encounter this property again for the channel with additive white Gaussian noise considered in section 1.6.2, but the relevant metric there will be Euclidean.

1.5.6. *Fundamental theorem: Gallager's proof*

This section is dedicated to the proof of the fundamental theorem, introduced by Gallager [13], simplified thanks to certain restrictive assumptions usual in coding. This proof does not have a spontaneous nature, in the sense that it starts with an increase of the probability of error chosen so as to lead to the already known result sought, but it has the merit of providing very interesting details on the possible performances of block codes. In addition, it implements *random coding*, a basic technique introduced by Shannon for the proof of the theorem and already discussed in section 1.5.4. Here is what the simplifying assumptions consist of:

– The source connected to channel input is without memory and of equal probability, i.e. it chooses with an equal probability and independently of others one of the M possible messages. These messages may be the $M = q_s^k$ symbols of the kth extension of a q_s-ary source, itself deduced from the initial source by ideal source coding, in accordance with the standardization of blocks of the Shannon paradigm from section 1.2.4.

– To each message selected by the source in this manner, the encoder associates in a unique manner a codeword of n symbols belonging to the alphabet of channel input, whose size is noted q. In other codewords, coding consists of an application of integers from 1 to $M = q_s^k$ to the set of codewords with n symbols of this alphabet, with $q^n > q_s^k$ since coding must be redundant. This type of coding is called *block coding*.

– Less essentially, the channel is supposed to be without memory, this assumption being introduced during the proof.

1.5.6.1. *Upper bound of the probability of error*

Let X_n be the set of possible codewords of length n at the channel input and Y_n be the set of codewords that can be received at its output. We suppose finite X_n and Y_n. The channel is characterized by the set of its transition probabilities, $\{\Pr(\underline{y}|\underline{x})\}$, each one of them being the probability that to $\underline{x} \in X_n$ transmitted there must correspond $\underline{y} \in Y_n$ received.

A code of M words is used, defined by a bijective mapping of the set of message indices, i.e. integers from 1 to M, in that of M codewords, of length n, that is, $\{\underline{x}_1, \underline{x}_2, \dots, \underline{x}_M\}, \underline{x}_i \in X_n$ regardless of $i, 1 \leq i \leq n$. The emission of the mth codeword represents the mth message. We suppose that the reception occurs with maximum probability, i.e. the decoder associates the number m identifying the decoded message to the received sequence $\underline{y}$, if:

$$\Pr(\underline{y}|\underline{x}_m) > \Pr(\underline{y}|\underline{x}_{m'}) \ \forall\, m' \neq m, 1 \leq m' \leq M. \qquad [1.23]$$

An error occurs if for a transmitted m this decision rule leads to m' which is different from m. We can write the probability of an error by introducing the function $\phi_m(\underline{y})$ defined as follows:

$$\phi_m(\underline{y}) = \begin{cases} 1 & \text{if there exists } m' \neq m \text{ such that } \Pr(\underline{y}|\underline{x}_m) \leq \Pr(\underline{y}|\underline{x}_{m'}), \\ 0 & \text{if not,} \end{cases}$$

that is:

$$P_{em} = \sum_{\underline{y} \in Y_n} \Pr(\underline{y}|\underline{x}_m)\phi_m(\underline{y}), \qquad [1.24]$$

where P_{em} is the probability of an error when m is transmitted.

We now introduce an upper bound of $\phi_m(\underline{y})$, that is:

$$\phi_m(\underline{y}) \leq \left[\frac{\sum_{m' \neq m} \Pr(\underline{y}|\underline{x}_{m'})^{1/(1+s)}}{\Pr(\underline{y}|\underline{y}_m)^{1/(1+s)}} \right]^s, \quad s > 0, \qquad [1.25]$$

where s is a positive parameter, which is arbitrary for the moment. It is indeed an upper bound of $\phi_m(\underline{y})$, since:

– if $\phi_m(\underline{y}) = 0$, the right-hand side of the inequality is always positive;

– if $\phi_m(\underline{y}) = 1$, the numerator is by definition larger that the denominator, since it is the sum of positive terms including one, which according to [1.23] is larger than or equal to it. The expression between brackets is thus larger than 1, a property that remains after the expression is taken to the positive power of s.

According to [1.24], replacing $\phi_m(\underline{y})$ by its upper bound [1.25], we obtain:

$$P_{em} \leq \sum_{\underline{y} \in Y_n} \left\{ \Pr(\underline{y}|\underline{x}_m)^{1/(1+s)} \left[\sum_{m' \neq m} \Pr(\underline{y}|\underline{x}_{m'})^{1/(1+s)} \right]^s \right\}, \quad s > 0. \qquad [1.26]$$

Naturally, the upper bound [1.25] does not come from a spontaneous idea but from prior knowledge (by other means) of the result, at which we are trying to arrive. It leads to [1.26], an upper bound dependent on the parameter s, which one can thus adjust to obtain the tightest possible bound.

1.5.6.2. *Use of random coding*

This upper bound is valid for any code, but it is generally too complicated to be useful directly. To go further it is necessary to draw on the method of *random coding* already mentioned. We no longer consider a single code (how could code be specified *a priori* which would minimize the probability of error?) but a *probabilistic set* of codes. It will be easy to calculate an upper bound of its average probability of error. It will then be shown that if the inequality [1.17] is satisfied, the upper bound of this average error probability can be made lower than a certain positive ε. From that we will deduce that this set contains at least one code whose error probability is in turn lower than ε. That also means, less rigorously, that a code "close" to the average result of random coding must be "good" in the sense of the fundamental theorem, which legitimates the use of a quasi-random code introduced in section 1.5.4.

According to this method, random coding consists of choosing each codeword independently of the others with a probability $P(\underline{x})$ defined over the set of sequences of input X_n. We will calculate an upper bound of the average $\overline{P}_{em}$ of P_{em} with respect to the set of codes constructed in this manner.

Conditional probabilities in the right-hand side of [1.26] then become *random variables* in the sense that they depend on the codewords $\underline{x}_m$ and $\underline{x}_{m'}$ that have become random. They have been selected independently of each other as belonging to the code, so that these conditional probabilities are independent random variables. We will represent the averages by superscript bars in the rest of this section.

We can then note that in [1.26]:

– the average of the sum with respect to $\underline{y}$ is equal to the sum of averages of the terms between the curly brackets;

– as each one of these terms is the product of two independent factors, its average is equal to the product of averages;

– restricting ourselves to $s \leq 1$, the convexity $\cap$ of the z^s function implies $\overline{z^s} \leq \overline{z}^s$;

– finally, we may still invert sum and average, after replacing the term in the form $\overline{z^s}$ by its upper bound $\overline{z}^s$.

We thus deduce from [1.26]:

$$\overline{P}_{em} \leq \sum_{\underline{y} \in Y_n} \left\{ \overline{\Pr(\underline{y}|\underline{x}_m)^{1/(1+s)}} \left[\sum_{m' \neq m} \overline{\Pr(\underline{y}|\underline{x}_{m'})^{1/(1+s)}} \right]^s \right\}, \quad 0 < s \leq 1. \qquad [1.27]$$

The codewords $\underline{x}_m$ being chosen by random coding with the probability $P(\underline{x})$, by the definition of the average we have:

$$\overline{\Pr(\underline{y}|\underline{x}_m)^{1/(1+s)}} = \sum_{\underline{x} \in X_n} P(\underline{x}) \Pr(\underline{y}|\underline{x})^{1/(1+s)}, \qquad [1.28]$$

an expression independent of $\underline{x}_m$ and thus also valid for $\underline{x}_{m'}$. After substitution according to [1.28], [1.27] becomes:

$$\overline{P}_{em} \leq (M-1)^s \sum_{\underline{y} \in Y_n} \left[\sum_{\underline{x} \in X_n} P(\underline{x}) \Pr(\underline{y}|\underline{x})^{1/(1+s)} \right]^{1+s}, \quad 0 < s \leq 1. \qquad [1.29]$$

This bound is very general; it is valid regardless of the probability $P(\underline{x})$ and for channels "with memory" where errors for successive symbols are not independent. In the case of a channel without memory, in the sense that the successive errors in it are independent, this limit can be simplified. Let $x_1, x_2, \ldots, x_n$ be the symbols of the codeword at channel input and $y_1, y_2, \ldots, y_n$ be their corresponding symbols at the output. The independence of the successive transitions in the channel leads to:

$$\Pr(\underline{y}|\underline{x}) = \prod_{i=1}^{n} \Pr(y_i|x_i), \forall\, x \in X_n, \forall\, y \in Y_n.$$

Restricting ourselves to codes where the successive codeword symbols are chosen by random coding independently of each other, following the same law $p(x_i)$, we have:

$$P(\underline{x}) = \prod_{i=1}^{n} p(x_i), \; \underline{x} \triangleq (x_1, x_2, \ldots, x_n)$$

and the expression in brackets in [1.29] can be transformed. The sum then relates to the products in the form:

$$\prod_{i=1}^{n} p(x_i) \Pr(y_i \mid x_i)^{1/(1+s)}$$

corresponding to all the possible choices of $\underline{x}$ in X_n, i.e. with all the possible codewords with n symbols in the channel input alphabet: the sum of all the products of n terms is equal to the nth power of the sum of the terms written for all the symbols of this alphabet.

Indicating by $a_1, a_2, \ldots, a_q$ the symbols of the channel input alphabet (of size q), and by $b_1, b_2, \ldots, b_J$ those of the output alphabet (of size J), we thus deduce from [1.29] the limit:

$$\overline{P}_{em} \leq (M-1)^s \left\{ \sum_{j=1}^{J} \left[\sum_{k=1}^{q} p(a_k) \Pr(b_j|a_k)^{1/(1+s)} \right]^{s+1} \right\}^n, \quad 0 < s \leq 1.$$

Increasing $M - 1$ by $M = 2^{nR}$, where $R = (k/n) \log_2 q$ is the quantity of information per symbol in shannons, we obtain:

$$\overline{P}_{em} < 2^{-n[-sR+E_0(s,\underline{p})]}, \quad 0 < s \leq 1, \qquad [1.30]$$

where we have posed:

$$E_0(s,\underline{p}) \triangleq -\log_2 \left\{ \sum_j \left[\sum_k p(a_k) \Pr(b_j|a_k)^{1/(1+s)} \right]^{1+s} \right\}. \qquad [1.31]$$

This function depends, on the one hand, on the parameter s and, on the other hand, on the vector $\underline{p}$ having the components $p(a_1), p(a_2), \ldots, p(a_q)$. We note that the bound obtained is independent of the transmitted message m.

1.5.6.3. *Form of exponential limits*

The bound [1.30] is important, because within the limits of its conditions of validity it leads to the existence of a code such that its word error probability is limited for a given R by:

$$P_e < 2^{-nE(R)}, \qquad [1.32]$$

where n is the length of the codewords and where:

$$E(R) \triangleq \max_{s,\underline{p}}[-sR + E_0(s,\underline{p})], \quad 0 < s \leq 1. \qquad [1.33]$$

$E(R)$ is called the *reliability function*.

Without getting into detail of the discussion of [1.33] (which the reader will find in [14]), we can say that the curve representing the exponent $E(R)$ as a function of R appears as the envelope of the straight lines of slope $-s$ and ordinate at the origin $\max_{\underline{p}}[E_0(s,\underline{p})]$ in the interval of variation of $s, 0 < s \leq 1$ ($E_0(s,\underline{p})$ was defined in [1.31]). Apart for the teratological exception, this envelope is decreasing and convex

∪. For the smallest values of R, it merges with the straight line of slope -1, of the equation $E(R) = R_0 - R$, where $R_0 = \max_{\underline{p}}[E_0(1, \underline{p})]$. For example, in the case of the binary symmetric channel, the largest upper bound of the probability of error is obtained [1.32] for $s = 1$ (i.e. in the range of values of R where the envelope of the straight lines $E = -sR + E_0$ merges with the straight line of slope -1 and ordinate at the origin R_0) is written:

$$P_e < 2^{-n(R_0 - R)},$$

where, according to [1.31], $R_0 = 1 - \log_2(1 + 2\sqrt{p(1-p)})$. This exponential upper bound is much tighter than the one that can be deduced from [1.22].

Beyond a certain value of R, the absolute value s of the slope of the tangent to the curve representing $E(R)$, initially equal to 1, decreases and tends towards 0, the curve becoming tangent with the x-axis at the point $R = C$ for $S = 0$, where:

$$C = \max_{\underline{p}} I(X;Y)$$

is the capacity of the channel, [13,14]. In the case of a binary symmetric channel with probability of error p, an easy calculation shows that this capacity is equal to $1 - \mathcal{H}_2(p)$, where the function $\mathcal{H}_2(\cdot)$ has been defined by [1.21].

We see that the factor of n in the exponent of [1.32] is negative only if $R < C$, which is thus necessary in order to obtain a probability of error tending towards 0 when n tends towards infinity. It is the statement of the fundamental theorem (without the notations, but $R = (k/n)\log_2 q$ is still entropy of the channel input variable and, therefore, of the source). Moreover, in addition to this asymptotic result, it shows how the word error probability varies according to their length n. It is clear that to obtain the same word error probability (or rather the same bound [1.32] of this probability) n needs to be larger as R becomes closer to C. However, the length n of the codewords measures the complexity of the coding and decoding operations, which for block random coding is an increasing exponential function.

We can improve the bound [1.32] for the majority of channels, but only for the smallest values of R, by operating a selection of the codes resulting from random coding to preserve only the best (we are said to "expurgate" the set of codes). Then the curve representing the function $E(R)$ obtained is still decreasing and convex ∪. It does not deviate from the straight line of the equation $E(R) = R_0 - R$ apart from beyond a certain point where it is tangential to it and grows quicker than this straight line when R decreases to, finally, tangentially reach the y-axis at $R = 0$.

1.6. Channels with continuous noise

1.6.1. *Introduction*

Up until now we could satisfy ourselves with the description of a discrete channel by its transition probabilities and took the example of the binary symmetric channel, which is the simplest model there is. We will now consider some more realistic channel models.

One of the principal restrictions made above relates to the finite character of the channel input and output alphabets. For the input alphabet it stems naturally from the choice made to limit ourselves to finite discrete, i.e. digital, sources. The omnipresence of thermal noise, modeled well by the addition of white Gaussian noise, makes the assumption of a finite output alphabet exaggeratedly restrictive. We will thus devote a particular development to the channel with additive Gaussian noise. The noise will be initially supposed to be white, but this restriction will be easily raised. More briefly, we will also consider the channel with fadings where the received signal undergoes fluctuations represented by a Rayleigh process multiplying its amplitude before adding white Gaussian noise.

1.6.2. *A reference model in physical reality: the channel with Gaussian additive noise*

Capacity of a channel with additive white Gaussian noise

Before examining the case of a finite number of input signals disturbed by addition of white Gaussian noise, which interests us mainly, we will make a detour by the case where the channel input variables are themselves continuous. Let X be a continuous random variable with probability density function $p_X(x)$, i.e. the probability that the value taken by X belongs to the infinitesimal interval $(x, x+\mathrm{d}x)$ is equal to $p_X(x)\mathrm{d}x$. The definition [1.2] of the entropy of a discrete random variable is no longer usable in this case but, by analogy, for the continuous variable X we define the differential entropy:

$$H_{\mathrm{d}}(X) \triangleq -\int p_X(x)\log[p_X(x)]\mathrm{d}x, \qquad [1.34]$$

where the integral is calculated for the set of values taken by X, for example, the set of real numbers. This value has some but not all of the properties of the entropy of a discrete variable. Thus, it can be negative and loses certain properties of invariance. Its principal interest lies in the fact that the mutual information generalized to continuous variables is still expressed, as in [1.5] or [1.6], by a difference between two entropies which are now differential.

By analogy with the discrete case where the capacity of a binary symmetric channel is equal to mutual information for the probability distribution of the input symbols making their entropy maximal, we will admit that the capacity of the channel is obtained for the distribution of input variables, which is here continuous, which renders the differential entropy maximal. It is demonstrated that for a given finite variance σ^2 this distribution is Gaussian. To reach the capacity we will admit that the distribution of the channel input variables must be such.

If X is a zero-mean Gaussian random variable σ^2, by definition it has a probability density function:

$$p_X(x) = \frac{1}{\sigma\sqrt{2\pi}} \exp(-x^2/2\sigma^2). \quad [1.35]$$

The calculation by [1.34] of its differential entropy (in shannons) yields the result:

$$H_{\mathrm{d}}(X) = \log_2 \sigma + (1/2) \log_2(2\pi \mathrm{e}), \quad [1.36]$$

where e is the base of the Napierian logarithms. It is thus equal, to the nearest constant, to the logarithm of the standard deviation σ of X, i.e. to the half of the logarithm of its variance σ^2. The mutual information between the input and output variables (both Gaussians) expressed by the difference between differential entropies at output, respectively not conditional and conditional at input, is thus equal to half of the logarithm of the ratio of the corresponding variances. However, the addition of Gaussian noise to the channel input variable X gives a variable Y which is also Gaussian with variance equal to the sum of those of channel noise and input signal. Indeed, the sum of two Gaussian variables is a Gaussian, the noise and the signal are independent so that their variances are added and, in addition, the differential entropy of Y conditionally to the input variable X is equal to the differential entropy of noise because it is additive.

The assumption of a source without memory has its equivalent here in the limitation of the band to a certain value B, so that the sampling theorem makes it possible to represent exactly and reversibly any signal pertaining to the set of functions in a band limited to B by the sequence of values which it takes with periodic intervals, known as *samples*. The period of sampling must be $T = 1/2B$. These samples are random statistically independent variables and, with our assumptions, Gaussian, centered and of variance $P/2B$, where P is the power of the received signal. Thus, thanks to the discretization of time realized by the sampling of interval T, we find the same diagram of communication as in our introduction, with the difference that the channel input alphabet has become the entire set of real numbers. Its symbols, the samples, undergo the sole disturbance of the addition of noise present in the band B. If we suppose that the one-sided power spectral efficiency of this noise has a constant value N_0 (the noise is then referred to as "white") these are, as free-noise samples, Gaussian variables, centered and mutually independent. Their variance is $N_0/2$.

The input samples have a variance of $P/2B$, those of additive noise $N_0/2$ and the output samples the sum of these variances or $(P + N_0B)/2B$. The capacity of this channel is thus equal to:

$$C = \frac{1}{2}\log_2\frac{P + N_0B}{N_0B} = \frac{1}{2}\log_2(1 + \frac{P}{N_0B}) = \frac{1}{2}\log_2(1 + \frac{P}{N}). \qquad [1.37]$$

In this expression, the signal to noise ratio P/N appears in the argument of the logarithm since $N \triangleq N_0B$ is the total noise power in the band B. This is the capacity by symbol (or sample); the capacity in shannons by second stems from that by multiplying it by the frequency $2B$ of samples, that is:

$$C' = B\log_2\left(1 + \frac{P}{N}\right) = B\log_2\left(1 + \frac{E_bR}{N_0B}\right) \qquad [1.38]$$

where the notation C' is employed to indicate that the capacity is expressed here as information flow, i.e. a quantity of information per unit of time.

Expression [1.38] of the capacity of an additive Gaussian channel is justly famous but sometimes erroneously interpreted. It has paradoxical consequences for the role of the bandwidth in communications through a channel with additive white Gaussian noise. Indeed, [1.38] shows that, C' being an increasing function of B for P and N_0 kept constant, it is necessary to increase the band to increase the capacity, although that involves a reduction of the signal to noise ratio P/N_0B. This conclusion is radically opposed to the dominant trend in traditional radio-electronics, where apparent common sense suggests limiting the bandwidth as much as possible in order to reduce the noise entering the receiver. It is true that the channel coding function[2] was then unknown, although it alone can exploit the increase in capacity due to band widening.

Generalization to the case where the noise power spectral density varies in the signal band is easy (the noise is then known as colored). It suffices to divide the band into infinitesimal intervals, each considered as defining a channel with additive white Gaussian noise and to treat these channels in parallel. The total capacity is equal to the sum of the capacities of the constituent channels. The calculation of variations indicates how to maximize it when the total power of the useful signal is given: the spectral density of this signal must be such that we obtain a constant by adding it to that of the noise. This result can be expressed by an image: if we assimilate the noise spectral density, variable in the band, to the thickness of the bottom of a container, all occurs as if the optimal spectral density of the useful signal were to compensate for the variable thickness of the bottom so that the total spectral density is represented

2. At least explicitly. The "modulation gain" brought by certain systems, such as frequency modulation, at the cost of widening the band, in fact, results from a form of channel coding, misunderstood by radio-electricians prior to the birth of information theory.

by a horizontal, which we could obtain by pouring a liquid whose volume would represent the total power of the signal (a result introduced by Shannon, often called *water-filling*).

1.6.3. *Communication via a channel with additive white Gaussian noise*

We note that the capacity [1.37] or [1.38] is finite, although the alphabet of the channel is continuous. Therefore, communicating via an additive white Gaussian noise channel with a flow of information of R' shannons per second ($R' < C'$), implies the use of a repertory of $M = 2^{R'T}$ signals of band B and duration τ, where τ is large compared with the signal interval. Each signal is associated by a bijective relation to one of the M source messages. The set of samples (with real values) present in the time interval τ can be regarded as a codeword which can itself be represented by a point in a space with $\mathcal{D} = 2B\tau$ dimensions, having for coordinates all its samples in the interval τ. We show that the relevant metric is then Euclidean [15], with an energy meaning. The rule of optimal decision stays the choice of the codeword represented by the point nearest to that representing the received signal, for this metric. Reasoning homologous to that in section 1.5.5, with the exception that the Euclidean metric replaces that of Hamming, leads to similar conclusions. In fact, this geometrical representation of Shannon makes it possible to directly prove that a negligible probability of decoding error can be achieved with random coding, when the number of dimensions $\mathcal{D}$ tends towards infinity, if the flow of information remains smaller than the capacity given by [1.38]. Thus we prove the fundamental theorem for this particular channel [16].

1.6.3.1. *Use of a finite alphabet, modulation*

The process of communication through a channel with additive white Gaussian noise, which we have just considered, has only a theoretical interest, because its implementation would be exaggeratedly complicated. Indeed, it is necessary to employ a repertory of M signals where, for a given flow of information R', M varies as an exponential function of the duration τ, which must be large so that a small probability of error is obtained. In practice it is necessary to use an alphabet comprising q symbols, the M necessary messages being obtained by combinations of n of them. To satisfy the condition of redundancy we take $M = q^k$ with $k < n$, i.e. a block code of the type that has been considered in section 1.5.4 for $q = 2$. Each symbol of the alphabet is separately represented by a specific signal, which is the modulation operation[3]. It makes it possible to use a code built on the basis of a finite alphabet in a channel receiving continuous signals. The capacity of such a channel is obviously limited to

3. Even if the signal thus obtained does not rigorously conform with the assumption of band limitation, the geometrical representation of signals in an Euclidean space with a finite number of dimensions remains, at the cost of a redefinition of the bandwidth; see, for example [1], p. 135.

$\log_2 q$ shannons per sample, which is the asymptotic value for large P/N. Coding is only useful if the signal to noise ratio is small enough so that the capacity [1.37] is definitely lower than this value. We will note that the capacity calculated supposing an alphabet of finite size q is very close to [1.37] for the smallest values of the signal to noise ratio; the curves representing it according to P/N are indeed tangential at the origin. More refined means of distributing points in Euclidean space using codes defined for finite alphabets will be shown in Chapter 4.

1.6.3.2. *Demodulation, decision margin*

The continuous character of the received signals requires detailed attention. Indeed, let us suppose, for example, that the alphabet is binary ($q = 2$) and that the process of modulation (known as "antipodal", optimal for this alphabet) consists of transmitting a signal of a certain form compatible with the properties of the channel, in order to represent one of the symbols, for example 0, and the opposed signal to represent the other, i.e. 1. Let $s(t)$ and $-s(t)$ be the corresponding received signals with energy $E = \int s^2(t)\,dt$, integration taking place on the support of $s(t)$. The optimal reception in the sense of maximum probability using a correlator or a matched filter results in a real number called sufficient statistics which is a Gaussian variable Y of probability density function $p_Y(y) = g[y - (-1)^x\sqrt{E}; \sigma^2]$, where x is the binary value 0 or 1 taken by the transmitted symbol X and where $g(\cdot; \sigma^2)$ denotes here the Gaussian probability density function [1.35]. Its variance is that of the additive noise, that is $\sigma^2 = N_0/2$. For input symbols of equal probability, the ratio of the probability that $x = 0$ has been transmitted to that of $x = 1$ being transmitted, called the probability ratio, conditionally to the observation y of the channel output (including the correlator or adapted filter), is equal to $\exp(4y\sqrt{E}/N_0)$, so that its logarithm is proportional to the observation y. The optimal decision for the transmitted symbol is thus $\hat{x} = 0$, if y is positive, and $\hat{x} = 1$ if y is negative (the case $y = 0$ makes a decision impossible; we can merely arbitrarily choose a value of $\hat{x}$ with a probability of error equal to $1/2$).

With regard to decoding, two unequally effective and complex strategies are possible:

– make a *hard decision* regarding each binary symbol transmitted according to the sign of received variable y. We are then brought back to the problem considered in Paragraph 1.5.4, since the initial channel with continuous output is converted into a binary symmetric channel;

– preserve and exploit in the decoder the real value y of the sufficient statistic which, as we saw, is equal to the nearest positive factor to the logarithmic probability ratio $\log \frac{\Pr(X=0)}{\Pr(X=1)} = \log \frac{\Pr(X=0)}{1-\Pr(X=0)}$, i.e. provides information on the probability of the transmitted symbol. In this case we speak of a *soft decision*, although this is rather a case of an absence of an explicit decision.

The second strategy exploits information that the first strategy lacks (carried by the margin $|y|$ of each decision providing information on its reliability). It is thus more effective in principle: the calculation of the corresponding channel capacity indeed shows an advantage in its favor by a factor that varies according to the signal to noise ratio, from $\pi/2$ (if E/N_0 is very small) to 2, an asymptotic value when E/N_0 tends towards infinity. In practice, the same decoding error probability is obtained, with the same code, for a difference of the signal to noise ratios of about 2 dB (for small values of this ratio), in conformity with the ratio of the capacities ($10\log_{10}(\pi/2) = 1.96$ dB). However, the implementation of this second strategy is more difficult, because it requires the decoder to deal with real numbers: the logarithmic probability ratios as they are given by the demodulator before any binary decision, and not the symbols of the input alphabet; this is referred to as decoding with soft or balanced decisions. However, an important branch of the studies of channel coding is based on the algebraic properties of finite bodies. The use of soft decisions prohibits treating decoding as an algebra problem, even though the construction of the code makes use thereof.

1.6.4. *Channel with fadings*

We suppose now that the signal is received through a channel "with fadings", where its amplitude is multiplied by a random stationary variable A that follows the Rayleigh law before the addition of a white Gaussian noise with one-sided spectral density N_0. A random variable A of unitary variance following the Rayleigh law has as a probability density function:

$$p_A(a) = \begin{cases} 2a\exp(-a^2), & a > 0, \\ 0, & a \leq 0. \end{cases} \qquad [1.39]$$

It is the probability density function of the absolute value of a complex signal whose real and imaginary parts are independent Gaussian random variables, centered and with the same variance 1/2.

We suppose that the signal is transmitted with constant amplitude. It is, therefore, modulated only in phase. The received average power is equal to P and we admit that the signal band remains limited to B (this assumption is not rigorously exact, but can be allowed by way of an approximation). An interleaving and disinterleaving device provides the successive samples of the received signal, after disinterleaving sufficiently distant in time in the channel to be regarded as independent.

These samples are none other than those of one of the two components in signal quadrature. Since the amplitude of the signal follows the Rayleigh law [1.39], they have a centered Gaussian probability density function and, since they are made independent by interlacing, we are brought back to the problem of a sequence of independent Gaussian samples with power P received at the frequency $2B$ in the presence of additive Gaussian noise with variance $N_0/2$, i.e. to the same problem as that of a Gaussian signal received in the presence of additive white Gaussian noise. We have already calculated the capacity [1.38] of this channel.

The presence of fadings of Rayleigh thus does not modify the capacity, although it complicates the reception notably and, in practice, often degrades the result. The conservation of the capacity of the channel with additive Gaussian noise in the presence of Rayleigh fadings suggests that it must be possible to arbitrarily reduce the degradation which they cause. In fact, rotation operators in Euclidean space $\mathcal{R}^n$ achieve that for sufficiently large n thanks to an effect of "diversity"[4] [17]. When auxiliary devices, in particular, those of interleaving and diversity, make it possible to effectively employ error correcting codes, we will note that the benefit of coding, measured by the increase in the signal to noise ratio needed without coding to obtain the same probability of error as with coding, is much more important for the channel with fadings than where the only disturbance is the addition of Gaussian noise. Indeed, the probability of error in the absence of coding decreases when the signal to noise ratio increases a lot less quickly in the presence of signal fadings, and the benefit brought by the system of coding is a reduction of the probability of error.

$$C = E_A\left\{\frac{1}{2}\log_2\left(1 + A^2\frac{P}{N}\right)\right\}$$

1.7. Information theory and channel coding

The presence of disturbances in the channel limits the possible flow of information, but not the quality with which the message can be restored: this lesson of information theory created the basis for channel coding. The limitation of the flow of information to a value smaller than the channel capacity is achieved by the introduction of redundancy, but it is only one of the conditions necessary to control the error rate upon decoding. The entire problem of channel coding lies in the manner of doing it.

The ultimate possible limit in information theory, i.e. channel capacity, has long appeared inaccessible and the assertion "*All codes are good, except those we can think off*" expresses in humorous form an opinion that was until recently dominant. Turbo-codes and the extraordinary torrent of research that they unleashed have contradicted this assertion and now it is in tenths of a decibel that we express the variation of the best experimental results with respect to capacity, for a channel with additive white Gaussian noise.

4. Which consists of jointly exploiting several supports of the same information.

The key question of channel coding is the complexity of decoding. If we could get rid of it, almost any code would be satisfactory. Indeed, not only random coding is good on average, but it is known that almost all codes are good. Unfortunately, the complexity of decoding for a random code increases exponentially with the length of the code, which must be large in order to obtain small probabilities of error. Efficient use of random coding is thus absolutely out of the question; we can only hope to employ codes provided with a structure which facilitates their decoding. Coding being incomparably easier than decoding, two extreme manners of undertaking the study of channel coding were conceivable and both were actually tried out:

– to seek at first to build codes provided with good distance properties, deferring to a later stage the more difficult problem of decoding them;

– to first resolve the problem of decoding, risking the properties of the codes to remain unexploited; it is, of course, necessary that they have a minimum of structure, but the linearity, whether they are block or convolutional codes, is enough for general decoding algorithms to be designed.

Very schematically, the first tendency gave rise to algebraic codes and the second led to the development of convolutional codes, for which the principal results are, in fact, not families of codes, but decoding algorithms.

The results of these studies were initially confronted with reality in space communication applications, where the channel is well modeled by the addition of white Gaussian noise, and where the improvement of coding and decoding devices that the immense progress of electronic technology now allows with reliability and economy costs much less than the improvement of the energy cost of the connection. We saw that the weighting of decoding avoids a costly loss of information. The ease of its implementation in algorithms stemming from the second tendency is the main reason for its success.

Research in channel coding in the simplest cases (binary symmetric channel and channel with additive white Gaussian noise with a weak signal to noise ratio) have produced an impressive arsenal of tools. Apart from the important exception of the Reed-Solomon codes, these are mainly binary codes. For other, often much more complicated, channels in general we still employ the means created in this manner, but auxiliary techniques (interleaving diversity ...) are needed to adapt them to the characteristics of the channel.

Regarding the channel with additive white Gaussian noise, the capacity C' given by [1.38] exceeds the limit of $2B$ shannons per second, intrinsic for the binary alphabet, when the signal to noise ratio is large. Non-binary codes, or means of combining binary codes with modulation processes with more than two states, such as "multilevel" coding or lattice-coded modulations that will be seen in Chapter 4, must then be used to increase the flow of information beyond this limit.

1.8. Bibliography

[1] BATTAIL G., *Théorie de l'information. Application aux techniques de communication*, Masson, 1997.

[2] SHANNON CE., "A mathematical theory of communication", *BSTJ*, Vol. 27, pp. 379–457 and pp. 623–656, July and October 1948. These articles have been reprinted with a commentary of W. Weaver in the form of a book entitled *The mathematical theory of communication*, University of Illinois Press, 1949.

[3] WYNER A.D., "Another look at the coding theorem of information theory — A tutorial", *Proc. IEEE*, Vol. 58, No. 6, pp. 894–913, June 1970.

[4] BERGER T., *Rate-distorsion theory : a mathematical basis for data compression*, Prentice Hall, 1971.

[5] MOREAU N., *Techniques de compression des signaux*, Masson, 1995.

[6] BERROU C., A. Glavieux, "Near optimum error-correcting coding and de-coding: turbo-codes", *IEEE Trans. Corn.*, Vol. 44, No. 10, pp. 1261–1271, Oct. 1996.

[7] KOLMOGOROV A.N., "Logical basis for information theory and probability theory", *IEEE Trans. on Inf. Th.*, Vol. IT-14, No. 5, pp. 662–664, Sept. 1968.

[8] CHAITIN G.J., *Algorithmic Information Theory*, Cambridge University Press, 2nd revised edition, 1988.

[9] HUFFMAN D.A., "A method for the construction of minimum redundancy codes", *Proc. IRE*, Vol. 40, pp. 1098–1101, 1952.

[10] RISSANEN J.J., LANGDON G.G., Jr. "Arithmetic coding", *IBM J. Res. & Dev.*, Vol. 23, No. 2, pp. 149–162, March 1979.

[11] GUAZZO M., "A general minimum-redundancy source-coding algorithm", *IEEE Trans. on Inf. Th.*, Vol. IT-26, No. 1, pp. 15–25, Jan. 1980.

[12] PETERSON W.W., WELDON E.J., Jr., *Error-correcting codes*, 2nd edition, MIT Press, 1972.

[13] GALLAGER R.G., "A simple derivation of the coding theorem and some applications", *IEEE Trans. on Inf. Th.*, Vol. IT–13, No. 1, pp. 3–18, Jan. 1965.

[14] GALLAGER R.G., *Information theory and reliable communication*, Wiley, 1968.

[15] SHANNON CE., "Communication in the presence of noise", *Proc. IRE*, pp. 10–21, Jan. 1949.

[16] WOZENCRAFT J.M., JACOBS I.M., *Principles of communication engineering*, Wiley, 1965.

[17] BOUTROS J., VITERBO E., RASTELLO C., BELFIORE J.C., "Good lattice constellations for both Rayleigh fading and Gaussian channels", *IEEE Trans. on Inf. Theory*, Vol. 42, No. 2, pp. 502–518, March 1996.

Chapter 2

Block Codes

2.1. Unstructured codes

2.1.1. *The fundamental question of message redundancy*

We wish to transmit messages from point A to point B through space (transmission channel), or from point A to point A through time (recording channel). Any transmission of information is a voluntary energy modulation. The channel which allows the transmission is traversed by random energy impulses. This parasitic energy produces transmission errors: noise. In a binary transmission, 1 is transformed into 0, and conversely. When we have difficulties transmitting a word or a message because of the noise, we naturally tend to repeat the word or the message. It is then said that we add redundancy to the information. Now, let us consider that the message to be transmitted is coded into binary, i.e. it consists of a sequence of 1 and 0.

One of the first problems that had to be dealt with during World War II was how to contact the American spies in hostile German territory. The spies could not ask for retransmission for fear of being discovered. If the message was short it was completely destroyed by jamming if the jamming affected it. If the message was reinforced with redundancy, there was more chance than it would be affected, but it was less susceptible. The question that would then arise was the following: was a lot or little redundancy necessary for the security of these transmissions?

The answer was provided by C. Shannon (1948). He created the information theory, which led him to formalize the problem, to define a mathematical measure of

---

Chapter written by Alain POLI.

information, and to give an answer to the question at hand in the form of very fine theorems. His answer was: it is necessary to add "enough" redundancy to be statistically sure of the effectiveness of protection. The lower bound of this quantity is determined on the basis of a channel characteristic: capacity. C. Shannon proved that if we added "enough" redundancy, there was a coding which made it possible to have a statistically reliable transmission. The empirical proof was provided very quickly (before 1950) by Hamming, Golay, and others who offered examples of codes constructions.

2.1.2. *Unstructured codes*

In the rest of this section we restrict ourselves to binary codes, i.e. with coefficients in $F_2 = \{0, 1\}$, unless otherwise mentioned. Each codeword has the same length n. These are "block" codes.

We wish to code a set of messages. Each message, or information word, is coded by a binary word (codeword). The set of codewords is called the *code*.

DEFINITION 2.1 (HAMMING DISTANCE). *Let there be two n-tuples $x = (x_1, x_2, \dots, x_n)$ and $y = (y_1, y_2, \dots, y_n)$. The Hamming distance between x and y is the number of positions where these two vectors are different. It is noted $d_H(x, y)$.*

DEFINITION 2.2 (HAMMING WEIGHT). *The Hamming weight of x is equal to the number of non-zero components. It is noted $w_H(X)$. It is also equal to $d_H(x, 0_n)$, where 0_n indicates the vector $(0, \dots, 0) \in (F_2)^n$.*

DEFINITION 2.3 (SPHERE WITH A CENTER x AND RADIUS ρ). *The sphere with its center at x and radius ρ, noted $B_\rho(x)$, is defined by: $B_\rho(X) = \{y \in (F_2)^n / d_H(x, y) \leq \rho\}$.*

DEFINITION 2.4 (EUCLIDEAN DISTANCE). *Let there be two n-tuples $x = (x_1, x_2, \dots, x_n)$ and $y = (y_1, y_2, \dots, y_n)$ whose components are real values in the interval $[-1, 1]$. The Euclidean distance between them is equal to:*

$$\left(\sum_{i=1,n} (x_i - y_i)^2 \right)^{1/2}$$

2.1.2.1. *Code parameters*

A code is a set of codewords, characterized by a family of parameters:

1) the length n of each codeword. It is also said that the code has a length n,

2) the number M of codewords. It characterizes the transmission capacity of the code,

3) the minimum distance d of the code. It is related to the capacity of correction of the code,

4) the maximum correction capacity per codeword, noted t,

5) the minimum weight of a code, noted w.

We speak of a code (n, M, d), or a (n, M, d) code.

EXAMPLE 2.1. In $(F_2)^5$ the family $\{10101, 00010, 01111, 11000\}$ is a code (5,4,3). Find w and d.

EXAMPLE 2.2. In $(F_3)^3$ the family $\{102, 110, 200, 121\}$ is a (3, 4, *d*) code. Find w and d.

EXAMPLE 2.3. Is the family $\{01101, 10120, 11012, 00000, 11111\}$ a (5, 4, *d*) code in $(F_3)^5$? If yes, find w and d.

2.1.2.2. *Code, coding and decoding*

In this section we introduce the concepts of code, coding and decoding with maximum probability.

DEFINITION 2.5 (CODE). *An unstructured binary code of length n is a family of vectors included in* $(F_2)^n$.

DEFINITION 2.6 (CODING). *Coding consists of associating a codeword, element of* $(F_2)^n$*, to an information word taken in* $(F_2)^k$*, for* $k < n$*. The most elementary coding is done using a coding table.*

DEFINITION 2.7 (MAXIMUM LIKELIHOOD DECODING). *It will be admitted that, in the usual cases we have:*

$$\begin{aligned}\text{Prob}(0 \textit{ error in a transmitted word}) &\leq \text{Prob}(1 \textit{ error}) \\ &\leq \text{Prob}(2 \textit{ errors}) \\ &\leq \cdots\end{aligned}$$

This assumption means that the channel is not too bad with respect to the length of codewords. We note that the distance from an transmitted codeword c *to the received word* $\mathrm{r} = \mathrm{c} + \mathrm{e}$ *is equal to the Hamming weight of the error vector* e*. The assumption made is thus equivalent to supposing that it is more probable that the distance between the transmitted word and the received word is* 0 *rather than* 1, 1 *rather than* 2, *etc. Maximum likelihood decoding thus implies decoding the word received by the nearest codeword. If this codeword is not unique we do not decode.*

2.1.2.3. *Bounds of code parameters*

The parameters n, M and d are connected with each other by various constraints. If two are fixed values, then the value of the third is limited by certain traditional inequalities. In general, we cannot calculate the best possible value for this third parameter.

A bound on M is as follows. What is the largest number of words M of length n in a code allowing the correction of t errors per word? The disjoint spheres are counted and total volume is calculated. It is necessary to have:

$$2^n \geq \left[1 + \binom{1}{n} + \binom{2}{n} + \cdots + \binom{t}{n}\right] M$$

EXAMPLE 2.4 ($n = 5, t = 1$). We have $32 \geq M[1 + 5] = 6M$. Thus, M is less than or equal to 5.

It should be noted that it may not be possible to obtain the value M.

Similarly, M being fixed, the best error correcting capability can be much lower than the value of t obtained from the previous formula.

2.2. Linear codes

As we can see it from the exercises in the preceding section, it is very difficult to construct unstructured codes. A code is equivalent to the data of a family of spheres with radius ρ, disjoint two by two. The number of spheres is M, and the code corrects ρ errors per word (with maximum likelihood decoding). The best possible code is equivalent to the best packing of spheres, which is a very complex problem. In order to be able to build codes more easily, we agree to lose some freedom by imposing an algebraic structure on the code. We will thus consider the binary codes with a particular property: stability during addition.

2.2.1. *Introduction*

These codes have a structure of vector subspaces of $(F_2)^n$. If the code C is a vector subspace of dimension k, it is said that the dimension of the linear code C is k. The number of words in C is then 2^k. From now on we will speak of a linear code (n, k, d) instead of $(n, M = 2^k, d)$.

2.2.2. *Properties of linear codes*

These codes have properties used for their decoding or construction.

2.2.2.1. *Minimum distance and minimum weight of a code*

The following proposition makes it possible to simplify obtaining the minimum distance when the code is linear.

PROPOSITION 2.1. *The minimum weight of a code is equal to its minimum distance.*

Proof. Indeed, the difference between two codewords of a linear code is a codeword of this code, and in addition we have: $d_H(x, y) = w_H(x - y)$. □

To know the correction capacity of a linear code of dimension k, it is enough to explore the weights of 2^k codewords instead of the $2^{k-1}(2^k - 1)$ distances between the codewords taken two by two.

2.2.2.2. *Linear code base, coding*

Let us simply demonstrate on an example a particular basic form of a code (systematic form).

EXAMPLE 2.5. In $(F_2)^5$ we take $e_1 = 10110, e_2 = 00101, e_3 = 11011$. It is a free family. We note that $\{e_1 + e_2, e_1 + e_2 + e_3, e_2\}$ is another base of $L(e_1, e_2, e_3)$.

In *systematic form* this base is described as:

10011
01000
00101

DEFINITION 2.8 (GENERATOR MATRIX). *We call a linear code generator matrix* $C(n, k, d)$ *any matrix whose rows are vector representations of a base of* C.

This matrix is in systematic form when it is written in the form $G = (I_k R)$, *where* (I_k) *is the identity matrix of rank k, or when* $G = LI_k$

DEFINITION 2.9 (SYSTEMATIC CODING). *Systematic coding corresponds to the following operation:*

$$(i_1, i_2, \ldots, i_k)G = (i_1, i_2, \ldots, i_k)(I_k R)$$
$$= (i_1, i_2, \ldots, i_j, \ldots, i_k, r_{k+1}, \ldots, r_l, \ldots, r_n)$$

The i_j *are information bits, and the* r_l *are redundancy symbols.*

EXAMPLE 2.6 ($n = 6$, $e_1 = 101101$, $e_2 = 111011$, $e_3 = 101100$). Construct G. Put G in systematic form. Encode (101) with G. Encode (101) with (I_3R). Encode (a, b, c) with the two matrices, and compare.

2.2.2.3. *Singleton bound*

The following proposition introduces the Singleton inequality and the Singleton bound.

PROPOSITION 2.2. *Let* (n, k, d) *be a linear code* $C(n, k, d)$. *We have the inequality (called Singleton inequality):* $d \leq n - k + 1$.

Proof. Consider a generator matrix in systematic form. □

2.2.3. *Dual code*

The code C being a vector subspace it admits an orthogonal, noted $C^{\perp}$.

PROPOSITION 2.3. *If* $(I_k R)$ *is a generator matrix of* C*, then* $H = (-R^T I_{n-k})$ *is a generator matrix of* $C^{\perp}$*, known as a parity check matrix of* C.

Proof. The verification is direct. □

A generator matrix of $C^{\perp}$ is referred to as *a parity check matrix* of C.

2.2.3.1. *Reminders of the Gaussian method*

To pass from a generator matrix of C to a parity check matrix (and reciprocally) we often use the method of Gaussian pivots.

EXAMPLE 2.7. In F_2 we take:

$$G = \begin{pmatrix} 111000 \\ 011101 \\ 011101 \\ 100111 \end{pmatrix}$$

We find:

$$H = \begin{pmatrix} 110100 \\ 110001 \end{pmatrix}$$

with 1 permutation of columns.

EXAMPLE 2.8. In F_3 we take:

$$G = \begin{pmatrix} 21012 \\ 12101 \\ 20212 \end{pmatrix}$$

We find:

$$H = \begin{pmatrix} 21210 \\ 20001 \end{pmatrix}$$

without permutation of columns.

EXAMPLE 2.9. In F_3 we takes:

$$G = \begin{pmatrix} 22021 \\ 22101 \\ 11022 \end{pmatrix}$$

We find:

$$H = \begin{pmatrix} 21000 \\ 10001 \end{pmatrix}$$

with 2 permutations of columns.

2.2.3.2. *Lateral classes of a linear code C*

We note by C_u the set $\{u + c/c \in C\}$. The element u is called a representative of the class C_u.

PROPOSITION 2.4. *If* $b \in C_a$ *then* $C_b = C_a$.

Proof. It is enough to prove the inclusion $C_b \subseteq C_a$ (due to cardinals). If $u = b + c$, since $b = a + c'$ then $u = a + b + c' = a + b + c'' \in C_a$ □

We can thus take as representative of each class the one whose weight is minimum in its class. This is used in certain decodings.

PROPOSITION 2.5. *The set of lateral classes of C forms a partition of* $(F_2)^n$*, in parts of the same cardinal.*

Proof. Any u of $(F_2)^n$ is in its own class. The set of classes is thus a repetition of $(F_2)^n$. It remains to prove that two distinct classes do not have common elements, which stems from the previous proposition. □

EXAMPLE 2.10. Let C be a code of length $n = 5$, and generator matrix:

$$G = \begin{pmatrix} 10111 \\ 01110 \\ 11101 \end{pmatrix}$$

The set of classes is the following (each line is a class, the first is the code, on the left is a representative):

00000 10111 01110 11101 11001 01010 10011 00100

10000 00111 11110 01101 01001 11010 00011 10100
01000 11111 00110 10101 10001 00010 11011 01100
00001 10110 01111 11100 11000 01011 10010 00101

We note that the 3rd line is equal to the following line:

00010 10101 01100 11111 11011 01000 10001 00110

This observation is important for decoding.

2.2.3.3. *Syndromes*

Now we introduce the concept of a vector syndrome.

PROPOSITION 2.6. *Two elements a and b are in the same class if $a - b \in C$.*

Proof. The proof bears on the necessity and the sufficiency of the condition:

1) Let us suppose $a \in C_v$ for a certain v, and $a - b \in C$. Then $b = a - c = (v + c') - c = v + c'' \in C_v$;

2) If a and b are in C_v for a certain v then $a = v + c_1$, $b = v + c_2$, $a - b = c_1 - c_2 \in C$. □

EXAMPLE 2.11. $p = 2$, and a code with generator matrix:

$$G = \begin{pmatrix} 10111 \\ 01110 \\ 11101 \end{pmatrix}$$

generates the code C:

1) Using the Gaussian method we find the parity check matrix $H = \begin{pmatrix} 11010 \\ 10001 \end{pmatrix}$ without changing the columns;

2) Let $v = 11111$ be a received word. Calculate $H[v]^T$ ($[v]^T$ is v transposed).

PROPOSITION 2.7. *Let H be a parity check matrix of the code C:*

1) If $b \in C_a$, then we have $H[b] = H[a]$,

2) If $H[d] = H[a]$, then we have $d \in C_a$.

Proof.

1) $b = a + c, c \in C$. Then $H[b] = H[a] + H[c] = H[a]$,

2) $H[d - a] = [0]$, which is equivalent to $d - a \in C$. □

In conclusion, the syndrome of a vector u characterizes the class to which u belongs. This makes it possible to simplify the decoding practice.

2.2.3.4. *Decoding and syndromes*

Let v be a received word. The equality $H[v]^T = [s]^T$ defines the vector $[s]^T$ called the *syndrome* of v.

2.2.3.5. *Lateral classes, syndromes and decoding*

We use maximum likelihood decoding. Thus we decode by a codeword of C which is the closest to the received word in the sense of Hamming distance.

PROPOSITION 2.8. *If v is the received word, then the error is any element of the class of v.*

Proof. For any u of C_v we have $v - u \in C$. □

EXAMPLE 2.12. $p = 2$, and:

$$G = \begin{pmatrix} 10111 \\ 01110 \\ 11101 \end{pmatrix}$$

Let us suppose receiving 11110. The error can be 10000, 00111, 11110, 01101, 01001, 11010, 00011, 10100. We will suppose maximum likelihood decoding and, thus, that the error was 10000. The decoded word will then be 01110.

In fact, we calculate the syndrome of the received word v. This syndrome is the same for any element of C_v. We then suppose that the error is the word with the smallest weight in C_v (this is maximum likelihood decoding).

2.2.3.6. *Parity check matrix and minimum code weight*

The following proposition expresses a property of minimum code distance and parity check matrix.

PROPOSITION 2.9. *The minimum distance of a code is greater than or equal to d, if there is no zero linear combination of $d - 1$ columns of a parity check matrix of C.*

Proof. Let $(c_0, c_1, \ldots, c_{n-1})$ be a codeword. Writing:

$$H \times \begin{pmatrix} c_0 \\ c_1 \\ \vdots \\ c_{n-1} \end{pmatrix}$$

is equivalent to making a linear combination of the columns of H. If there is no zero linear combination with less that $d-1$ columns of H, then the kernel of H (i.e the code C) does not have a word with weight lower than d. □

2.2.3.7. *Minimum distance of C and matrix H*

The study of the columns of H gives the minimum distance of C.

EXAMPLE 2.13. We take as code C the code (known as *the Hamming code*) (7,4,3). Its parity check matrix:

$$H = \begin{pmatrix} 1010101 \\ 0110011 \\ 0001111 \end{pmatrix}$$

has neither a zero column, nor two equal columns. The minimum distance of C is 3.

EXAMPLE 2.14. We take as code C the Hsiao code (8,4,4) whose parity check matrix is:

$$H = \begin{pmatrix} 10000111 \\ 01001011 \\ 00101101 \\ 00011110 \end{pmatrix}$$

It is a code that corrects 1 error and detects 2 of them.

2.2.4. *Some linear codes*

The best known linear codes are the Hamming codes and the Reed-Muller codes (known as *RM codes*). Hamming codes have a parity check matrix formed by all the non-zero r-tuples. They are the $(2^r-1, 2^r-1-r, 3)$ codes. An RM code with a length of 2^m and order r is built on the basis of vectors $v_0, v_1, \ldots, v_m$, where $v_0 = (11\cdots1)$ and v_i has 2^{i-1} "0" then 2^{i-1} "1" as components from left to right, in alternation. The codewords of an RM code of length 2^m and order r are all the products (component by component) of a maximum of r codewords v_i. An RM code of the order r has a length q, a dimension $1 + \binom{m}{1} + \binom{m}{2} + \cdots + \binom{m}{r}$, and a minimum distance 2^{m-r}.

EXAMPLE 2.15 ($m = 3$, $r = 2$). We have $v_0 = 11111111$, $v_1 = 01010101$, $v_2 = 00110011$, $v_3 = 00001111$. The code has 11111111, 01010101, 00110011, 00001111, 00010001, 00000101, 00000011, 00000001, 00000000 as words.

2.2.5. *Decoding of linear codes*

There are various more or less complex decodings possible, such as, for example, lattice decoding, studied by S. Lin and T. Kasami amongst others.

Step by step decoding

Let us now introduce a very easy algorithm that can be used for all linear codes.

Let there be a linear code C, of length n, corrector of t errors by word, for which we take a generator matrix G. We will suppose that C is binary, although this decoding extends directly to non-binary codes.

Preparation of decoding

The following steps must be performed before proceeding to decoding:

1) construction of the parity check matrix H on the basis of G;

2) construction of the table of pairs (weight, syndromes):

– we will take as vector x any vector whose Hamming weight (noted $w_H(x)$) is less than or equal to t,

– we will pose $H[x]^t = [z_x]^t$,

– it is necessary to memorize in a table all the pairs $(w_H(x), z_x)$.

Decoding

Let c be an transmitted codeword, which is supposed to have been altered by an error x satisfying $w_H(x) \leq t$. For each received word $c + x$ we have an initialization phase and an iterative phase.

Initialization phase

The initialization phase comprises three stages:

1) calculation of $H[c + x]^t$ (equal to $H[x]^t$), which we will call $[z_x]^t$,

2) search for z_x in the table of pairs, from which we deduce $w_H(x)$,

3) initialization of a variable P to the found value of $w_H(x)$.

Iterative phase, for $i = 1$ with n

Let us use l_i to indicate the binary vector of Hamming weight equal to 1, where 1 is in position i. The iterative stage comprises two stages:

1) calculation of $H[c + x + l_i]^t$, and search for $w_H(x + 1_i)$ in the table. If it is not found, the error cannot be corrected. We pass to 5);

2) analysis of $w_H(x + 1_i)$:
– if $w_H(x + 1_i) \geq P$, we do nothing,
– if $w_H(x + 1i) \leq P$, then : $c + x \leftarrow c + x + 1_i$ and $P \leftarrow w_H(x + 1_i)$.

REMARK. We may stop the iterations as soon as $w_H(x + 1_i) = 0$, since the error is then corrected.

EXAMPLE 2.16. Let C be a BCH code (see section 2.4), with $n = 15, t = 2$, and $g(X) = 1 + X^4 + X^6 + X^7 + X^8$. Its generator matrix is:

$$G = \begin{pmatrix} 100010111000000 \\ 010001011100000 \\ 001000101110000 \\ 000100010111000 \\ 000010001011100 \\ 000001000101110 \\ 000000100010111 \end{pmatrix}$$

We find the parity check matrix:

$$H = \begin{pmatrix} 110000010000000 \\ 011000001000000 \\ 001100000100000 \\ 000110000010000 \\ 000011000001000 \\ 000001100000100 \\ 000000110000010 \\ 000000011000001 \end{pmatrix}$$

The table of pairs $(w_H(x), z_x)$ contains 121 elements. Let us suppose that the transmitted codeword is $c = 0$ and the received codeword is $x = (000100010000000)$.

Initialization phase, we have $H(c + x)^t = z_x = (10111011)^t$. We go through the table of pairs and find $w_H(x) = 2$. We pose $P = 2$.

Iterative phase:

– $i = 1, 2, 3$:

– we find nothing in the table,

– therefore, we do nothing;

– $i = 4$:

– $H[c+x+1_4]^t = [11100011]^t$, and the $w_H(x+1_4)$ equals 1, lower than P,

– we thus replace P by 1 and the received vector by $c + x + 1_4$, i.e. (000000010000000);

– $i = 5, 6, 7$: nothing changes;

– $i = 8$: $H[c + x + 1_8]^t = [00000000]^t$ and the $w_H(x + 1_8)$ equals 0. The corrected word is thus $c + x + 1_8$.

2.3. Finite fields

2.3.1. *Basic concepts*

We presume that the reader is already familiar with the notions of *modulo* n calculations, $\mathcal{Z}/(p)$ field, p prime (also noted $\mathbb{F}_p$) and Euclid and Bezout equalities. We also presume that the concept of ring of polynomials on the $\mathbb{F}_p$ field is also known. An important result concerning the ring of polynomials is the following.

PROPOSITION 2.10. *Any non-zero polynomial of degree n has at most n roots in a field.*

Proof. The proof is outside the scope of this book. □

A useful result for us is provided in the following proposition.

PROPOSITION 2.11. *If β is a root of a polynomial $f(X)$ of $\mathbb{F}_2[X]$, then β^2 is also a root.*

Proof. Let us pose $f(X) = f_0 + f_1X + \cdots + f_nX^n$, $f_i \in \mathbb{F}_2$. Since $f_i^2 = f_i$, we have the equalities $f(\beta^2) = f_0 + f_1\beta^2 + \cdots + f_n\beta^{2n} = (f_0 + f_1\beta + \cdots + f_n\beta^n)^2 = 0^2 = 0$. □

2.3.2. *Polynomial modulo calculations: quotient ring*

Let us suppose a polynomial $a(X) \in \mathbb{F}_2[X]$. The set noted $\mathbb{F}_2[X]/(a(X))$ is the set of polynomial expressions in X, with coefficients in $\mathbb{F}_2$, where we add and multiply two elements calculating in $\mathbb{F}_2[X]$ then taking the remainder of the division of the result by $a(X)$. We easily prove that it is a ring.

EXAMPLE 2.17. Let us consider $A = \mathbb{F}_2[X]/(a(X))$, with $a(X) = 1+X+X^2+X^3$. Let us pose $u_1(X)=1+X+X^2$ and $u_2=X+X^3$. In $\mathbb{F}_2[X]$ we have $u_1(X)u_2(X) = X + X^2 + X^4 + X^5$, the remainder of whose division by $a(X)$ is $1 + X^2$, which is the result of $u_1(X)u_2(X)$ in A.

EXAMPLE 2.18. Let us pose $a(X) = X^5 + 1$. Let us pose $u_1(X) = 1+X+X^2$. Calculate $u_1(X) = 1+X+X^2, Xu_1(X), X^2u_1(X), X^3u_1(X), X^4u_1(X)$, and examine the representation in the form of binary vectors. We see that we obtain a circular *shift* with each multiplication by X.

The ring $\mathbb{F}_2[X]/(a(X))$ is called the *quotient ring*.

2.3.3. *Irreducible polynomial modulo calculations: finite field*

When $p(X)$ is irreducible, of degree n, we demonstrate that $\mathbb{F}_2[X]/(p(X))$ is a (finite) field with 2^n elements. The field $\mathbb{F}_2[X]/(p(X))$ is also noted $\mathbb{F}_q$, with $q = 2^n$. It is said that $\mathbb{F}_2[X]/(p(X))$ is a representation of $\mathbb{F}_q$. If there are two irreducibles of the same degree n, then we have two representations of the same $\mathbb{F}_q$ field.

It is sometimes necessary (for example for certain decodings) to seek the roots of a polynomial in a given field. Let us give an example of such a search for the roots of a polynomial $b(Y)$ in a finite field.

EXAMPLE 2.19. In $\mathbb{F}_2[X]/(1 + X + X^4)$ we seek the roots of $b(Y) = 1 + Y + Y^2$:

1) is it $1 + X$? We have $(1 + X)^2 + (1 + X) + 1 = 1 + X + X^2 \neq 0$. It is not a root;

2) is it $X + X^2$? We have $(X + X^2)^2 + (X + X^2) + 1 = 1 + X + X^4 = 0$. It is a root;

3) is it $1 + X + X^2$? We have $(1 + X + X^2)^2 + (1 + X + X^2) + 1 = 0$. It is a root.

We can write $1 + Y + Y^2 = (Y - (X + X^2))(Y - (1 + X + X^2))$ (verify it). We will also verify that $(X + X^2)^2 = 1 + X + X^2$ (see proposition 2.11).

2.3.4. *Order and the opposite of an element of* $\mathbb{F}_2[X]/(p(X))$

We can study the order and the opposite of an element of a ring, but here we place ourselves in a finite field. Let $\beta \in \mathbb{F}_{2^n}$, non-zero. Let us note that it is invertible because it is in a field. We consider the family $E = \{\beta, \beta^2, \beta^3, \ldots\}$ of distinct successive powers of β.

2.3.4.1. *Order*

The order of β is the smallest positive integer e such that $\beta^e = 1$ (e depends on β).

PROPOSITION 2.12. $|E| = e$.

Proof. E is finite because it is part of a finite field. Let us pose $E = \{\beta, \beta^2, \beta^3, \ldots, \beta^r\}$. This means that β^{r+1} was already obtained in the form of β^i, with $i \leq r$. Let us suppose $\beta^{r+1} = \beta^t$, with $t \geq 2$. We then have $\beta\beta^r = \beta\beta^{t-1}$, and since β is invertible, we have $\beta^r = \beta^{t-1}$, which means that β^{t-1} has already been obtained, which contradicts the definition of E. Thus $\beta^{r+1} = \beta$, from where we directly deduce that $\beta^r = 1$. The order of β is thus equal to r. □

EXAMPLE 2.20. In $\mathbb{F}_2[X]/(1 + X + X^2)$ the order of $1 + X$ is 3, the opposite of $1 + X$ is X.

EXAMPLE 2.21. In the field $\mathbb{F}_2[X]/(1 + X + X^4)$ let us pose $\beta = X^3$. We find that its order is 5.

EXAMPLE 2.22. In the field $\mathbb{F}_2[X]/(1 + X + X^4)$ let us pose $\beta = 1 + X$. We find that its order is 15.

2.3.4.2. *Properties of the order*

The three following propositions express the properties of the order.

PROPOSITION 2.13. *The order e of β divides 2^{n-1}.*

Proof. The set of $q - 1$ invertibles of the field forms a multiplicative group. The set of powers of β forms a multiplicative subgroup. We know that the cardinal of a subgroup divides the cardinal of the group which it contains. Lastly, e is the cardinal of the subgroup. Thus, e divides $q - 1$. □

PROPOSITION 2.14. *If x is of the order e, then $x^u = 1$ involves $e|u$.*

Proof. If $u = \lambda e$ then $x^u = (x^e)^\lambda = 1$. If $x^u = 1$, then by the Euclid equality $u = q_e + r, r < e$, and thus $1 = x^u = x^{qe}x^r = (x^e)^q x^r = x^r$. Since e is the order of x we must have $r = 0$. □

PROPOSITION 2.15. *If x is of the order e, then x^r is of the order $e/gcd(e, r)$.*

Proof. Let us note (a, b) for $\text{pgcd}(a, b)$. We have: $(x^r)^{e/(e,r)} = (x^e)^{r/(e,r)} = 1$. Thus, the order of x^r divides $e/(e, r)$ (see proposition 2.14). If we have $(x^r)^E = 1$, then $e|rE$, i.e. $rE = \lambda e$ for a certain λ, from where $[r/(e, r) \times E = e/(e, r)] \times \lambda$. Since we see that $(r/(e, r), e/(e, r)) = 1$, E is a multiple of $e/(e, r)$. □

Let us provide a method to compute the order of a β of $\mathbb{F}_q$, $q = 2^n$.

1) make the lattice of the divisors of $q - 1$;

2) to test β^i where i is a maximum divisor of $q - 1$;

3) if for a maximum divider k we have $\beta^k = 1$, then start again with the lattice of dividers of k;

4) if $\beta^k \neq 1$ for any maximum divisor k of $q - 1$, then the order of β is $q - 1$ (see proposition 2.13).

EXAMPLE 2.23. Let $\mathbb{F}_{2^6}$ be represented by $\mathbb{F}_2[X]/(1 + X + X^6)$. We seek the order of $\beta = 1 + X$:

1) the lattice of divisors of $2^6 - 1 = 63$ is as follows;

2) we must calculate β^9 and β^{21}. We find $\beta^9 = \beta + \beta^2 + \beta^4 + \beta^5 \neq 1$ which proves that its order does not divide 9. Moreover, we find $\beta^{21} = 1$. Thus, the order of β is 21 or 7. The calculation yields $\beta^7 = \beta + \beta^3 + \beta^4 + \beta^5 \neq 1$. Therefore, the order of β is 21.

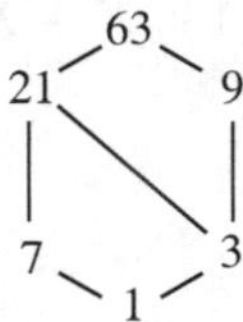

Figure 2.1. *Lattice of divisors of* $2^6 - 1 = 63$

2.3.4.3. *Primitive elements*

An element of $\mathbb{F}_q$ is called primitive if its order is $q - 1$. We will see that there always exists such an element in a field. We will prove the existence, then give the number of such elements in $\mathbb{F}_q$.

2.3.4.3.1. Existence

Propositions 2.16, 2.17 and 2.18 prove the existence of primitive. Let us pose $q - 1 = p_1^{m_1} \cdots p_k^{m_k}$ (primary decomposition of $q - 1$).

PROPOSITION 2.16. *There exists an element y_1 whose order is of the form $p_1^{m_1} p_2^{i_2} \cdots p_k^{i_k}$.*

Proof. If not, the order of all $x \neq 0$ of the field would be the root of $X^{(q-1)/p_1} - 1$, which is not possible, because of the degree (see proposition 2.10). □

Of course, there is also an element y_2 whose order is of the form $p_1^{i_1} p_2^{m_2} p_3^{i_3} \cdots p_k^{i_k}$ and so on. There thus exist particular elements $y_1, y_2, \ldots, y_k$.

PROPOSITION 2.17. *Let* $z_1 = y_1^{p_2^{m_2} p_3^{i_3} \ldots p_k^{i_k}}$. *Its order is* $p_1^{m_1}$.

Proof. Applying proposition 2.15 we see that the order of the element z_1 is equal to $(p_1^{m_1} p_2^{i_2} \cdots p_k^{i_k})/(p_1^{m_1} p_2^{i_2} \cdots p_k^{i_k}, p_2^{m_2} p_3^{i_3} \cdots p_k^{i_k})$. □

Using the same argument we also obtain the elements $z_2, \ldots, z_k$ that have respective orders $p_2^{m_2}, \ldots, p_k^{m_k}$.

PROPOSITION 2.18. *The element* $t = z_1 \cdots z_k$ *has as an order of* $q-1$.

Proof. Let E be the order of t. E is of the form $p_1^{r_1} \cdots p_k^{r_k}$, (see proposition 2.13), with $r_i \leq m_i$ for all i. We have $t^{p_1^{r_1} p_2^{r_2} \ldots p_k^{r_k}} = 1$. Let us raise to the power of $p_2^{m_2 - r_2} \cdots p_k^{m_k - r_k}$. We have: $(t^{p_1^{r_1} p_2^{r_2} \ldots p_k^{r_k}})^{p_2^{m_2-r_2} \ldots p_k^{m_k - r_k}} = t^{p_1^{r_1} p_2^{m_2} \ldots p_k^{m_k}} = z_1^{p_1^{r_1} p_2^{m_2} \ldots p_k^{m_k}} = 1$. Thus (see proposition 2.14) $p_1^{m_1} | p_1^{r_1} p_2^{m_2} \cdots p_k^{m_k}$, and then $m_1 | r_1$, which means that $r_1 = m_1$. By symmetry we also obtain the equalities $r_2 = m_2, \ldots, r_k = m_k$, and thus the order of t is equal to $q-1$. □

We cannot formally construct a primitive, but if we know one of them we can find them all, as indicated by the following proposition.

PROPOSITION 2.19. *Let* x *be a primitive. The element* x^i *is primitive, if* $(i, q-1) = 1$.

Proof. The order of x^i is $(q-1)/(q-1, i)$ (see proposition 2.15). □

If we are not in a field there may not exist a primitive for the group of invertibles, as the following examples show. Let us recall that φ is the Euler indicator. The number of integers smaller than m, and relatively preceding this m, is equal to $\varphi(m)$.

EXAMPLE 2.24. In $\mathcal{Z}/(9)$ we have $\varphi(9) = 6$. The group of invertibles thus has 6 elements. Element 2 has an order 6. It is a primitive from the group of invertibles.

EXAMPLE 2.25. In $\mathcal{Z}/(8)$ there are 4 invertibles. The invertibles 1, 3, 5, 7 have the respective orders 1, 2, 2, 2. Thus, there are no primitives.

2.3.4.3.2. The number of primitives

The number of primitives is specified by the following result.

COROLLARY 2.1. *The number of primitives in* $\mathbb{F}_q$ *is* $\varphi(q-1)$.

Proof. By definition of the Euler function φ, and by proposition 2.19. □

2.3.4.4. *Use of the primitives*

The primitive elements are often used in the application of error correcting codes.

2.3.4.4.1. The use of a primitive to represent the elements of a field

Any element β of $\mathbb{F}_2[X]/(p(X))$, with irreducible $p(X)$ of nth degree, is a polynomial expression in X with binary coefficients of a degree no more than $n-1$. The product of two elements β_1 and β_2 is thus a product of two *modulo* $p(X)$ polynomials. It is a rather complex operation, both time and power consuming. Therefore, in practice it is interesting to change the representation of the field. We choose a primitive α and express any non-zero element of the field as a power of this primitive. The advantage is as follows. Let $\beta_1 = \alpha^i$ and $\beta_2 = \alpha_j$. The product is α^{i+j} where $i+j$ is calculated *modulo* $q-1$, which is very easy and fast. Let us note that this representation renders the sum $\beta_1 + \beta_2$ more difficult to calculate than with the polynomial expression of the elements. This disadvantage can be mitigated by using a Zech table.

2.3.4.4.2. Zech's log table to calculate the sum of two elements

If $\beta_1 = \alpha^i$ and $\beta_2 = \alpha^j$, with $i > j$, then $\beta^i + \beta^j = \alpha^j(\alpha^{i-j}+1)$. The Zech table has $1+\alpha^k$ as input and α^m as output with $1+\alpha^k = \alpha^m$.

EXAMPLE 2.26. In $\mathbb{F}_2[X]/(1+X+X^4)$ we take as primitive $\alpha = X$. We have the following representation:

$$
\begin{array}{lll}
1 = 1 & \alpha^5 = \alpha + \alpha^2 & \alpha^{10} = 1 + \alpha + \alpha^2 \\
\alpha = \alpha & \alpha^6 = \alpha^2 + \alpha^3 & \alpha^{11} = \alpha + \alpha^2 + \alpha^3 \\
\alpha^2 = \alpha^2 & \alpha^7 = 1 + \alpha + \alpha^3 & \alpha^{12} = 1 + \alpha + \alpha^2 + \alpha^3 \\
\alpha^3 = \alpha^3 & \alpha^8 = 1 + \alpha^2 & \alpha^{13} = 1 + \alpha^2 + \alpha^3 \\
\alpha^4 = 1 + \alpha & \alpha^9 = \alpha + \alpha^3 & \alpha^{14} = 1 + \alpha^3
\end{array}
$$

Let $\beta_1 = \alpha^2 + \alpha^3$ and $\beta_2 = 1 + \alpha + \alpha^2 + \alpha^3$. The product is equal to $\alpha^{6+12} = \alpha^3$. The Zech table is presented as follows:

$$
\begin{array}{llll}
1 + \alpha = \alpha^4 & 1 + \alpha^2 = \alpha^8 & 1 + \alpha^3 = \alpha^{14} & 1 + \alpha^4 = \alpha \\
1 + \alpha^5 = \alpha^{10} & 1 + \alpha^6 = \alpha^{13} & 1 + \alpha^7 = \alpha^9 &
\end{array}
$$

This is sufficient because we have the equality $1 + \alpha^{i+(q/2)} = \alpha^{i+q/2}(1 + \alpha^{(q/2)-i-1})$. We have $\beta_i = \alpha^2(1+\alpha) = \alpha^2(\alpha^4) = \alpha^6$, as well as $\beta_2 = 1 + \alpha(1+\alpha) + \alpha^3 = 1 + \alpha^5 + \alpha^3 = 1 + \alpha^3\alpha^8 = \alpha^{12}$.

2.3.4.5. *How to find a primitive*

We cannot find a primitive formally, but we can use the following algorithm:

1) create the lattice of the divisors of 2^{n-1};

2) choose a non-zero element β;

3) use the maximum divisors. If no maximum divisor d yields $\beta^d = 1$, then the element β is primitive.

EXAMPLE 2.27. In $\mathbb{F}_{64}$ represented by $\mathbb{F}_2[X]/(1 + X + X^6)$ let us consider the non-zero element $\beta = X$. It is primitive. We finds $\beta^9 = X^3 + X^4 \neq 1$ and $\beta^{21} = 1 + X + X^3 + X^4 + X^5 \neq 1$. Thus, the order of β is 63. It is primitive.

2.3.4.6. *Exponentiation*

We saw how to search for the order of an element, and how find out if it is primitive. For large fields (i.e large q) we are led to calculate β^i for very large i. One of the best methods is to proceed as follows:

1) break up i in base 2;

2) calculate the exponentiations by 2, i.e. $\beta, \beta^2, \beta^{2^2}, \beta^{2^3}$, *etc.*;

3) calculate the necessary products (see example 2.28).

We prove that the complexity is in $O(\log i)$ instead of $O(i)$.

EXAMPLE 2.28. Calculation of β^{21}, with the notations of example 2.9:

1) $21 = 16 + 4 + 1$;

2) $\beta \to \beta^2 \to \beta^4 \to \beta^8 \to \beta^{16}$ which yields $1 + X \to 1 + X^2 \to 1 + X^4 \to 1 + X^2 + X^3 \to X + X^4$;

3) $\beta^{21} = \beta^{16}\beta^4\beta^1 = (X + X^4)(1 + X^4)(1 + X) = 1$.

This method is used, for example, for calculations necessary for the use of RSA in cryptography.

2.3.5. *Minimum polynomials*

Let $\beta \in \mathbb{F}_{2^n}$. Let us consider the part $C_\beta = \{\beta, \beta^2, \beta^{2^2}, \beta^{2^3}, \beta^{2^4}, \ldots\}$.

PROPOSITION 2.20. *There exists a polynomial with binary coefficients, which admits all the elements of this part as the set of its roots. This polynomial is irreducible.*

Proof. Let us examine C_β. It is a finite part, because it is included in a finite field. Let us pose: $C_\beta = \{\beta, \beta^2, \beta^{2^2}, \ldots, \beta^{2^{t-1}}\}$. This means that β^{2^t} is an element of the form β^{2^i}, with $0 \leq i \leq t-1$.

Let us suppose $i \neq 0$. We then have $(\beta^{2^{t-1}})^2 = (\beta^{2^{i-1}})^2$. Thus, $(\beta^{2^{t-1}}/\beta^{2^{i-1}})^2 = 1$. However, the polynomial $Z^2 - Z$ has only two roots (see proposition 2.10), which are 0 and 1. This leads to $\beta^{2^{t-1}} = \beta^{2^{i-1}}$. Thus, 2^{t-1} has been already obtained, which goes against the definition of C_β. Therefore, $\beta^{2^{t-1}} = \beta$. A consequence of this equality is that the class C_β is stable under exponentiation by 2.

Now, let us consider the polynomial $(Y-\beta)(Y-\beta^2)(Y-\beta^{2^2})\cdots(Y-\beta^{2^{t-1}})$. It has the symmetrical functions of its roots as coefficients. Thus, each of its coefficients is invariant under exponentiation by 2. Each coefficient is, therefore, binary. This polynomial is irreducible, since otherwise it would have a divisor of a strictly smaller degree than it does. Moreover, this divisor would have at least one element of C_β as root. As this class is invariant under exponentiation by 2, and according to proposition 2.11, this polynomial should have all the elements of C_β as roots. This is impossible according to proposition 2.10. Thus, this divisor strictly cannot exist. □

It is said that this irreducible polynomial is the minimum polynomial of β, and we note it by $M_\beta(Y)$. The part C_β is called the *cyclotomic class* of β.

EXAMPLE 2.29. $\mathbb{F}_2[X]/(1+X+X^3), \beta = 1+X$:

1) the cyclotomic class of β is $\{1+X, 1+X^2, 1+X+X^2\}$;

2) we have $M_\beta(Y) = (Y-(1+X))(Y-(1+X^2))(Y-(1+X+X^2)) = 1+Y^2+Y^3$.

When β is primitive, the polynomial $M_\beta(Y)$ is referred to as *irreducible primitive*, or simply *primitive*.

2.3.6. *The field of nth roots of unity*

When we study a cyclic code of length n, we are led to seek the smallest field $\mathbb{F}_q(q = 2^r)$ containing the nth roots of unity (i.e. x such that $x^n = 1$). If x has as an order n, it is said that it is a nth primitive root of unity.

PROPOSITION 2.21. *The smallest field $\mathbb{F}_q(q = 2^r)$, which contains the nth roots of unity, is such that r is the order of 2 modulo n.*

Proof. $\mathbb{F}_q$ has $q-1$ non-zero elements, which form a multiplicative group. The set of nth roots of unity forms a subgroup thereof. Thus, n divides $2^r - 1$. Written differently, we have $2^r - 1 = \lambda n$, or otherwise $2^r = 1$ *modulo* n, which shows that r is the order of 2 *modulo* n. □

PROPOSITION 2.22. *Let γ be an element of $\mathbb{F}_q$ ($q = 2^r$, r is of the order 2 modulo n), which is a nth root of unity, primitive or not. We have:*

1) $1+\gamma+\gamma^2+\cdots+\gamma^{n-1} = 0$ *if* $\gamma \neq 1$,

2) $1+\gamma+\gamma^2+\cdots+\gamma^{n-1} = 1$ *if* $\gamma = 1$.

Proof. Indeed:

1) γ is a root of the polynomial $\frac{1+X^n}{1+X}$, since the group of the nth roots is the set of roots of the polynomial $1 + X^n$;

2) n is odd, since it divides $2^r - 1$.

2.3.7. *Projective geometry in a finite field*

We consider $\mathbb{F}_{q^{m+1}}$ as space vector of dimension $m + 1$ over $\mathbb{F}_q$, with $q = 2^s$. Let α be a primitive of $\mathbb{F}_{q^{m+1}}$ and β be a primitive of $\mathbb{F}_q$. We have $\beta = \alpha^{\frac{q^{m+1}-1}{q-1}}$. We can build a particular geometry, known as projective geometry. We define the "points" in $\mathbb{F}_{q^{m+1}}$, then the projective subspaces of the dimensions $1, 2, \ldots$ in the following way.

2.3.7.1. *Points*

A point, noted (α^i), is the subspace of $\mathbb{F}_{q^{m+1}}$ generated by α^i, deprived of 0. We have $(\alpha^i) = \{\alpha^i, \beta\alpha^i, \ldots, \beta^{q-2}\alpha^i\} = L(\alpha^i)\backslash\{0\}$. It is a subspace of dimension 1 deprived of 0.

2.3.7.2. *Projective subspaces of order* 1

If $\alpha^j \notin (\alpha^i)$, then a projective subspace of order 1, denoted (α^j, α^i), equals $L(\alpha^j, \alpha^i)\backslash\{0\}$. It is often called a "projective straight line".

2.3.7.3. *Projective subspaces of order t*

It is a subspace of dimension $t + 1$ deprived of 0, in other words it is $L(\alpha^{i_1}, \ldots, \alpha^{i_{t+1}})\backslash\{0\}$. For $t = 2$, it often called a "projective plane".

2.3.7.4. *An example*

Let us take $q = 2, s = 1, m = 2, \mathbb{F}_8 = \mathbb{F}_2[X]/(1 + X + X^3), \alpha = X$. The points are as follows: $(\alpha^0) = \{\alpha^0\}$ because $\beta = 1, (\alpha^1), (\alpha^2), (\alpha^3), (\alpha^4), (\alpha^5), (\alpha^6)$.

The projective straights are as follows:

$$(\alpha^0, \alpha^1) = \{(\alpha^0), (\alpha^1), (\alpha^3)\}, \text{ because } \alpha^0 + \alpha^1 = \alpha^3$$

$$(\alpha^0, \alpha^2) = \{(\alpha^0), (\alpha^2), (\alpha^5)\}$$

$$(\alpha^0, \alpha^4) = \{(\alpha^0), (\alpha^4), (\alpha^5)\}, \text{ because } \alpha^0 + \alpha^4 = \alpha^5$$

$$(\alpha^1, \alpha^2) = \{(\alpha^1), (\alpha^2), (\alpha^4)\}$$

$$(\alpha^1, \alpha^5) = \{(\alpha^1), (\alpha^5), (\alpha^6)\}$$

$$(\alpha^2, \alpha^3) = \{(\alpha^2), (\alpha^3), (\alpha^5)\}$$

$$(\alpha^3, \alpha^4) = \{(\alpha^3), (\alpha^4), (\alpha^6)\}, \text{ because } \alpha^3 + \alpha^4 = \alpha^6$$

The only projective plane is the private field of 0.

2.3.7.5. *Cyclic codes and projective geometry*

We note that we can pass from a point to another by multiplication by α. Indeed, the number n of points in the geometry is $\frac{q^{m+1}-1}{q-1}$, and 1 and $\alpha^{\frac{q^{m+1}-1}{q-1}}$ belong to the same point (1). The set of the points can thus be arranged like a cyclic sequence. This suggests considering cyclic codes in $\mathbb{F}_q[X]/(X^n - 1)$, which is what we will return to in the description of PG-codes (see section 2.4).

2.4. Cyclic codes

After the theoretical results of C. Shannon and the first linear code constructions (Hamming, Golay) American engineers were required to be able to obtain codes stable not only under addition (linear codes), but also stable under circular sliding (or *shift*). The codes obtained (cyclic codes) are linear codes with additional properties.

This new requirement led the mathematicians to exploit the structure of $A = \mathbb{F}_2[X]/(X^n - 1)$, and, in particular, to study the ideal A. An ideal A is a non-empty part, stable under addition, and stable under multiplication by any element of A. It is a cyclic code of length n. Everywhere hereinafter n is odd.

2.4.1. *Introduction*

The following results express the properties of a cyclic code.

PROPOSITION 2.23. *Any code C, stable under addition and circular shift may be represented as an ideal A.*

Proof. The circular shift on the right represents the multiplication by X in A. The code C is thus stable under addition and multiplication by X. It is therefore stable under addition and multiplication by any polynomial: thus, it is an ideal A. Conversely, an ideal A is clearly a code stable under addition and circular shift. □

PROPOSITION 2.24. *Any cyclic code has the form $(g(X))$ (i.e. the set of multiples of $g(X)$), with $g(X)$ dividing $X^n - 1$. More precisely, there is between the cyclic codes of length n and the set of divisors of $X^n - 1$.*

Proof. We know that the ring $A = \mathbb{F}_2[X]/(X^n - 1)$ is principal, i.e. it is the set of the multiples of one of its elements.

Let C be a cyclic code in A. Let $g(X)$ be a polynomial of minimum degree in the code. In $\mathbb{F}_2[X]$ we have: $X^n - 1 = q(X)g(X) + r(X)$. In A we deduce $r(X) = q(X)g(X)$, and, thus, $r(X)$ is in C. Due to the minimality of the degree of $g(X)$ it necessarily follows that $r(X) = 0$. Thus $g(X)$ divides $X^n - 1$.

Reciprocally, let (X) be a divisor of $X^n - 1$. It is straightforward to prove that $(g(X))$ is a cyclic code.

It remains to be shown that two divisors distinct from $X^n - 1$ generate two distinct codes C_1 and C_2. Let us suppose $C_1 = C_2$. Then in $\mathbb{F}_2[X]$, we have $g_2(X) = q(X)g_1(X) + \lambda(X)(X^n - 1)$, for a certain $\lambda(X)$. Thus $g_1(X)$ divides $g_2(X)$, since $g_1(X)$ divides $X^n - 1$. Using symmetry, we prove the equality $g_1(X) = g_2(X)$. □

PROPOSITION 2.25. *Any cyclic code $(a(X))$ where $a\ (X)$ is unspecified is still generated by $PG(X)$, where $PG(X) = (a(X), X^n - 1)$.*

Proof. Let us pose $PG(X) = (a(X), X^n - 1)$. Using the Bezout equality we obtain $PG(X) = \lambda(X)a(X) + \mu(X)(X^n - 1)$, for certain $\lambda(X), \mu(X)$. This proves that $PG(X)$ is in the code $(a(X))$. Any multiple of $a(X)$ is thus a multiple of $PG(X)$. In $(a(X))$ there exists a generator of minimum degree. Because of the degrees, it must necessarily be $PG(X)$. □

2.4.2. *Base, coding, dual code and code annihilator*

We now develop the basic ideas on the cyclic codes.

2.4.2.1. *Cyclic code base*

Let there be a cyclic code of length n, with a generator $g(X)$ of degree $n - k$, and $g(X)|X^n - 1$.

PROPOSITION 2.26. *Let $h(X) = (X^n - 1)/g(X)$, that is to say k is the degree of $h(X)$. The family $\mathcal{F} = \{g(X), Xg(X), \dots, X^{k-1}g(X)\}$ is one of the basis of the code $(g(X))$.*

Proof. Let the word $a(X)g(X)$ belong to the code. Let us pose that $a(X) = q(X)h(X) + r(X)$, with $r(X) = 0$ or $\deg(r(X)) < \deg(h(X))$. In A we have $a(X)g(X) = r(X)g(X)$ since $g(X)h(X) = 0$, and thus the family $\mathcal{F}$ is a generating one. Let us prove that it is of rank k.

Let us suppose $\alpha_0 g(X) + \alpha_1 X_g(X) + \cdots + \alpha_{k-1}X^{k-1}g(X) = 0$, in A. In $\mathbb{F}_2[X]$ we deduce it $(\alpha_0 + \alpha_1 X + \cdots + \alpha_{k-1}X^{k-1})g(X) = \lambda(X)(X^n - 1)$, but the degree of the first member is at the most $n - 1$. Thus two members are zero, and we have $\alpha_0 = \alpha_1 = \cdots = \alpha_{k-1} = 0$. □

EXAMPLE 2.30 ($n = 7$, $g(X) = 1 + X + X^3$). A base of $(g(X))$ is $\{1 + X + X^3, X + X^2 + X^4, X^2 + X^3 + X^5, X^3 + X^4 + X^6\}$. The code generator matrix g associated to this base is:

$$G = \begin{pmatrix} 1101000 \\ 0110100 \\ 0011010 \\ 0001101 \end{pmatrix}$$

To build the code we make all the linear combinations of the lines of this generator matrix G. We find 2^4 words.

2.4.2.2. *Coding*

The first coding can be derived from proposition 2.26.

We suppose to have information blocks of length k. Each block will be encoded by means of a length n code. We thus add $n - k$ symbols of redundancy to k bits of information.

We will describe how two of classical codings for cyclic codes are performed. Thus, let us suppose that the code considered here has a length n, and is generated by a polynomial $g(X)$ of degree $n - k$. Information is a block of length k, which is represented by a binary sequence, let's say $i_0, i_1, \ldots, i_{k-1}$. We will associate the polynomial to this sequence (known as *information polynomial*) according to: $i(X) = i_0 + i_1X + \cdots + i_{k-1}X^{k-1}$.

The first coding consists of calculating the polynomial $i(X)g(X)$. It is clearly in the code, since it is a multiple of $g(X)$: it is a codeword. This coding is referred to as non-systematic.

The second coding consists of calculating first $X^{n-k}i(X)$. Then we calculate the remainder $r(X)$ of the division of this new polynomial $X^{n-k}i(X)$ by $g(X)$. The polynomial obtained is $r(X) + X^{n-k}i(X)$. The sequence of its coefficients is sent through the transmission channel. Generally, we first send the largest degree.

The following proposition proves that we have indeed carried out a coding.

PROPOSITION 2.27. *If $i(X)$ is an information polynomial, then the polynomial $r(X)+X^{n-k}i(X)$ is the corresponding codeword.*

Proof. The polynomial $r(X) + X^{n-k}i(X)$ is divisible by $g(X)$, thus it belongs to the code. □

This second coding is known as *systematic*, because information appears in it clearly. We sometimes speak of a "systematic code". It is not correct, because a cyclic code is an ideal in A. It does not depend at all on the performed coding.

EXAMPLE 2.31 ($n = 7, g(X) = 1+X+X^3$). Let us take the information block equal to 1011. The polynomial $i(X)$ is equal to $1 + X^2 + X^3$. The polynomial $X^{n-k}i(X)$ is $X^3 + X^5 + X^6$. The remainder of the division of this polynomial by $g(X)$ is 1. The coded polynomial is, therefore, $1 + X^3 + X^5 + X^6$. As the length of the code is 7, we will send the following sequence of 7 binary symbols through the channel 1001011 (the first sent is on the right).

EXAMPLE 2.32 ($n = 15$, $g(X) = 1 + X^3 + X^4$). Let us take the information block equal to 10111110001. The polynomial $i(X)$ is equal to $1 + X^2 + X^3 + X^4 + X^5 + X^6 + X^{10}$. The polynomial $X^{n-k}i(X)$ is $X^4 + X^6 + X^7 + X^8 + X^9 + X^{10} + X^{14}$. The remainder of the division of this polynomial by $g(X)$ is $X^2 + X^3$. The coded polynomial is thus $X^2 + X^3 + X^4 + X^6 + X^7 + X^8 + X^9 + X^{10} + X^{14}$. We will therefore send 001110111110001 (the first sent is on the right).

2.4.2.3. *Annihilator and dual of a cyclic code C*

Let there be a code C equal to $(g(X))$.

DEFINITION 2.10 (ANNIHILATOR). *The annihilator of $(g(X))$ is the set of polynomials $v(X)$ with a zero product with all the words of the code C. It is written Ann(C).*

Let $h(X) = (X^n - 1)/g(X)$.

PROPOSITION 2.28. *The annihilator of C is the cyclic code $(h(X))$.*

Proof. $Ann(C)$ is clearly a cyclic code and is therefore generated by a $h(X)$ which divides $X^n - 1$, which is of minimal degree, but $h(X)$ is in the code *Ann* (C). Thus, $H(X)$ divides $h(X)$. Since $H(X)g(X) = 0$, it means that the degree of $H(X)$ is equal to or higher than that of $H(X)$. Thus, $H(X) = H(X)$. □

DEFINITION 2.11 (ORTHOGONAL). *The orthogonal (i.e. dual) of $(g(X))$ is the set of all the polynomials v (X) with zero scalar product with all the codewords of the code C. It is noted by $(g(X))^{\perp}$, or $C^{\perp}$.*

PROPOSITION 2.29. *The dual of C is the cyclic code $((h(X^{-1}), X^n - 1))$.*

Proof. Let τ be the application of A in A which sends all $a(X)$ over $a(X^{-1})$. We prove (and we will admit it) that τ is an automorphism. In addition we prove (and we will also admit it) the equality:

$$a(X)b(X) = \sum_{i=0}^{n-1} < a(X), X^i\tau(b(X)) > X^i$$

This equality implies that $a(X)b(X) = 0$, if we have $\angle a(X), X^i\tau(b(X))\rangle = 0$ for all i. We proceed in two stages:

1) a polynomial $b(X)$ is thus in $Ann(C)$ if $\tau(b(X)) \in (g(X))^{\perp}$. Thus, $h(X^{-1}) \in (g(X))^{\perp}$ and, thus, $(h(X^{-1})) \subseteq (g(X))^{\perp}$;

2) in addition, let $u(X) \in (g(X))^{\perp}$. Then $\langle g(X), u(X)X^i\rangle = 0$ and $\sum_{i=0,n=1} \langle g(X), u(X)X^i\rangle X^i = 0$. This is equivalent to $\tau(u(X)) \in Ann(g(X))$. Thus, $u(X) \in (h(X^{-1}))$, and finally $(h(X^{-1})) \supseteq (g(X))^{\perp}$, which proves the equality. Thus, $\tau(u(X))$ is a multiple of $h(X)$, and consequently, according to proposition 2.25, the code $(h(X^{-1}))$ is equal to the code $(h(X^{-1})), X^n - 1)$. □

2.4.2.4. *Cyclic code and error correcting capability: roots of $g(X)$*

We consider a cyclic code of length n generated by a polynomial $g(X)$ with minimum distance d.

One of the large advantages of cyclic codes is that we can have information on their minimum distance, i.e. on their error correcting capability. More precisely, the error correcting capability of a code is linked to the roots of the generator.

Let α be a primitive n^{th} root of unity.

2.4.2.5. *The Vandermonde determinant*

The following result expresses a property of the Vandermonde determinant.

PROPOSITION 2.30. *Let us consider the determinant D with $d-1$ undetermined $X_0, X_1, \ldots, X_{d-2}$:*

$$D = \begin{vmatrix} 1 & 1 & 1 & 1 \cdots & 1 \\ X_0 & X_1 & X_2 & X_3 \cdots & X_{d-2} \\ X_0^2 & X_1^2 & X_2^2 & X_3^2 \cdots & X_{d-2}^2 \\ \vdots & \vdots & \vdots & \vdots \ddots & \vdots \\ X_0^{d-2} & X_1^{d-2} & X_2^{d-2} & \cdots \cdots & X_{d-2}^{d-2} \end{vmatrix}$$

It is equal to the product $\prod_{j>i,i=0,\ldots,d-3}(X_i - X_j)$.

Proof. A determinant is an alternate form of its columns. We observe here that if it is supposed that two variables are equal, the determinant D is zero. It is thus divisible by the product $P = \prod_{j>i,i=0,\ldots,d-3}(X_i - X_j)$.

Let us consider the coefficient of X_{d-2}^{d-2} in P. It is equal to $\prod_{j>i,i=0,\ldots,d-4}(X_i - X_j)$. By a recurrence on the size of the determinant we easily prove that this coefficient is equal to the determinant:

$$\begin{vmatrix} 1 & 1 & 1 & 1 \cdots & 1 \\ X_0 & X_1 & X_2 & X_3 \cdots & X_{d-3} \\ X_0^2 & X_1^2 & X_2^2 & X_3^2 \cdots & X_{d-3}^2 \\ \vdots & \vdots & \vdots & \vdots \ddots & \vdots \\ X_0^{d-3} & X_1^{d-3} & X_2^{d-3} & \cdots \cdots & X_{d-3}^{d-3} \end{vmatrix}$$

This proves that the determinant D is equal to the product P. □

COROLLARY 2.2. *Let $j_1, \ldots, j_{d-1}$ be distinct integers in $\{0, \ldots, n_1\}$. Let there be the determinant D defined by:*

$$D = \begin{vmatrix} (\alpha^i)^{j_1} & (\alpha^i)^{j_2} & \cdots & (\alpha^i)^{j_{d-1}} \\ (\alpha^{i+1})^{j_1} & (\alpha^{i+1})^{j_2} & \cdots & (\alpha^{i+1})^{j_{d-1}} \\ \vdots & \vdots & \ddots & \vdots \\ (\alpha^{i+d-2})^{j_1} & (\alpha^{i+d-2})^{j_2} & \cdots & (\alpha^{i+d-2})^{j_{d-1}} \end{vmatrix}$$

It is not zero.

Proof. One of the properties of the determinants is to be a multi-linear function of their columns. We can thus write: $D = \alpha^{i+2i+3i+\cdots+(d-2)i} D'$, with:

$$D' = \begin{vmatrix} 1 & 1 & \cdots & 1 \\ \alpha^{j_1} & \alpha^{j_2} & \cdots & \alpha^{d_{d-1}} \\ \vdots & & \ddots & \vdots \\ \alpha^{(d-2)j_1} & \alpha^{(d-2)j_2} & \cdots & \alpha^{(d-2)j_{d-1}} \end{vmatrix}$$

Since α is of the order n and we have $d - 2 < n$, the elements $\alpha^{j_1}, \ldots, \alpha_{j_{d-1}}$ are all different from each other. We may thus apply proposition 2.30 and D' is non-zero, as well as D. □

2.4.2.6. *BCH theorem*

We can provide a lower bound of the minimum distance from a cyclic code. This result is based on corollary 2.2.

PROPOSITION 2.31 (BCH THEOREM). *Let there be a code C of length n admitting among its roots the following elements:* $\alpha^i, \alpha^{i+1}, \ldots, \alpha^{i+d-2}$, *where* α *is a* n^{th} *root of unity, whose order is greater than* $d-2$. *Then the code has a minimum distance of at least d.*

Proof. A parity check matrix of the code is clearly:

$$\begin{pmatrix} 1 & \alpha^i & \alpha^{2i} & \alpha^{3i} & \cdots & \alpha^{(n-1)i} \\ 1 & \alpha^{i+1} & \alpha^{2(i+1)} & \alpha^{3(i+1)} & \cdots & \alpha^{(n-1)(i+1)} \\ 1 & \alpha^{i+2} & \cdots & \cdots & \cdots & \alpha^{(n-1)(i+2)} \\ 1 & \cdots & \cdots & \cdots & \cdots & \cdots \\ 1 & \alpha^{i+d-2} & \cdots & \cdots & \cdots & \alpha^{(n-1)(i+d-2)} \end{pmatrix}$$

Based on the corollary 2.2, for any choice of $d-1$ columns of this matrix we obtain a determinant resembling the one studied in corollary 2.2. Every element of the kernel of H (that is, of the code considered) thus has a Hamming weight that cannot be less or equal to $d-1$. □

2.4.3. *Certain cyclic codes*

We provide here some of the most used classical codes. We will not speak of the generalized RS codes, of the alternating codes, or of the Goppa codes. Among the latter we find codes resulting from algebraic geometry, which is outside the scope of our subject matter. Often we will present only binary codes, although they also exist in $\mathbb{F}_q$.

2.4.3.1. *Hamming codes*

Cyclic Hamming codes are equivalent to linear Hamming codes. They are very simple codes, with error correcting capability equal to 1. Let there be the field $\mathbb{F}_q$, $q = 2^r$. Let α be a primitive of this field.

PROPOSITION 2.32. *The binary code C having for roots the elements of the class of* α *is a cyclic code* $(n = q-1, k = n-r, 3)$*, called the Hamming code.*

Proof. We take as a generator the minimum polynomial of α. Since α is of the order $q-1$, no polynomial in the form $1 + X^1 (i < q-1)$ can have α as root. The minimum weight of the code is thus 3. □

There exists a generalization of these Hamming codes. Let there be the field $\mathbb{F}_{q^m}$, $q = 2^r$. Let α be a primitive of $\mathbb{F}_{q^m}$. We pose $\beta = \alpha^s$. We seek a cyclic code with β for root, with a error correcting capability of 1, i.e. a code $(n, k, 3)$.

PROPOSITION 2.33. *Such a code verifies the following properties:*

1) we have $n = (q^m - 1)/(q^m - 1, s)$, *and* $k = q - 1 - w$ *where* w *is the cardinal of the class of* β*;*

2) such a code exists if $((q^m - 1)/(q^m - 1, s), q - 1) = 1$.

Proof. Indeed:

1) the length is the order of β. As the order of α is $q^m - 1$, we directly have the order of β;

2) since we want $d = 3$ no polynomial of the form $1 + \mu X^i (\mu \in \mathbb{F}_q, i < n)$ should admit β as a root. Thus β^i should only belong to $\mathbb{F}_q$ if it equals 1. Multiplicative groups (β) and $\mathbb{F}_q \backslash \{0\}$ must have an intersection reduced to $\{1\}$. As their respective cardinals are $(q^m - 1)/(q^m - 1, s)$ and $q - 1$, the proposition follows directly. □

PROPOSITION 2.34. *The parameter* s *must be a multiple of* $q - 1$.

Proof. So that $((q^m - 1)/(q^m - 1, s), q - 1) = 1$, it is necessary that $(q^m - 1, s)$ be a multiple of $q - 1$, since $q^m - 1$ is divisible by $q - 1$. □

According to this proposition we see that the length of such a code cannot exceed $(q^m - 1)/(q - 1)$. We will now demonstrate that this length can be reached.

PROPOSITION 2.35. *There exists such a code of length* $(q^m - 1)/(q - 1)$, *if* $(m, q - 1) = 1$.

Proof. We pose $\beta = \alpha^{q-1}$. The length is indeed $(q^m - 1)/(q - 1)$. The multiplicative groups (β) and $\mathbb{F}_q \backslash \{0\}$ have the respective cardinals $(q^m - 1)/(q^m - 1, q - 1)$ and $q - 1$, i.e. again $(q^m - 1)/(q - 1)$ and $q - 1$. But we have $(q^m - 1)/(q - 1) = q^{m-1} + q^{m-2} + \cdots + q + 1$. Since $q^i - 1 = s_i(q - 1)$ for certain s_i, we have $(q^m - 1)/(q - 1) = \lambda(q - 1) + m$. It follows that such a code exists if $((q - 1), m) = 1$. □

2.4.3.2. *BCH codes*

Let there be $\mathbb{F}_q, q = 2^r$ and a primitive element α. A binary BCH code of length n is a code admitting as roots the cyclotomic classes of elements $\alpha^i, \alpha^{i+1}, \ldots, \alpha^{i+2\delta-1}$ for any i.

PROPOSITION 2.36. *This code has a minimum distance equal to at least* $2\delta + 1$. *Moreover, its dimension is at least* $n - \delta r$.

Proof. The result for the minimum distance is a consequence of the BCH theorem. The dimension stems from the fact that all the cyclotomic classes have a cardinal that divides r (see proposition 2.46). □

2.4.3.3. *Fire codes*

These are binary codes directly defined by their generator $g(X) = p(X)(X^c - 1)$, where $p(X)$ is an irreducible polynomial of degree m, not dividing $X^c - 1$. The length n of the code is the least common multiple (LCM) of c and of the exponent e of $p(X)$.

Such a code C can correct any packet of errors (or *burst*) of length b and detect all *bursts* of length d, if the following conditions are verified:

1) $d \geq b$,

2) $b + d \leq c + 1$.

To prove this capacity it is enough to demonstrate that this code cannot contain the sum of a *burst* of length b and a *burst* of length d.

A *burst* of length b can be represented by a polynomial $B(X)$ of degree $b - 1$ and constant 1.

PROPOSITION 2.37. *C does not contain the sum of a burst of length b and of a burst of length d.*

Proof. Reduction and absurdum: let us suppose that C contains a polynomial $X^i B_1(X) + X^j B_2(X)$. By the cyclicity of code this is equivalent to saying that C contains a polynomial in the form of $B_1(X) + X^k B_2(X)$ (with $k = j - i$ *modulo* n). Since $g(X)$ is divisible by X^c– 1, the latter must divide $B_1(X) + X^k B_2(X)$. Let us pose $k = qc + r$ (Euclidean division). We deduce that $X^c - 1$ must divide $B_1(X) + X^r B_2(X) + (X^{qc} - 1)B_2(X)$, therefore, $B_1(X) + X^r B_2(X)$. We may write $B_1 X(X) + X^r B_2(X) = (X^c - 1)M(X)$. We proceed in two stages.

Let us suppose that $M(X)$ is not zero. We observe that we have $r + d - 1 > b - 1$, and that, therefore, $r + d - 1 = c + u$, where u is the degree of $M(X)$. Thus $r = c - d + 1 + u \geq b + u$. From this we deduce that $r \geq b > b - 1$ and that $r > u$. From this we see that in the right-hand side term there exists a monomial X^r, but that this monomial cannot exist in the right-hand side term. The contradiction is obvious.

Thus, $B_1(X) + X^r B_2(X) = 0$, which involves (because of the constant of $B_1(X)$) that $r = 0$ and $B_1(X) = B_2(X)$. Thus we are led to suppose that C contains a polynomial $B_1(X)(1 + X^{qc})$. Since $g(X)$ is divisible by $p(X)$, the latter must divide $B_1(X)(1 + X^{qc})$. Due to the degrees it cannot divide $M_1(X)$. It thus divides $1 + X^{qc}$. Thus, $1 + X^c$ and $1 + X^n$ must divide $1 + X^{qc}$, which is impossible since $qc < n$ and n is the LCM of e and c. □

2.4.3.4. *RM codes*

Let α be a primitive of $\mathbb{F}_q, q = 2^m$. Binary RM codes are defined by the expression of the powers of α, which are roots of the code. Their roots is the α^i such that the Hamming weight of the binary decomposition of i is strictly lower that $m - r$. It is an RM code of the order r.

PROPOSITION 2.38. *An RM code of the order r has length* $q-1$, *dimension* $1 + \binom{m}{1} + \binom{m}{2} + \cdots + \binom{m}{r}$, *and minimum distance* $2^{m-r} - 1$.

Proof. The proof is outside the scope of this work. □

2.4.3.5. *RS codes*

RS codes are codes whose coefficients are in $\mathbb{F}_{2^r}$, with $r \geq 2$, of length $2^r - 1$. There roots are $\alpha^i, \alpha^{i+1}, \ldots, \alpha^{i+\delta-1}$ where α is a primitive.

PROPOSITION 2.39. *Such a code is a* $(q-1, q-1-\delta, \delta+1)$ *code. Its BCH distance is its true distance.*

Proof. We have a $\deg g(X) = \delta$, and $W_H(g(X)) = \delta + 1$. □

2.4.3.6. *Codes with true distance greater than their BCH distance*

In an exhaustive article[1] "One the minimum distance of cyclic codes", J.H. van Lint and R.M. Wilson provide all the binary codes with length not exceeding 61 that have a true distance strictly larger than their BCH distance. Here are some examples; each code is described in the form (length, dimension, true distance, BCH distance):

$$(31, 20, 6, 4), (31, 15, 8, 5), (39, 12, 12, 8), (45, 16, 10, 8),$$

$$(47, 23, 12, 6), (55, 30, 10, 5), (55, 20, 16, 8)$$

They are decodable by the FG-algorithm which we provide hereafter.

2.4.3.7. *PG-codes*

2.4.3.7.1. Introduction

We recall (see section 2.3) that we regard $\mathbb{F}_{q^{m+i}}$ as a vector space of dimension $m + 1$ in $\mathbb{F}_q$, with $q = 2^s$. Let α be a primitive of $\mathbb{F}_{q^{m+i}}$ and β a primitive of $\mathbb{F}_q$. We have $\beta = \alpha^{\frac{q^{m+1}-1}{q-1}}$. We have a projective geometry with $n = \frac{q^{m+1}-1}{q-1}$ points.

We will construct codes in $\mathbb{F}_2[X]/(X^n - 1)$, but prior to this we will provide two definitions.

1. VAN LINT J.H., WILSON R.M., "On the minimum distance of cyclic codes".

DEFINITION 2.12. *Let there be 2 integers i and j, and their respective writings in base 2 be $(i_0, \ldots, i_u)$ and $(j_0, \ldots, j_u)$. It will be said that j is under i, if $(j_0, \ldots, j_u)$ is less than or equal to $(i_0, \ldots, i_u)$ for the produced order.*

For example, $j = 25$ and $i = 37$. Then j is not under i. With $j = 25$ and $i = 29$, j is under i.

DEFINITION 2.13. *We use $W_s(t(2^s - 1))$ to indicate the maximum number of integers in the form $i(2^s - 1)$ disjoint two by two that are under $t(2^s - 1)$.*

For example $s = 2, t = 5$. Since 3 and 12 are under 15 and are disjoint, we have $W_2(5(2^2 - l)) = 2$.

2.4.3.7.2. PG-codes

We consider the code C such that its orthogonal $C^\perp$ contains all the projective subspaces of the order r of the field $\mathbb{F}_{q^{m+i}}$. This code C is a code called a *PG-code of the order* r. The code $C^\perp$ is characterized by the following proposition.

PROPOSITION 2.40. *The $C^\perp$ code has for zeros all the elements in the form $\alpha^{t(2s-1)}$ where $W_s(t(2^s - 1)) \leq r, t \neq 0$.*

Proof. The proof is outside the scope of this work. □

There does not exist a formula giving the dimension of this code. It should be constructed, so that the sought code C can be deduced from it.

The minimum distance of C is given in the following proposition.

PROPOSITION 2.41. *The BCH distance of the code C is given by:*

$$d_{\text{BCH}} = \frac{p^{s(m-r+1)} - 1}{p^s - 1} + 1$$

Proof. The proof is outside the scope of this work. □

The error correcting capability of the code C stems from the following proposition.

PROPOSITION 2.42. *We have the following results:*

1) the number J of projective subspaces of rank r, containing a projective subspace of a fixed rank r - 1, verifies: $J = d_{\text{BCH}} - 1$;

2) the J projective subspaces have a two by two intersection which is reduced to a projective subspace of the order $r-1$;

3) we can correct up to $J/2$ errors by majority decoding.

Proof. Indeed:

1) it is the number of subspaces of the field that contains a fixed subspace of dimension r;

2) two projective subspaces of order r cannot have an intersection of order $r-1$ since they are distinct;

3) see section 2.6. □

PROPOSITION 2.43. *The length of C is equal to the number of points.*

Proof. This is straightforward. □

Since $d_{BCH} = J+1$, we can correct up to $J/2$ errors. The J projective subspaces are disjoint two by two, apart from the projective subspace of order $r-1$.

2.4.3.7.3. An application

Majority decoding makes it possible to carry out a cheap and fast electronic operation, especially when decoding is in one stage. If decoding has more than three stages, the complexity becomes very high.

The Japanese needed to find a powerful code with a cheap decoder. They wanted to use it for their Teletext. Constraints: length of information 81, number of errors to be corrected: 8. The solution found uses an information length of 82. The code is then shortened by one position. The respective values of the parameters are: $p = 2$, $r = 1$, $m = 2$. Decoding has 1 stage. We deduce from it the dimension of C: 82, the length of the code: 273, its error correcting capability: 8 ($s = 4$, because $273 = 2^{4\times2} + 2^4 + 1$). It is a (273, 82, 18) code shortened by 1 position, decodable by majority vote with 1 level. The price of an encoder/decoder was 175 FF in 1995.

2.4.3.8. *QR codes*

Binary quadratic residue codes (QR codes) have a length p, where p is a prime number in the form $8m+1$ or $8m-1$. For each such p there are 4 QR codes. One has all the *modulo p* squares as roots, another has all these squares and 1, another has all the non squares, and the last one has all non squares and 1.

PROPOSITION 2.44. *If $p = 8m+1$, then $d^2 > p$, and if $p = 8m-1$, then $d(d-1) \geq p-1$.*

Proof. The proof is outside the scope of this work. □

These codes have an important group of automorphisms. We can then think of finding good algorithms of trapping isolated errors.

2.4.4. *Existence and construction of cyclic codes*

Faced with a list of tasks proposed by an industrialist, we are often led to seek if there exists a code which fulfills the requirements. If one does exist, it then has to be constructed. There are tables of known codes, for a certain number of values of the parameters n, k and d.

2.4.4.1. *Existence*

It is often useful to simply know if there exists a cyclic code with the given parameters. It is the case when we are trying to satisfy a list of tasks. The first stage consists of testing the existence of a code. More precisely, we are led to examine whether there exists a polynomial $g(X)$, divisor of $X^n - 1$, with a given degree.

PROPOSITION 2.45. *There exists a generator $g(X)$, divisor of $X^n - 1$, with a given degree s, and only if there exists in $\mathcal{Z}/(n)$ a set of cyclotomic classes under multiplication by 2, whose cardinal is equal to s.*

Proof. Let $\mathbb{F}_{2^r}$ be the smallest field containing the nth roots of unity. Let α be a primitive of this field, and β be a nth primitive root of unity.

Let us suppose that the polynomial $g(X)$ has a degree s. Its s roots are powers of β. The corresponding exponents are elements of $\mathcal{Z}/(n)$. According to proposition 2.11 (see section 2.3), these roots are grouped by cyclotomic classes, and, therefore, by the powers.

The inverse is straightforward. The polynomial that admits cyclotomic classes as a set of roots divides $X^n - 1$ and is binary (see proposition 2.20). □

PROPOSITION 2.46. *In $\mathcal{Z}/(2^n - 1)$ there exists a cyclotomic class with a cardinal s if s divides n.*

Proof. If there exists a class with a cardinal s, then s divides n. Let there be x in $\mathcal{Z}/(2^n - 1)$, whose class has a cardinal s. By definition of cyclotomic classes we have $2^s x = x$. In addition, there is also $2^n x = x$. We use the Euclidean equality between x and n : $n = qs + r, 0 \leq r < s$. From there we obtain $(2^s)^q \times 2^r x = x$, then $2^s x = x$, which implies $r = 0$, otherwise the class of x would contain less than s elements. Thus, s divides n.

If there exists an s such that s divides n, then there exists a class with a cardinal s. Let $x = (2^n - 1)/(2^s - 1)$. We have $(2^s - 1)x = 0$, and the cardinal of the class of x is thus at most equal to s. Let us suppose that the cardinal is $t (0 < t < s)$.

We then have: $(2^t-1)x = 0$, i.e. in $\mathcal{Z}$: $(2^t-1)x = \mu \times n$. This implies: $(2^t-1)((2^n-1)/(2^s-1)) = \mu(2^n-1)$, from where $2^t - 1 = \mu(2^s-1)$, which is impossible. □

2.4.4.2. *Construction*

There exist various possibilities to construct a binary cyclic code with a given length n:

– we can use the cyclotomic classes of $\mathcal{Z}/(n)$, then construct minimum polynomials;

– we can directly seek $g(X)$ by factorizing $X^n - 1$;

– we may also be led to seek a code which contains given words.

2.4.4.2.1. Use of classes of $\mathcal{Z}/(n)$

As soon as we ensure the existence of the generator of a cyclic code $g(X)$ with length n, we construct it using the following proposition.

PROPOSITION 2.47. *Let $g(X)$ be the generator of a cyclic code of length n with a degree $n-k$. Let $\{C_{i_1}, C_{i_2}, \dots, C_{i_r}\}$ be the family of cyclotomic classes found in $\mathcal{Z}/(n)$. The polynomial $g(X)$ has the elements of the forms α^j as roots, where j traverses the joining of the classes $\{C_{\alpha i_1}, C_{\alpha i_2}, \dots, C_{\alpha i_r}\}$.*

Proof. This is straightforward. This proposition simply indicates the link between classes in $\mathcal{Z}/(n)$ and the roots of $g(X)$. We will note that the cardinal of the joining of classes must be equal $n-k$. □

2.4.4.2.2. Factorization by the Berlekamp method

We use a linear algebra method introduced by E. Berlekamp. This method is based on the following propositions describing and justifying the factorization of a polynomial $f(X)$ of the r degree. In the case of cyclic codes we are led to factorize polynomials of the form $X^n - 1$, for odd n.

PROPOSITION 2.48. *In $A = \mathbb{F}_2[X]/(f(X))$ the elevation to the square, which we will note h, is a linear endomorphism.*

Proof. We have successively:

$$a(X) \to h(a(X)) = a(X)^2 = r_a(X) + q_a(X) + q_a(X)f(X)$$

$$b(X) \to b(X)^2 = r_b(X) + q_b(X)f(X)$$

From this we deduce:

$$\begin{aligned} h(a(X) + b(X)) &= (a(X) + b(X))^2 = r_a^2(X) + r_b^2(X) + q(X)f(X) \\ &= r_a(X) + r_b(X) \end{aligned}$$

□

In $\mathbb{F}_2[X]$ let us suppose that $f(X)$ divides $a^2(X) - a(X)$, for a certain $a(X)$ of a degree strictly smaller than the degree of $f(X)$.

PROPOSITION 2.49. *The GCD $(f(X), a(X))$ is a non-trivial factor of $f(X)$.*

Proof. We have $a^2(X) - a(X) = a(X)(a(X) - 1) = \lambda(X)f(X)$, for a certain $\lambda(X)$. Any irreducible factor $p(X)$ of $f(X)$ divides either $a(X)$ or $a(X) - 1$, but not both, because otherwise it would divide their difference 1. Thus this PGCD $(f(X), a(X))$ is formed by a family of primary factors of $f(X)$. It can be equal neither to $f(X)$ nor to 1, because of the hypotheses regarding the degree of $a(X)$. □

To factorize $f(X)$ it is enough to find $a(X)$, which the Berlekamp method gives us. The identical application is noted id.

PROPOSITION 2.50. *Any element of A, different from 1, which is in the kernel of $h - id$, is such a polynomial $a(X)$.*

Proof. Indeed, $a(X) \in Ker(h - id)$ is equivalent to $a^2(X) - a(X) = 0$ in A, i.e. $a^2(X) - a(X) = \lambda(X)f(X)$ in $\mathbb{F}_2[X]$. □

To find the kernel of $h - id$ we proceed as follows:

1) considering the base $\{1, X, X^2, \dots, X^{r-1}\}$ (r is the degree of $f(X)$) we construct the matrix M of the endomorphism h, then we construct $M - I$;

2) using the Gaussian method we seek a base of the kernel of $M - I$;

3) if the only polynomial is 1, we cannot factorize. The polynomial $f(X)$ has only one primary factor. Otherwise, take a polynomial different from 1. It is the sought after $a(X)$;

4) we calculate $(f(X), a(X))$. We obtain a factor $f_i(X)$ of $f(X)$, then the second one by simple division of $f(X)$ by $f_1(X)$. We then have $f(X) = f_1(X)f_2(X)$, and we reiterate with these two new polynomials.

EXAMPLE 2.33 ($f(X) = 1 + X^2 + X^3 + X^4$). We have:

$$M - I = \begin{pmatrix} 0010 \\ 0101 \\ 0101 \\ 0010 \end{pmatrix}$$

from where successively, by the Gaussian algorithm:

$$\begin{pmatrix} 0010 \\ 1100 \\ 1100 \\ 0010 \end{pmatrix} \rightarrow \begin{pmatrix} 1100 \\ 0010 \\ 1100 \\ 0010 \end{pmatrix} \rightarrow \begin{pmatrix} 1100 \\ 0010 \\ 0000 \\ 0010 \end{pmatrix} \rightarrow \begin{pmatrix} 1010 \\ 0010 \\ 0000 \\ 0100 \end{pmatrix} \rightarrow \begin{pmatrix} 1010 \\ 0100 \\ 0000 \\ 0000 \end{pmatrix}$$

which yields the kernel matrix: $\binom{0101}{1000}$. It provides $a(X) = X + X^3$, and we easily find $(f(X), a(X)) = 1 + X$. The second factor is $1 + X + X^3$. Neither of these two factors can be factorized further.

We can prove that the factorization of $X^{2^r} - X$ gives all the irreducibles with a degree dividing r.

EXAMPLE 2.34. Let us factorize $X^7 - 1$ in $\mathbb{F}_2[X]$. With the same notations as in the previous example we have:

$$M = \begin{pmatrix} 1000000 \\ 0000100 \\ 0100000 \\ 0000010 \\ 0010000 \\ 0000001 \\ 0001000 \end{pmatrix} \text{ and } M - I = \begin{pmatrix} 0000000 \\ 0100100 \\ 0110000 \\ 0001010 \\ 0010100 \\ 0000011 \\ 0001001 \end{pmatrix}$$

We notice the simplicity of the construction of this matrix. Using the Gaussian method we obtain the needed matrix:

$$M = \begin{pmatrix} 0110100 \\ 0001011 \\ 1000000 \end{pmatrix}$$

We take $a(X) = X + X^2 + X^3$ (first line) and we obtain $f_1(X) = 1 + X + X^3$, then $f_2(X) = 1 + X + X^2 + X^4$. We factorize $f_2(X)$. The new matrix $M - I$ equals:

$$M - I = \begin{pmatrix} 0011 \\ 0111 \\ 0100 \\ 0000 \end{pmatrix}$$

The kernel matrix is $\binom{0011}{1000}$.

We take $a(X) = X^2 + X^3$ and obtain $1 + X + X^2 + X^4 = (1 + X)(1 + X^2 + X^3)$. We factorize $f_1(X)$. The new matrix $M - I$ equals:

$$M - I = \begin{pmatrix} 000 \\ 011 \\ 010 \end{pmatrix}$$

The kernel matrix is (1000).

We cannot factorize further. It can be easily verified that $1 + X + X^3$ is irreducible. Finally we have $X^7 - 1 = (1 + X)(1 + X + X^3)(1 + X^2 + X^3)$. Thus, there are $2^3 - 2$ non-trivial cyclic codes (the trivial ones have a 0 or 1 generator).

2.4.4.2.3. Construction of a cyclic code generated by given words

We may sometimes have to find the smallest cyclic code, which contains one or more given codewords. Let $m(X)$ be a given binary codeword of length n. We consider it as an element of $A = \mathbb{F}_2[X]/(X^n - 1)$. We seek the smallest cyclic code of A containing this codeword.

PROPOSITION 2.51. *Let $\lambda(X)$ be the polynomial of the smallest degree, such that we have $\lambda(X)m(X) = 0$. The required code is the ideal $((X^n - 1)/\lambda(X))$.*

Proof. Indeed:

1) the set of polynomials $u(X)$, such that $u(X)m(X) = 0$ is an ideal of A. As any ideal is principal, this ideal is generated by a polynomial with the smallest possible degree. Thus, it is the polynomial $\lambda(X)$ of the statement;

2) the polynomial $\lambda(X)$ divides $X^n - 1$. Thus, $m(X)$ is in the code $((X^n - 1)/\lambda(X))$. We pose $g(X) = (X^n - 1)/\lambda(X)$;

3) everything under the strict code of the code $(g(X))$ is generated by a polynomial of the form $u(X) \times g(X)$ (with $u(X) \neq 1$). If $m(X)$ is in such a subcode, then $m(X)$ must be canceled by $(X^n - 1)/u(X)g(X)$, which is impossible, because its degree is strictly smaller than that of $\lambda(X)$.

Thus, the required code is $((X^n - 1)/\lambda(X))$. □

Using the Gaussian method pivots we easily find the required code. Let us note that the pivots may be in any column.

EXAMPLE 2.35. Find the smallest cyclic code containing the following codeword: 110010100001110. By the Gaussian method we find, for example:

$$110010100001110\,1$$

$$011001010000111\,X$$

$$011110001001101\,1 + X^2$$

$$101111000100110\,X + X^3$$

$$100111000011101\,1 + X^2 + X^4$$

$$000000000000000\,1 + X + X^3 + X^5 = \lambda(X)$$

Each binary vector-row is followed on its right by a polynomial $v(X)$. This translates the fact that the row is equal to $v(X)m(X)$. These polynomials $v(X)$ appear during the application of the Gaussian method.

We find g $(X) = (X^{15} - 1)/\lambda(X) = 1 + X + X^2 + X^4 + X^5 + X^8 + X^{10}$. The pivots are in columns 7, 8, 9, 10 and 11, and the first column is on the right with number 1.

In the general case we want to determine the smallest cyclic code containing the codewords $m_1(X), m_2(X), \ldots, m_s(X)$. Using the previous construction we construct the polynomials $\lambda(X), \lambda_2(X), \ldots, \lambda_s$. The polynomial with the smallest degree canceling the $m_i(X)$ is clearly the LCM of the $\lambda_i(X)$. Another method is to seek the PGCD $(m_1(X), m_2(X), \ldots, m_s(X))$. It is the generator of the required code.

2.4.4.3. *Shortened codes and extended codes*

2.4.4.3.1. Shortened codes

We remove the s first components of information from each codeword of the code C. This amounts to considering only those codewords in C, which have these s components equal to 0. A shortened cyclic code is a linear code.

2.4.4.3.2. Extended codes

We add a parity symbol to each codeword, which is such that the sum of the symbols of each extended codeword is even. The following is an interesting question: is it possible for an extended cyclic code to be cyclic? That is one of the suggested exercises.

2.4.4.4. *Specifications*

A specification is a set of constraints that the system of coding must satisfy. The principal parameters, which industry specialists make a point of taking into account, are as follows:

- length L of the information string to be coded,
- maximum redundancy rate,
- maximum length of the codewords,
- gross flow (i.e. in terms of binary symbols),
- net flow (i.e. in terms of information bits),
- residual error rate for an input error rate (i.e. p_r for p_e),
- average space without errors between two badly decoded consecutive words,
- electronic constraints (delicate).

2.4.4.5. *How should we look for a cyclic code?*

There is no general method. We can start by looking for the possible values of k: those that divide the length of information strings. Then one can try associating the possible values of n to each possible value of k. We will study the values of k and n by ascending values, in order to have the shortest possible code (thus, *a priori*, the most economic). Then for fixed n and k we must examine whether there exists a cyclic code. To that end we study the distribution of cyclotomic classes in $\mathcal{Z}/(n)$.

On the basis of this study we find what can be the error correcting capability of the code. It is then necessary to use the formulae connecting p_r to p_e, as well as the BCH theorem. We can have an idea of the decoding power of the code from the following proposition.

PROPOSITION 2.52. *Let there be a cyclic code of length n, dimension k, with error correcting capability of t errors per word. Let p_c be probability of channel error, and p_r be the residual probability per corrected word. We have the inequality:*

$$np_r \leq \sum_{i=t+1}^{n} (t+i)\binom{n}{i} p_c^i (1-p_c)^{n-i}$$

Proof. Since p_r is the probability of error per symbol of a received word, the expectation of the number of residual errors per corrected word is np_r (binomial distribution). We will provide an increase of this expectation.

Let us consider the event "the word has been decoded incorrectly". This event is included in the following event E "for any value i ($i = t+1, t+2, \ldots, n$) of the number of transmission errors occurring, the decoding algorithm decodes using likelihood decoding". This means that the number of errors in the "corrected" word is at most equal to $i+1$ (i comes from the channel, t comes from decoding). The expectation of the number of errors in this event E is:

$$\sum_{i=t+1}^{n} (t+i)\binom{n}{i} p_c^i (1-p_c)^{n-i}$$

which yields the result announced in the statement. □

It will be noted that p_r is also the probability of residual errors for the information block recovered after decoding, provided that this decoding is systematic. Otherwise the residual rate is much greater.

In practice, when p_c is not greater than 10^{-3}, we are able to take as a relation:

$$np_r = (2t+1)\binom{n}{t+1} p_c^{t+1} (1-p_c)^{n-t-1}$$

It has to be well noted that the value p_r obtained is the one provided by the code. If we require a residual probability of p', we must then check for the code considered, the inequality:

$$(2t+1)\binom{n}{t+1} p_c^{t+1} (1-p_c)^{n-t-1} \leq np'$$

If it is satisfied, the code is acceptable.

Lastly, we are able to take into account the more delicate constraints on electronics.

EXAMPLE 2.36 (EXAMPLE OF CYCLIC SEARCH FOR CODE). Specifications:

– channel (i.e. input) error rate: 10^{-4},

– maximum redundancy rate: 0.18,

– maximum acceptable residual error rate: 10^{-5},

– information strings of length 105.

We will look for a natural (i.e. not shortened) cyclic code:

1) The possible values for k are the divisors of 105:

$$\{1, 3, 5, 7, 15, 21, 35, 105\}$$

2) Using the constraint on the redundancy rate we find the inequality $n \leq k/(0.82)$. This yields the possible values for n for a value of k:

k	1	3	5	7	15	21	35	105
n	1	3	3	7	15	15	31	127

We see that there exists, perhaps, a natural cyclic code of length 127. We examine the classes of $\mathcal{Z}/(127)$. We will look for a joining of these classes, with a cardinal $127 - 105 = 22$. The cardinals of classes are divisors of 7 (because $127 = 2^7 - 1$). There are, thus, cardinal classes 1, 7. Since $22 = 3 \times 7 + 1$, we conclude that there exists a code whose roots contain $\{\alpha, \alpha^3, \alpha^5, \alpha^0\}$.

The apparent distance of the code is 8. It thus corrects 3 errors per codeword. If we approximate the member on the right of the formula linking p_r to p_e by $(t+1) \times \binom{n}{(t+1)} \times p^{t+1} \times (1 - p_e)^{n-t-1}$, we must verify the inequality:

$$(t+1) \times \binom{n}{t} \times p^t \times (1 - p_e)^{n-t} \leq 127 \times 10^{-5}.$$

We obtain: $(4) \times \binom{127}{4} \times 10^{-16} \times (0.9999)^{123}$, to compare with 127×10^{-5}. We also have: 4.084×10^9 to compare with 127×10^{-5}. It is acceptable, therefore there exists a natural cyclic code that satisfies the required constraints.

2.4.4.6. *How should we look for a truncated cyclic code?*

We conduct the same study as before, but we allow ourselves to truncate the considered codes, which yields a greater choice.

2.4.5. *Applications of cyclic codes*

Since the beginning of 1970s many applications of error correcting codes have been introduced. Let us cite a few.

– transmissions of images by remote spacecrafts,

– satellite transmissions,

– underwater transmissions,

– optical discs,

– Hubble,

– bar-codes,

– computer memory,

– mobiles,

– CD readers,

– cryptography.

2.5. Electronic circuits

The implementation of error correction on board a satellite, a remote spacecraft, in computer memory, on optical or magnetic discs, in CD readers, is carried using electronic circuits. These circuits primarily use shift registers, carrying out multiplications or divisions of polynomials with coefficients in $\mathbb{F}_2$ or $\mathbb{F}_{2^r}$.

In this section, circuits are drawn without taking traditional standards into account, as far as logical gates and oscillation are concerned. We will not represent connections with the clock.

2.5.1. *Basic gates for error correcting codes*

There is the flip-flop, represented as follows, which contains a binary value. This flip-flop is under the control of a clock. With each beat (or signal) of this clock the flip-flop transmits the value that it contained and receives the value presented at input. A flip-flop has an input and an output (see Figure 2.2).

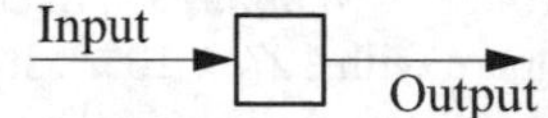

Figure 2.2. *Flip-flop (or oscillation)*

There are also logical gates, "OR", "AND", "exclusive OR" represented as follows (see Figure 2.3).

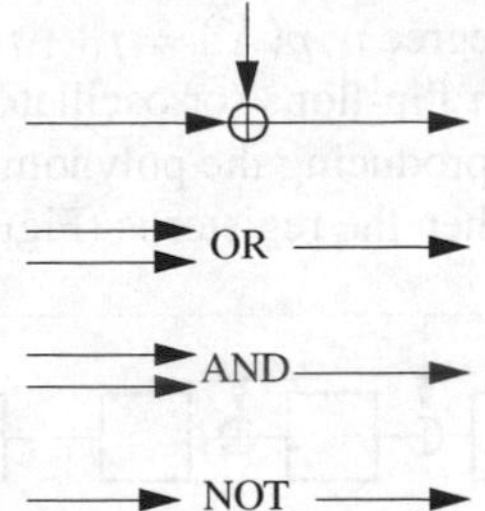

Figure 2.3. *Logical gates*

These logical gates are not under the control of the clock.

If two signals follow different sets of logical gates, the difference in propagation time is one of the limitations of certain algorithms.

In transmissions binary vectors often represent binary polynomials. These always circulate from the coefficient with the highest degree to the constant coefficient. When a polynomial enters a register it enters starting with the monomial coefficient of the highest degree. For example, if we take the polynomial $1 + X + X^4 + X^7$ the set of binary values associated to it is 11001001, and the first appearing at the input of the register will be the coefficient on the right.

2.5.2. *Shift registers*

A shift register (Figure 2.4) is a succession of *flip-flops* connected to each other in sequence. Such a register, in general, has one input and one output.

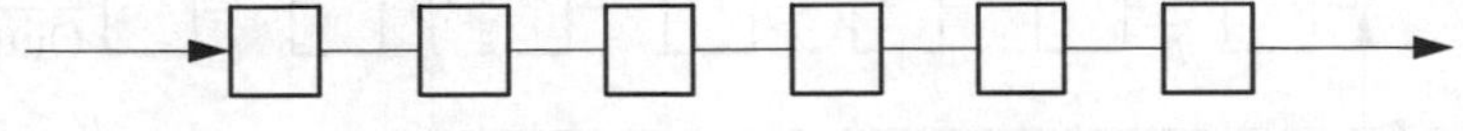

Figure 2.4. *Shift register*

2.5.3. *Circuits for the correct codes*

From shift registers we construct increasingly complex circuits, used in encoders and decoders.

2.5.3.1. *Divisors*

Such a circuit has an input and sometimes also an output. It divides the input by the polynomial corresponding to the *feedback* connections.

To a binary polynomial of degree n, $p(X) = p_0 + p_1X + \cdots + p_nX^n$ we associate a feedback register formed by n flip-flops (or oscillator), *modulo* 2 adders (with 2 or more inputs), and a feedback reproducing the polynomial coefficients. For example, if $p(X) = 1 + X^2 + X^3 + X^6$, then the register is (Figure 2.5):

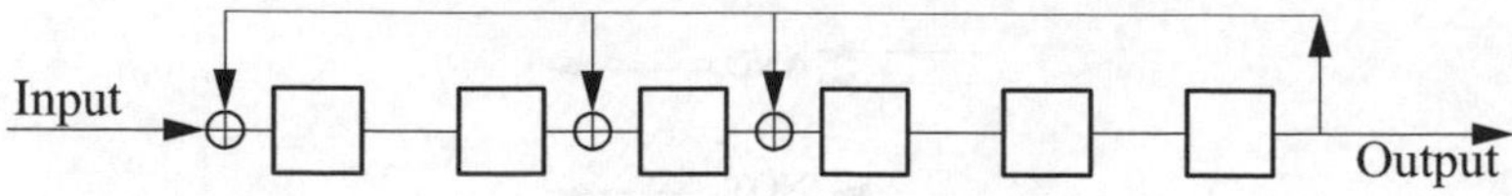

Figure 2.5. *Divisor*

PROPOSITION 2.53. *Let there be a register with shifts with connections that correspond to an irreducible polynomial p(X). Let us suppose that this register turns autonomous (i.e. without input), and that the initial contents are not zero. A shift corresponds to the multiplication by X in the field* $\mathbb{F}_2[X]/(p(X))$.

Proof. The register with shifts in fact translates into the inequality:

$$p_0 + p_1X + \cdots + p_{n-1}X^{n-1} = p_nX^n$$

which is equivalent to calculating *modulo* $p(X)$. Since a shift is equivalent to multiplying the contents of the register by X, the register with *feedback* multiplies the contents of the register by X *modulo* $p(X)$. □

2.5.3.2. *Multipliers*

The following circuit R in Figure 2.6 multiplies the entry by the polynomial associated to connections. If we input 1, the contents of R are equal to the polynomial of connections, i.e. $1 + X + X^5$.

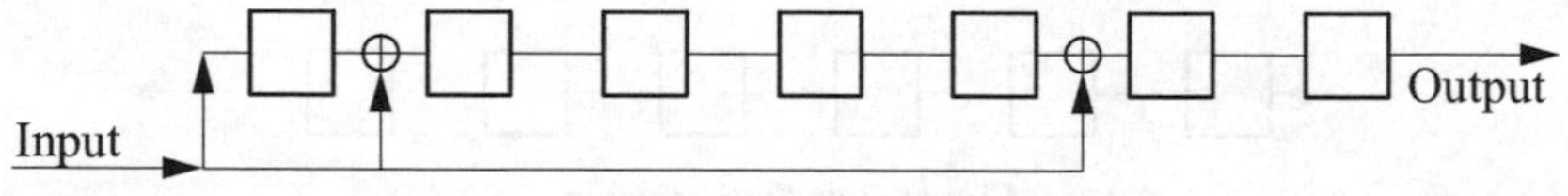

Figure 2.6. *Multiplier*

2.5.3.3. *Multiplier-divisors*

The following circuit in Figure 2.7 multiplies the input by the polynomial of the lower connections, and at the same time calculates the *modulo* result of the polynomial of the upper connections.

2.5.3.4. *Encoder (systematic coding)*

It is a particular case of the multiplier-divisors. For a systematic coding we multiply the information polynomial (at input) by $X^{\deg g(X)}$.

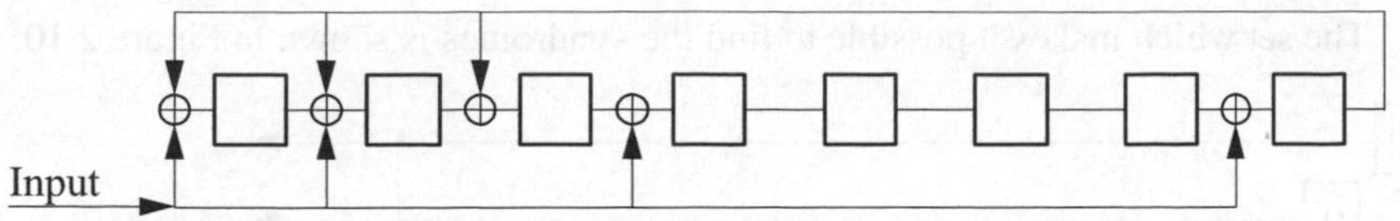

Figure 2.7. *Dividing multiplier*

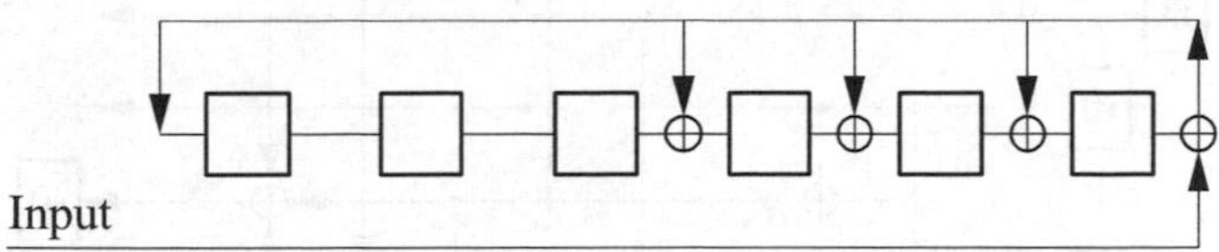

Figure 2.8. *Encoder*

2.5.3.5. *Inverse calculation in* $\mathbb{F}_q$

Let us consider the field $\mathbb{F}_q = \mathbb{F}_2[X]/(p(X))$, where $p(X)$ is an irreducible primitive (meaning that X is a primitive of the field). Let $a(X)$ be a non-zero element of this field. We can express $a(X)$ as a power of X : $a(X) = X^i$. Its reverse, let us say $b(X)$, is equal to X^{q-1-i}. If we multiply $a(X)$ by X^{q-1-i}, we then find 1. The calculation of the inverse of $a(X)$ can be carried out using the registers drawn hereunder in Figure 2.9. In this example they calculate the inverse of $1 + X + X^2 + X^3$ in the field $\mathbb{F}_2[X]/(1 + X^3 + X^5)$.

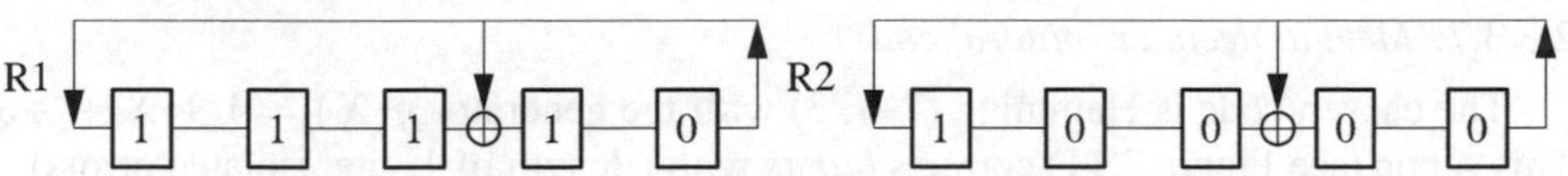

Figure 2.9. *Calculation of the inverse*

2.5.3.6. *Hsiao decoder*

This code is well adapted to computer read-write memories thanks to its decoding speed.

As a parity check matrix of the Hsiao code we take the following matrix M:

$$\begin{pmatrix} 10001110 \\ 01001101 \\ 00101011 \\ 00010111 \end{pmatrix}$$

We wish to transmit the following word $m = (i_0, i_1, i_2, i_3, r_0, r_1, r_2, r_3)$, where the i_j are information bits, and r are the redundancy symbols. With the receiver we calculate the syndromes (s_0, s_1, s_2, s_3), obtaining the product $M \times m^T$ (i_0 is on top).

The set which makes it possible to find the syndromes is shown in Figure 2.10.

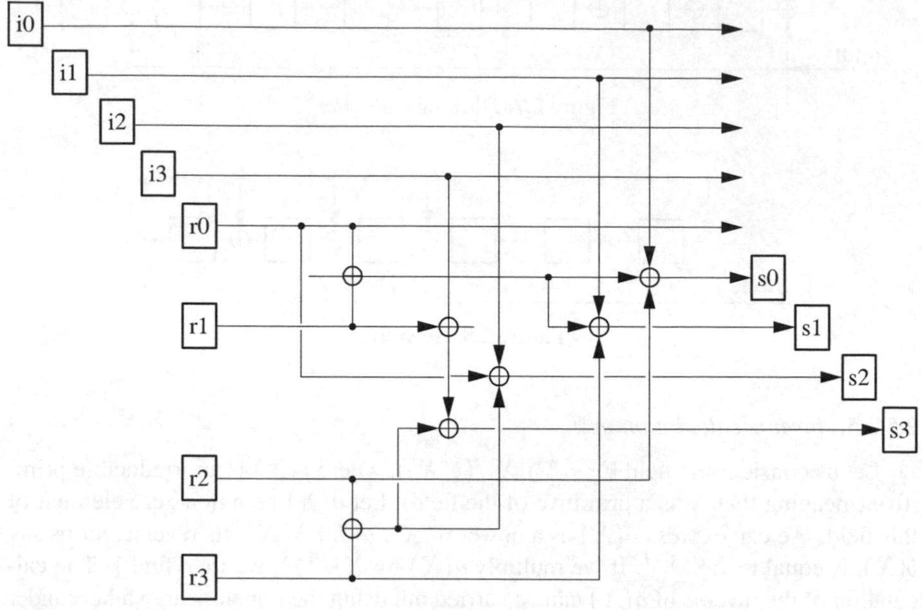

Figure 2.10. *Hsiao decoder*

2.5.3.7. *Meggitt decoder (natural code)*

The chosen code is Hamming (7, 4, 3) with the generator $g(X) = 1 + X + X3$. This set up (see Figure 2.11) corrects *bursts* with a length of 1, (i.e. isolated errors).

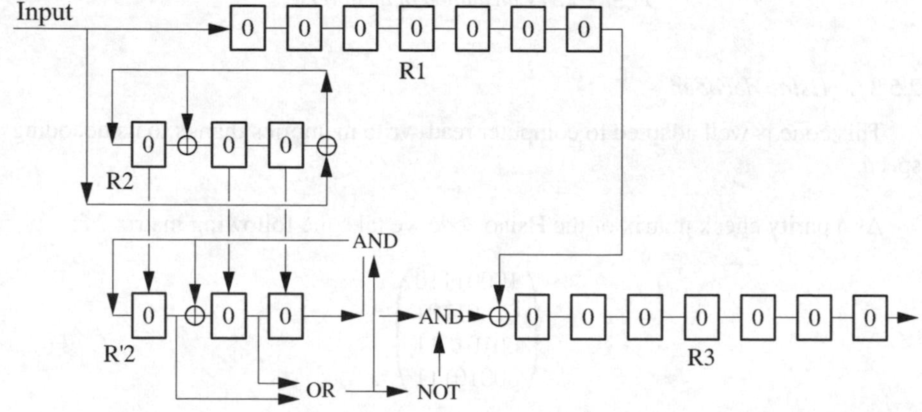

Figure 2.11. *Meggitt decoder (natural code)*

2.5.3.8. *Meggitt decoder (shortened code)*

The chosen code is the Hamming code, shortened by one position, i.e. the (6,3,3) code. Let us note that we "enter" the register of division by $g(X)$ by pre-multiplying by $X^{1+\deg g(X)} = X^4$, which is equal to $X + X^2$ *modulo* $g(X)$.

This set up (see Figure 2.12) corrects *bursts* with a length of 1 for each word of length 6.

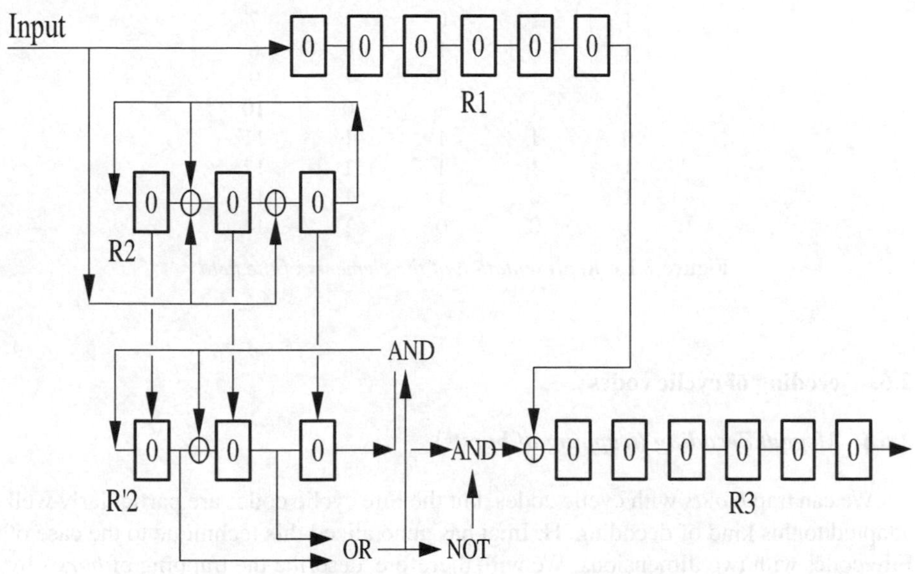

Figure 2.12. *Meggitt decoder (truncated code)*

2.5.4. *Polynomial representation and representation to the power of a primitive representation for a field*

We take a register whose *feedback* is an irreducible primitive of the degree n. We take $\alpha = X$ as the primitive element. We initialize the register with $(1, 0, 0, \dots, 0)$ and make it turn $2^n - 2$ times. We obtain two representations of the $\mathbb{F}_{2^n}$ field, one of them polynomial (it is the sequence of the register's contents), the other is the sequence of the powers of the primitive a (it is the sequence of clock signal numbers).

In the following example (Figure 2.13) we give two representations of the non-zero elements of $\mathbb{F}_{16}$ represented by $\mathbb{F}_2[X]/(1 + X + X4)$. The primitive selected is X. The column on the right provides the powers of X corresponding to the polynomial writing given on the left.

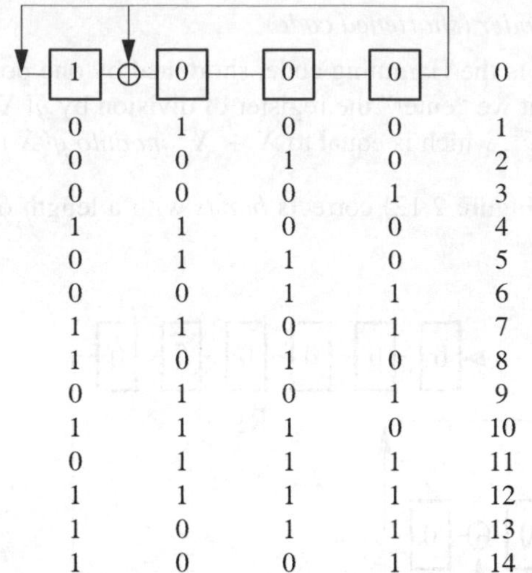

1	0	0	0	0
0	1	0	0	1
0	0	1	0	2
0	0	0	1	3
1	1	0	0	4
0	1	1	0	5
0	0	1	1	6
1	1	0	1	7
1	0	1	0	8
0	1	0	1	9
1	1	1	0	10
0	1	1	1	11
1	1	1	1	12
1	0	1	1	13
1	0	0	1	14

Figure 2.13. *Representation of the elements of the field*

2.6. Decoding of cyclic codes

2.6.1. *Meggitt decoding (trapping of bursts)*

We can trap *bursts* with cyclic codes, but the Fire cyclic codes are particularly well adapted to this kind of decoding. H. Imai has generalized this technique to the case of Fire codes with two dimensions. We will, therefore, describe the trapping of *bursts* by means of a binary Fire code, of length n and generator $g(X)$.

We will suppose that the *burst* $b(X)$ has a maximum corrigible length b. The other cases are directly derive from here. It is thus represented as a polynomial of the $b-1$ degree, and with a constant 1.

2.6.1.1. *The principle of trapping of bursts*

The received word $r(X)$ is equal to the transmitted word $c(X)$, to which a *burst* $X^i b(X)$ has been added during transmission. Simultaneously with calculating the remainder of $r(X)$ in an associated divisor register $g(X), R_g$ we memorize $r(X)$ in a register with shifts, R of length n. We wish to achieve that the *burst* also be at the output of the register associated with $g(X)$ when it is at the output of the register with shifts. By simple addition we then eliminate the *burst*. The two following propositions bring the solution.

2.6.1.2. *Trapping in the case of natural Fire codes*

The following proposition concerns the solution in the case of natural Fire codes.

PROPOSITION 2.54. *Pre-multiplying the input of R_g by $X^{\deg g(X)}$ we will obtain the burst at the output of R and the output of R_g simultaneously.*

Proof. Let $r(X)$ be the remainder of the division of the *burst* $X^i b(X)$ by $g(X)$. After $[n-(i+b)]$ clock beats the *burst* became $X^{n-b}b(X)$ (which is of the $n-1$ degree). It is wedged against the register with shifts R. In the register associated to $g(X)$ we have $b(X)$ after $n-i$ signals. In this register there is the wedged *burst* after $[n-(i+b)+\deg g(X)]$ clock beats. By pre-multiplying the input of R_g by $X^{\deg g(X)}$ we can correct the *burst*. □

2.6.1.3. *Trapping in the case of shortened Fire codes*

In general, Fire codes have a very large length. We are thus led to shorten them. We passes from n to a shortened length n'. The register R is shortened by a certain number of oscillations. The Meggitt decoding adapts well to this truncation.

PROPOSITION 2.55. *By pre-multiplying the input of R_g by $X^{n-n'+\deg g(X)}$ we will have the burst at the output of R and at the output of R_g at the same time.*

Proof. After $[n'-(i+b)]$ signals the *burst* became $X^{n'-b}b(X)$. It is wedged in R. In R_g we have $b(X)$ after $n-i$ signals. In this register we thus have the *burst* wedged after $[n-(i+b)+\deg g(X)]$ clock signals. To trap the *burst* we pre-multiply the input of R_g by $X^{n-n'+\deg g(X)}$. □

2.6.2. *Decoding by the DFT*

The Fourier transform is used in many fields, such as signal processing. The discrete Fourier transform (DFT) is used in finite bodies, for certain decodings.

2.6.2.1. *Definition of the DFT*

We consider the algebraic equation $A = \mathbb{F}_q[X]/(X^n-1)(q=2^r)$, odd n, where $\mathbb{F}_q$ is the smallest extension field of $\mathbb{F}_2$ containing the nth roots of unity (see section 2.3.6). Let β be a primitive n^{th} root of unity. For all $a(X)$ of A we pose $T(a(X)) = \sum_{i=0}^{n-1} a(\beta^i)X^i$ which defines the DFT, noted here T.

2.6.2.2. *Some properties of the DFT*

The five following propositions express certain properties of the discrete Fourier transform.

PROPOSITION 2.56. *T is a bijective linear application of A in A.*

Proof. Let us prove that T is linear. We have directly:

$$T(a(X) + b(X)) = \sum_{i=0}^{n-1}(a(\beta^i) + b(\beta^i))X^i = T(a(X)) + T(b(X))$$

Moreover:

$$T(\lambda a(X)) = \lambda T(a(X)), \text{ for all } \lambda \text{ of } \mathbb{F}_q$$

Let us prove that T is bijective. Since A is finite, it is enough to prove that T is injective, i.e. its kernel is reduced to $\{0\}$. Let there be $a(X)$ such that $T(a(X)) = 0$. Then $a(X)$ admits as roots the n distinct elements $\beta^0, \beta^1, \beta^2, \ldots, \beta^{n-1}$. Since its degree does not exceed $n-1$, $a(X)$ must be zero (see proposition 2.10). □

PROPOSITION 2.57. *We have the two following properties:*

1) $T^2 = \tau$,

2) $T^4 = id$.

Proof. Indeed:

1) Let us pose $a(X) = \sum_{i=0}^{n-1} a_j X^j$. We have:

$$\begin{aligned}
T^2(a(X)) &= T\left[\sum_{i=0}^{n-1} a(\beta^i)X^i\right] \\
&= \sum_{i=0}^{n-1} a(\beta^i)\sum_{j=0}^{n-1}\beta^{ji}X^j \\
&= \sum_j\left[\sum_i a(\beta^i)\beta^{ji}\right]X^j \\
&= \sum_j\left[\sum_i\sum_{k=0}^{n-1} a_k\beta^{ik}\beta^{ji}\right]X^j \\
&= \sum_j X^j\sum_k a_k\left[\sum_{i=0}^{n-1}\beta^{i(j+k)}\right]
\end{aligned}$$

However, $\sum_{i=0}^{n-1}\beta^{i(j+k)}$ equals 0 unless $k = -j$ (see proposition 2.22). We thus also have $T^2(a(X)) = \sum_j X^j a_{-j}$, which is equal to $\tau(a(X))$;

2) The proof of the second property is straightforward. □

Now we provide A with a second product, noted $*$, called *component by component product*, defined as follows. For $a(X) = \sum_{i=0}^{n-1} a_i X_i$ and $b(X) = \sum_{i=0}^{n-1} b_i X^i$ we pose:

$$a(X) * b(X) = \sum_{i=0}^{n-1} a_i b_i X^i$$

It is easily proven that A provided with the two laws $+$ and $*$ is a ring.

PROPOSITION 2.58. *T has the two following properties:*

1) $T(a(X)b(X)) = T(a(X)) * T(b(X))$,

2) $T(a(X) * b(X)) = T(a(X))T(b(X))$.

Proof. Indeed:

1) the proof of the first property is straightforward;

2) since T is surjective we have $a(X) = T(a'(X))$ and $b(X) = T(b'(X))$, for certain $a'(X)$ and $b'(X)$. We can write:

$$\begin{aligned} T(a(X) * b(X)) &= T(T(a'(X)) * T(b'(X))) \\ &= T(T(a'(X)b'(X))) \\ &= \tau(a'(X)b'(X)) \\ &= \tau(a'(X))\tau(b'(X)). \end{aligned}$$

But we also have $a'(X) = \tau(T(a(X)))$ and $b'(X) = \tau(T(b(X)))$. Thus $T(a(X)* b(X)) = T(a(X))T(b(X))$, because $\tau^2 = id$. □

PROPOSITION 2.59. *Let $s(X) \in A$. Let us note $W_H(s)$ its Hamming weight:*

1) the set E of the polynomials $L(X)$ such that $s(X) * T(L(X)) = 0$ is an ideal of A generated by a polynomial noted $L_s(X)$;

2) the degree of $L_s(X)$ is $W_H(s)$.

Proof. Indeed:

1) if $s(X) * T(L_1(X)) = s(X) * T(L_2(X)) = 0$, then $s(X) * T(L_1(X) + L_2(X)) = 0$. Moreover, $s(X) * T(XL_1(X)) = 0$, which proves this point;

2) if $s(X) * T(L(X)) = 0$ then $L(\beta^i) = 0$ as long as s_i is a non-zero coefficient of $s(X)$. The polynomial $L_s(X)$ is thus that which has as roots β^i such that s_i is a non-zero coefficient of $s(X)$. Its degree is thus the Hamming weight of $s(X)$. □

PROPOSITION 2.60. *Let $s(X) \in A$ and $S(X) = T(s(X))$. A polynomial $L(X)$ verifies $\langle S(X), X^i L(X)\rangle = 0, i = 0, 1, \ldots, n-1$, if $s(X) * T(L(X)) = 0$.*

Proof. By assumption we deduce $\sum_{i=0}^{n-1}\langle S(X), X^i L(X)\rangle X^i = 0$, i.e. $S(X)\tau(L(X)) = 0$ (see proof of proposition 2.29), or $T(S(X))T^2(L(X)) = 0$. We deduce from it:

$$\begin{aligned} 0 &= T\left(T(s(X))T^2(L(X))\right) \\ &= T^2(s(X)) * T^2\left(T(L(X))\right) \\ &= \tau\left(S(X) * T(L(X))\right) \end{aligned}$$

from where, finally, $s(X) * T(L(X)) = 0$. □

2.6.2.3. *Decoding using the DFT*

Let there be a cyclic code C, which we can suppose to be binary, length n. Let β a primitive n^{th} root of unity. Let us suppose that the set of the roots of code contain the sequence of powers $\beta, \beta^2, \beta^3, \dots, \beta^{2t}$, which is equivalent to saying that the code is a t corrector.

A word $c(X)$ of the code is transmitted. The received word $r(X)$ is equal to $c(X) + e(X)$ where $e(X)$ is the error word. We calculate the syndromes of $r(X)$, i.e. a part of the Fourier transform of $r(X)$ over the elements $\beta, \beta^2, \dots, \beta^{2t}$. Since $c(\beta^i) = 0$ for $i = 0, 1, 2, \dots, 2t$ we calculate, in fact, a part of the Fourier transform of $e(X)$. $T(e(X))$ is a polynomial whose coefficients from X to X^{2t} we know, which are the respective syndromes $S_1, S_2, \dots, S_{2t}$. We pose $S(X) = T(e(X))$. Since we want to calculate $e(X)$, we will seek the error positions. We are thus looking for the polynomial $L_e(X)$, of minimum degree, no more that t.

PROPOSITION 2.61. *Let us consider the following syndrome matrix:*

$$\begin{pmatrix} S_1 & S_2 & \cdots & S_{t+1} \\ S_2 & S_3 & \cdots & S_{t+2} \\ \vdots & \vdots & \ddots & \vdots \\ S_t & S_{t+1} & \cdots & S_{2t} \end{pmatrix}$$

There exists a linear combination of the r first columns (starting from the left) which is zero, if $L_e(X)$ is of the r degree.

Proof. If there exists such a zero linear combination, let us say $m_1 S_1 + m_2 S_2 + \cdots + m_{r-1} S_{r-1} + S_r = 0$ then the polynomial $m_1 + m_2 X + \cdots + m_{r-1} X^{r-1} + X^r$ is the polynomial of the lowest degree, which is orthogonal to $S(X)$ in the corresponding positions. It is thus the locator of $e(X)$. The reverse is direct. □

If the code is binary as we have supposed, it is enough to look for the roots of $L_e(X)$. Otherwise, we calculate the complete polynomial $S(X)$ using the relation of orthogonality $\langle S(X), X^i L_e(X)\rangle = 0$, then we calculate $T(S(X))$ that is equal to $\tau(e(X))$. Finally, we recover the error polynomial $e(X)$, and perform the correction.

We propose the following *decoding algorithm*, for a binary or non-binary code of length n:

1) calculate the $2t$ syndromes $r(\beta), r(\beta^2), \dots, r(\beta^{2t})$, which yield the $2t$ coefficients $S_1, \dots, S_{2t}$ of $S(X)$;

2) look for the kernel (straight) of the matrix:

$$\begin{pmatrix} S_1 & S_2 & \dots & S_{t1} \\ S_2 & S_3 & \dots & S_{t+1} \\ \vdots & \vdots & \ddots & \vdots \\ S_t & S_{t+1} & \dots & S_{2t} \end{pmatrix}$$

More precisely, we will look for the first zero combination of columns starting from the left. Only one vector appears then in the kernel, and it is the vector of the coefficients of $L_e(X)$;

3) calculate $S(X)$ completely finding all the scalar products $\langle S(X), X^i L_e(X)\rangle$. We obtain $\tau(e(X)) = T(S(X))$;

4) calculate the error polynomial $e(X)$ using $e(X) = \tau(\tau(e(X)))$;

5) correct the received word by cutting off $e(X)$ from it.

Let us give an example of decoding using the DFT.

EXAMPLE 2.37. Let there be a code of length $n = 15$, among the roots of which we have $\{\alpha, \alpha^2, \alpha^3, \alpha^4\}$, where α is a primitive of the $\mathbb{F}_{16}$ field represented by $\mathbb{F}_2[X]/(1 + X + X^4)$. The code thus corrects 2 errors.

We suppose that the error word is $e(X) = X + X^7$. We have $S_1 = \alpha + \alpha^7 = \alpha^{14}$, $S_2 = \alpha^{13}$, $S_3 = \alpha^2$, $S_4 = \alpha^{11}$. We look for the zero combination of the columns of the matrix:

$$\begin{pmatrix} S_1 & S_2 & S_3 \\ S_2 & S_3 & S_4 \end{pmatrix}$$

We find the relation:

$$\alpha^8 \begin{pmatrix} S_1 \\ S_2 \end{pmatrix} + \alpha^{14} \begin{pmatrix} S_2 \\ S_3 \end{pmatrix} + \begin{pmatrix} S_3 \\ S_4 \end{pmatrix} = (0)$$

The polynomial $L_e(X)$ is thus $\alpha^8 + \alpha^{14}X + X^2$. We deduce from it then, since $\langle L_e(X), X^j S(X)\rangle = 0, j = 0, 1, \dots, n-1$:

$$\alpha^8 S_3 + \alpha^{14} S_4 + S_5 = 0, \alpha^8 S_4 + \alpha^{14} S_5 + S_6 = 0, \dots$$

then, finally:

$$S(X) = \alpha^{14}X + \alpha^{13}X^2 + \alpha^2X^3 + \alpha^{11}X^4 + \alpha^4X^6 + \alpha^3X^7 + \alpha^7X^8 + \alpha X^9 + \alpha^9X^{11} + \alpha^8X^{12} + \alpha^{12}X^{13} + \alpha^6X^{14}$$

Applying T we find:

$$T(S(X)) = T^2(e(X)) = \sum_{j=0}^{14} S(\alpha^j)X^j = X^8 + X^{14}$$

Applying τ we find $X + X^7$, which is the error $e(X)$.

We observe that this algorithm works because we know $2t$ consecutive syndromes and that is sufficient to calculate $L_e(X)$.

2.6.3. *FG-decoding*

Just like in the previous decoding we will use the DFT but within a more general framework. We suppose to be using a binary code of length n, whose true distance d_{true} is strictly larger than its apparent distance d_{BCH}. We suppose that the sequence of $d_{\text{BCH}} - 1$ consecutive roots starts in β, where β is an nth primitive root of unity. FG-decoding can be also used if the sequence starts elsewhere than in β or if the code is not binary. For example, let us consider the code $(n = 21, k = 7, d_{\text{true}} = 8, d_{\text{BCH}} = 5)$. The 21st roots of unity are in $\mathbb{F}_{64}$. Let α be a primitive, such that $1 + \alpha + \alpha^6 = 0$. The element $\beta = \alpha^3$ is a primitive 21st root of unity. The cyclotomic classes containing the roots of the code are as follows:

$$C_\beta = \{\beta, \beta^2, \beta^4, \beta^8, \beta^{16}, \beta^{11}\}$$
$$C_{\beta^3} = \{\beta^3, \beta^6, \beta^{12}\}$$
$$\beta_{\beta^7} = \{\beta^7, \beta^{14}\}$$
$$C_{\beta^9} = \{\beta^9, \beta^{18}, \beta^{15}\}$$

The sequence of the $d_{\text{BCH}} - 1$ consecutive known syndromes is S_1, S_2, S_3, S_4. We note that it does not make it possible to determine the locator, which can have a degree of 3.

2.6.3.1. *Introduction*

The classical algorithms do not work because they require that the code has $2t$ consecutive roots to be able to correct at most t errors per word. There exists a table of the 147 codes whose length is less than 63. Many current publications describe the search for algorithms that make it possible to correct until t_{true}.

The Gröbner bases and the algorithm of B. Buchberger (a recent chapter in the theory of polynomials with n variables) gave birth to the famous algorithm of Chen *et al.*, but this algorithm is very complex. We have proposed a new algorithm, the FG-algorithm, which also uses Gröbner bases, but less complex. The name of this algorithm comes from Fourier (i.e. F) and Gröbner (G). We use the Fourier transform, with a system of polynomial equations with several variables solved by the algorithm of B. Buchberger.

The algorithm that we provide now is a simplified alternative of the general FG algorithm. This simplified algorithm applies to 139 of the 147 codes given in the table. The codes for which we must use the general FG are the following, each code being indicated by a sequence (length, dimension, d_{true}, d_{BCH}):

$$(33, 13, 10, 5)$$
$$(43, 15, 13, 7)$$
$$(47, 24, 11, 5)$$
$$(47, 23, 12, 6)$$
$$(51, 25, 10, 5)$$
$$(51, 17, 12, 6)$$
$$(55, 30, 10, 5)$$
$$(57, 21, 14, 6)$$

We can correct until t_{true} using a system of polynomial equations with several variables.

2.6.3.2. *Solving a system of polynomial equations with several variables*

To treat polynomial equations with n variables over a finite field $\mathbb{F}_q(q = 2^r)$, a traditional method is as follows:

1) define a total order for the set of monomials. We can take the lexicographical order for the exponents of the monomial. We will then pose $X_1^i 1 X_2^{i_2} \cdots X_n^{i_n} \leq X_1^{j_1} X_2^{j_2} \cdots X_n^{j_n}$, if $(i_1, i_2, \ldots, i_n) \leq (j_1, j_2, \ldots, j_n)$ for the chosen lexicographic order;

2) this total order makes it possible to order the sequence of the monomials of a polynomial in ascending order, let us say, from left to right. The largest monomial of f is called the *head term*, or Hterm (f). For $F = 1 + X_1X_2 + X_2^2 + X_1X_2^3$ we have Hterm $(f) = X_1X_2^3$.

2.6.3.3. *Two basic operations*

This concept of Hterm will make it possible to define two basic operations, which will be used in the B. Buchberger algorithm.

2.6.3.3.1. Reduce (f_1, f_2)

If Hterm $(f_1) = X_1^{i_1}X_2^{i_2}\cdots X_n^{i_n}$ for the produced order is lower or equal to a monomial $X_1^{j_1}X_2^{j_2}\cdots X_n^{j_n}$ of another polynomial f_2, then we replace f_2 by $f_2 - X_1^{j_1-i_1}X_2^{j_2-i_2}\cdots X_n^{j_n-i_n}f_1$. We remove the monomial of f_2 in question and thus modify the polynomial f_2. We make this operation for all the possible monomials in f_2. The polynomial obtained at the end is indicated by Reduce (f_1, f_2). It should be noted that this result depends on the sequence of simplifications which we have chosen, but that does not obstruct the algorithm of decoding that we deal with. We will say that we reduce f_2 with f_1.

EXAMPLE 2.38. Let $f_1 = 1 + X_1X_2 + X_2^2 + X_1^2X_2^3$ and $f_2 = 1 + X_1^2 + X_1^2X_2^2$. We have Reduce $(f_2, f_1) = f_1 + X_2f_2 = 1 + X_2 + X_1X_2 + X_1^2X_2 + X_2^2$. Let us note that we cannot modify f_2 using f_2.

2.6.3.3.2. Spol (f_1, f_2)

Let $f_1 = f_1' + \lambda_1\text{Hterm}(f_1)$, $\lambda_1 1 \in \mathbb{F}_q\backslash\{0\}$, $f_2 = f_2' + \lambda_2\text{Hterm}(f_2)$, $\lambda_2 \in \mathbb{F}_q\backslash\{0\}$. We pose ($[u, v]$ indicates the LCM of u and v):

$$\text{Spol}(f_1, f_2) = f_1\lambda_2\frac{[\text{Hterm}(f_1), \text{Hterm}(f_2)]}{\text{Hterm}(f_1)} + f_2\lambda_1\frac{[\text{Hterm}(f_1), \text{Hterm}(f_2)]}{\text{Hterm}(f_2)}$$

The result is determined clearly.

EXAMPLE 2.39 ($f1 = 1 + X_1X_2 + X_2^2 + X_1X_2^3$, $f_2 = 1 + X_1^2 + X_2 + X_1^2X_2$). We have $[\text{Hterm}(f_1), \text{Hterm}(f_2)] = [X_1X_2^3, X_1^2X_2] = X_1^2X_2^3$, as well as $\text{Spol}(f_2, f_1) = X_1 + X_1^2X_2 + X_2^2 + X_1X_2^2 + X_1^2X_2^2 + X_2^3$.

Let us note that the application $\text{Spol}(f_1, f_2)$ and the construction $\text{Reduce}(f_1, f_2)$ have a very simple significance in the case of a single variable. It is the calculation of a dividend in the division of f_2 by f_1 (if the degree of f_2 is the greater).

2.6.3.4. *The algorithm of B. Buchberger*

We start with a family of polynomials with n variables $\mathcal{F} = \{f_1, \ldots, f_R\}$:

1) calculate all the $\text{Spol}(f_i, f_j), i \neq j$. We associate all the $\text{Spol}(f_i, f_j)$ obtained to the family $\mathcal{F}$. We obtain a new family $\mathcal{F}' = \{g_1, \ldots, g_s\}$;

2) calculate all $\text{Reduce}(g_i, g_j)$, for $i \neq j$:

– if we transform g_j into g_j', then we preserve only g_j' and not g_j. We obtain a family $\mathcal{F}'' = \{h_1, \ldots, h_T\}$,

– if $\mathcal{F}'' = \mathcal{F}$, we stop and we preserve $\mathcal{F}$, which is the result of the Buchberger algorithm. There is a very difficult demonstration that proves that the final family is unique regardless of the transformations made,

– otherwise we return to 1) and reiterate.

We demonstrate that this algorithm is finite. The proof of this assertion is rather complicated and we will not provide it here. We now will apply these general results to the decoding of cyclic codes.

2.6.3.5. *FG-decoding*

2.6.3.5.1. Obtaining the initial system

To obtain the initial system we proceed as follows:

1) we pose $L(X) = Y_0 + Y_1X + Y_2X^2 + \cdots + Y_{t_{true}-1}X^{t_{true}-1} + X^{t_{\text{true}}}$. The polynomial $L(X)$ thus has its coefficients expressed as unknown functions of $t_{\text{true}}Y_0, Y_1, Y_2, \ldots, Y_{t_{\text{true}}-1}$;

2) from the nullity of the following scalar product ($i = 0, 1, \ldots, d_{\text{BCH}}-2-t_{\text{true}}$):

$$\langle (Y_0, Y_1, Y_2, \ldots, Y_{t_{\text{true}}-1}, 1), (S_{1+i}, \ldots, S_{t_{\text{true}}+1} + i) \rangle$$

we deduce the s variables $Y_{i_1}, \ldots, Y_{i_a}$, making it possible to express all the others by simple linear combinations. We obtain the following expression:

$$\begin{aligned} L(X) = {} & A_0(Y_{i_1}, \ldots, Y_{i_a}) + A_1(Y_{i_1}, \ldots, Y_{i_a})X + \cdots \\ & + A_{t_{\text{true}}-1}(Y_{i_1}, \ldots, Y_{i_a}) + X^{t_{\text{true}}-1} + X^{t_{\text{true}}} \end{aligned}$$

The number s of the remaining unknowns is known, as indicated in the following proposition.

PROPOSITION 2.62. *Let us suppose* $t_{\text{true}} + l \leq d_{\text{BCH}} - 1$. *The number s of necessary unknowns is equal to* $2t_{\text{true}} - d_{\text{BCH}} + 1$.

Proof. The proof is outside the scope of this work. □

2.6.3.5.2. The FG-decoding algorithm

Now let us provide the algorithm of this decoding:

1) we calculate all the coefficients of $S(X)$ which we know thanks to the roots of the code;

2) we pose $L(X) = Y_0 + Y_1X + Y_2X^2 + \cdots + Y_{t_{\text{true}}-1}X^{t_{\text{true}}-1} + X^t$ where the Y_i are the unknowns. This polynomial is a multiple of the locator $L_e(X)$ whose roots indicate here the positions in error (and not the reverse, as in the Berlekamp-Massey algorithm);

3) using the fact that $L(X)$ is orthogonal to any sequence of $t_{\text{true}} + 1$ consecutive syndromes, we eliminate variables using the $d_{\text{BCH}} - 1$ known consecutive syndromes. There remain s unknowns;

4) we calculate the following scalar product

$$P_i : Pi = \langle (A_0, A_1, A_2, \ldots, A_{t_{\text{true}}-1}), (S_{1+i}, \ldots, S_{t_{\text{true}}+i}) \rangle$$

for $i = d_{\text{BCH}} - 1 - t_{\text{true}}, \ldots, n-1$. We will note that P_i is a polynomial in the s remaining unknowns. At each step we deduce one of the two following results:

– if $S_{t_{\text{true}}+i}$ is known, an equation $S_{t_{\text{true}}+i} + P_i = 0$,

– if $S_{t_{\text{true}}+i}$ is not known, we pose $S_{t_{\text{true}}+i} = -P_i$;

5) at the end of stage 4), we have a family of equations $\{f_1{=}0,\ f_2{=}0, \ldots, f_R{=}0\}$. We apply the Buchberger algorithm to the family of polynomials $\{f_1, f_2, \ldots, f_R\}$;

6) we then find ourselves in one of the four cases described in the following propositions.

PROPOSITION 2.63. *We have a constant if the error is incorrigible.*

Proof. The proof is outside the scope of this work. □

In this case we cannot correct. We pass to the following word.

PROPOSITION 2.64. *We have a family of polynomials with one variable, all linear,* $\{Y_{i_1} - a_1, \ldots, Y_{i_s} - a_s\}$ *if the error is corrigible. In this case:*

$$L(X) = A_0(a_1, \ldots, a_s) + A_1(a_1, \ldots, a_s)X + \cdots$$
$$+ A_{t_{\text{true}}-1}(a_1, \ldots, a_s)X^{t_{\text{true}}-1} + X^{t_{\text{true}}}$$

is then the locator $L_e(X)$.

Proof. The proof is outside the scope of this work. □

In this case we obtain the required polynomial $L_e(X)$. We look for its roots and we can then correct directly (for a binary code).

PROPOSITION 2.65. *The family of equations may be empty, in which case we know the degree of the locator. In all other cases we have a family of polynomials with several variables. The error is corrigible, but* $L(X)$ *is not the locator. We have* $L(X) = X^i L_e(X)$ *for an* $i \geq 1$.

Proof. The proof is outside the scope of this work. □

In this case we add the constant coefficient of $L(X)$ to the family given by the algorithm of B. Buchberger. The algorithm is re-applied.

EXAMPLE 2.40. Let us consider the code (21, 7, 8, 5). Let us suppose that the error polynomial is $e(X) = 1 + X + X^4$. According to proposition 2.62 we will need 2 unspecified Y_0, Y_1. After stage 1 of the FG algorithm, we obtain $L(X) = \alpha^9 + \alpha^{24} Y_0 + \alpha^{30} Y_1 + Y_0 X + Y_1 X^2 + X^3$. At stage 6), we have $\{\alpha^{28} + \alpha^{39} Y0, \alpha^9 + \alpha^{48} Y_1\}$. This is the case where we are able to correct immediately. The obtained polynomial $L(X)$ is the error locator, $L_e(X)$. The weight of the error is 3. We directly find $Y_0 = \alpha^{52}, Y_1 = \alpha^{24}$. We obtain $L(X) = \alpha^9 + \alpha^{24}\alpha^{52} + \alpha^{30}\alpha^{24} + \alpha^{52} X + \alpha^{24} X^2 + X^3$. It has three roots, $\alpha^0, \alpha^3, \alpha^{12}$. Since $\beta = \alpha^3$, they correspond to the positions indexed by β^0, β, β^4, which is correct.

EXAMPLE 2.41. Let us consider the same code, but supposing that the error is $e(X) = 1 + X^4$. At stage 3) we have $L(X) = (\alpha^6 + \alpha^2 Y_0 + \alpha^{52} Y_1) + Y_0 X + Y_1 X^2 + X^3$. At stage 5), we obtain $\{\alpha^{30} + \alpha^{41} Y_0 + \alpha^{43} Y_1\}$. It is not a constant, therefore, the error is corrigible, but there we do not have 1st degree polynomials. The locator is thus of a degree lower than 3. Let us add the constant $\alpha^6 + \alpha^2 Y_0 + \alpha^{52} Y_1$ of $L(X)$ to the family obtained at stage 5), and re-use the B. Buchberger algorithm. We then obtain $\{\alpha^{54} + \alpha^{52} Y_1, \alpha^{20} + \alpha^8 Y_0\}$, from where we obtain $Y_0 = \alpha^{12}, Y_1 = \alpha^2$. We have $L(X)$ is $L(X) = X(\alpha^{12} + \alpha^2 X + X^2)$. The locator is thus $\alpha^{12} + \alpha^2 X + X^2$ whose roots are $1, \alpha^{12}$. The positions of the errors are thus indexed by β^1 and β^4, which is exact.

2.6.4. *Berlekamp-Massey decoding*

2.6.4.1. *Introduction*

We consider a cyclic code of length n, with apparent odd distance $d_{\text{BCH}} = 2t + 1$, where t is the apparent correction distance (i.e. t_{BCH}). Let β be an nth primitive root of unity. We will suppose that the zeros of the code are $1, \beta, \beta^2, \ldots, \beta^{d_{\text{BCH}}-2}$. The S_i syndromes are classically $S_i = r(\beta^i)$ $(i = 0, \ldots, d_{\text{BCH}} - 2)$, where $r(X)$ is the received word equal to the transmitted word $c(X)$ plus the error word $e(X)$. We will call the syndrome polynomial the polynomial $S(X) = \sum_{i=0}^{d-2} S_i X^i$. In this algorithm we call the error locator the polynomial $L(X)$ the inverses of whose roots give the positions of errors, and $EV(X)$ is the errors evaluator defined by:

$$EV(X) = \sum_i e_i \frac{L(X)}{(1 - \alpha^i X)} \quad \left(= \sum_i e_i U_i(X) \right)$$

We choose here to normalize the constant of $L(X)$, i.e. to agree that $L(0) = 1$.

These two polynomials make it possible to reconstitute the error word. Indeed, after factorizing $L(X)$ we have all the error positions. Let α^i be one of these positions with an error value e_i. We have $LV(\alpha^i) = e_i U_i(\alpha^i)$, which yields e_i.

2.6.4.2. *Existence of a key equation*

The polynomials $S(X), L(X)$ and $EV(X)$ are linked to each other.

PROPOSITION 2.66. *The polynomials $S(X), L(X)$ and $EV(X)$ satisfy an equation, known as the key equation:*

$$L(X)S(X) = EV(X) \pmod{X^{2t}}$$

Proof. We have:

$$\begin{aligned} S(X) &= \sum_{j=0}^{d_{\mathrm{BCH}}-2} \left[\sum_k e_k \alpha^{jk} X^j \right] \\ &= \sum_k e_k \left[\sum_j (\alpha^k X)^j \right] \\ &= \sum_k e_k \left[\frac{1-(\alpha^k X)^{2t}}{(1-\alpha^k X)} \right] \end{aligned}$$

From that we directly obtain $L(X)S(X) = \sum_k e_k \frac{L(X)}{(1-\alpha^k X)}$ *modulo* X^{2t}. □

PROPOSITION 2.67. *The couple of polynomials $(L(X), EV(X))$ is the only couple $(a(X), b(X))$ satisfying the key equation with the conditions:*

$$\begin{cases} (a(X), b(X)) = 1 \\ a(0) = 1 \\ \deg\, b(X) < \deg\,(a(X)) \leq t \end{cases}$$

Proof. Let us suppose to have the following two pairs $(a_1(X), b_1(X))$ and $(a_2(X), b_2(X))$. We have: $a_1(X)S(X) = b_1(X)(\text{mod } X^{2t})$, and $a_2(X)S(X) = b_2(X)(\text{mod } X^{2t})$. From that we deduce $a_2(X)b_1(X) = a_1(X)b_2(X)$ *modulo* X^{2t}, but the degrees of each member are at most equal to $2t - 1$. Thus, the congruence is, in fact, an equality: $a_2(X)b_1(X) = a_1(X)b_2(X)$. According to the assumption, $a_1(X)$ is relatively prime to $b_1(X)$, then $a_1(X)$ divides $a_2(X)$. By symmetry we deduce $a_1(X) = ka_2(X)$, where k is a constant. Since, by assumption, $a_1(X)$ and $a_2(X)$ have their constant equal to 1, k equals 1, $b_1(X) = b_2(X)$ and the solution pair is, therefore, unique. □

2.6.4.3. *The solution by successive stages*

We are going to build a pair which is the successive stages solution, for $j = 0, 1, \ldots, 2t$. At each stage we will build two pairs $(a_j(X), b_j(X))$ and $(a'_j(X), b'_j(X))$.

At each stage (or each level) j any couple $(u_j(X), v_j(X))$, such that we have $u_j(X)$ $S(X) = v_j(X)$ *modulo* X^j, will be referred to as a "solution". Moreover, it will be said that a solution is optimal if $\max(\deg\ u_j(X), 1 + \deg\ v_j(X))$ is the smallest possible among the solutions at level j. We will then indicate this maximum by d_j. The couple $(a_j(X), b_j(X))$ that we will build will be an optimal solution at the level j. The second couple $(a'_j(X), b'_j(X))$ will satisfy $a'_j(X)S(X) = b'_j(X) + X^j$ *modulo* $X^{j+1}, a'(0) = 0$, and $\max(\deg\ a'_j(X), 1 + \deg\ b'_j(X)) = j + 1 - d_j$. Initialization will be done posing $(a_0, b_0) = (1, 0); (a'_0, b'_0) = (0, 1), d_0 = 0$.

2.6.4.4. *Some properties of* d_j

The following results express some properties of d_j.

LEMMA 2.1. *We have* $d_{j+1} \geq d_j$.

Proof. If we have $a_{j+1}(X)S(X) = b_j(X + 1)$ *modulo* X^{j+1}, then we also have $a_{j+1}(X)S(X) = b_{j+1}(X)$ *modulo* X^j. Thus, an optimal solution at the level $j+1$ is a solution at the row j. □

PROPOSITION 2.68. *If we have* $2d_j \leq j$, *then there exists at most one optimal solution at stage* j.

Proof. Let us suppose having two optimal solutions at stage j, $(a_1(X), b_1(X))$ and $(a_2(X), b_2(X))$. We have $a_1(X)S(X) = b_1(X)$ *modulo* X^j, and $a_2(X)S(X) = b_2(X)$ *modulo* X^j. From that we deduce: $a_2(X)b_1(X) = a_1(X)b_2(X)$ *modulo* X_j. But $d_j = \max(\deg\ a_2(X), 1 + \deg\ b_2(X)) = \max(\deg\ a_1(X), 1 + \deg\ b_i(X))$. From that we obtain: $\deg\ a_2(X) \leq d_j$, $\deg\ b_1(X) \leq d_j - 1$, and $\deg\ a_1(X) \leq d_j$, $\deg\ b_2(X) < d_j - 1$. Thus, $\deg\ a_2(X) + \deg\ b_1(X) < 2d_j - 1$ and $\deg\ a_1(X) + \deg\ b_2(X) \leq 2d_j - 1$. Since, by assumption, we have $2d_j \leq j$ we obtain $\deg\ (a_2(X)b_1(X)) < j$ and $\deg\ (a_1(X)b_2(X)) < j$. Thus, the congruence is a polynomial equality $a_2(X)b_1(X) = a_1(X)b_2(X)$. Since $a_2(X)$ and $b_2(X)$ are relatively prime we see that the two solutions are the same as in the proof of proposition 2.67. □

2.6.4.5. *Property of an optimal solution* $(a_j(X), b_j(X))$ *at level* j

The following result expresses a property of an optimal solution $(a_j(X), b_j(X))$ at level j.

PROPOSITION 2.69. *If* $(a_j(X), b_j(X))$ *is an optimal solution at level* j *and if it is not a solution at the level* $j + 1$, *then we have* $d_{j+1} \geq j + 1 - d_j$.

Proof. By assumption we have $\max(\deg\ a_j(X), 1 + \deg\ b_j(X)) = d_j$, as well as the congruence $a_j(X)S(X) = b_j(X) + \delta X^j$ *modulo* $X^{j+1}, \delta \neq 0$. Let $(c_{j+1}(X), d_{j+1}(X))$ be an optimal solution at the level $j + 1$. We have a second congruence

$c_{j+1}(X)S(X) = d_{j+1}(X)$ *modulo* X^{j+1}. Multiplying the first by $c_{j+1}(X)$ and the second by $a_j(X)$ we obtain: $a_j(X)d_{j+1}(X) = b_j(X)c_{j+1}(X) + c_{j+1}(X)\delta X^j$ *modulo* X^{j+1}. Since $c_{j+1}(X)$, by assumption, has a constant equal to 1, we still have $a_j(X)d_{j+1}(X) = b_j(X)c_{j+1}(X) + \delta X^j$ *modulo* X^{j+1}.

We will now demonstrate by absurdity that we have:

$$\max(\deg\ a_j(X), 1 + \deg\ b_j(X)) + \max(\deg\ c_{j+1}(X), 1 + \deg\ d_{j+1}(X)) \geq j + 1$$

Let us suppose that we have:

$$\max(\deg\ a_j(X), 1 + \deg\ b_j(X)) + \max(\deg\ c_{j+1}(X), 1 + \deg\ d_{j+1}(X)) \leq j$$

It follows that:

$$\begin{aligned} &\deg\ a_j(X) + \deg\ d_{j+1}(X) \\ &\quad < \deg\ a_j(X) + 1 + \deg\ d_{j+1}(X) \\ &\quad \leq \max(\deg\ a_j(X), 1 + \deg\ b_j(X)) \\ &\qquad\qquad + \max(\deg\ c_{j+1}(X), 1 + \deg\ d_{j+1}(X)) \\ &\quad \leq j \end{aligned}$$

Thus, $\deg\ a_j(X) + \deg\ d_{j+1}(X) < j$, and we prove in the same way that $\deg\ b_j(X) + \deg\ c_{j+1}(X) < j$. This contradicts the existence of a non-zero δ. Thus:

$$\max(\deg\ a_j(X), 1 + \deg\ b_j(X)) + \max(\deg\ c_{j+1}(X), 1 + \deg\ d_{j+1}(X)) \geq j + 1$$

Since $d_{j+1} = \max(\deg\ c_{j+1}(X), 1 + \deg\ d_{j+1}(X))$ we obtain from it:

$$d_{j+1} \geq j + 1 - \max(\deg\ a_j(X), 1 + \deg\ b_j(X)) = j + 1 - d_j \qquad \square$$

2.6.4.6. *Construction of the pair* $(a'_{j+1}(X), b'_{j+1}(X))$ *at the* j *stage*

The issue of constructing the pair $(a'_{j+1}(X), b'_{j+1}(X))$ at the j stage is solved by the following result.

PROPOSITION 2.70. *If at the* j *stage we have an optimal solution* $(a_j(X), b_j(X))$, *and a pair* $(a'_j(X), b'_j(X))$, *then we can build a pair* $(a'_{j+1}(X), b'_{j+1}(X))$.

Proof. Indeed:

1) if $d_{j+1} = d_j$, then we pose $a'_{j+1}(X) = Xa'_j(X)$, $b'_{j+1}(X) = Xb'_j(X)$ and the pair $(a'_{j+1}(X), b'_{j+1}(X))$ is an appropriate pair;

2) if $d_{j+1} > d_j$ (see proposition 2.69) then $d_{j+1} \geq j + 1 - d_j > d_j$, which implies $2d_j \leq j$, and thus the optimal solution at the j level is unique. Let $(a_j(X), b_j(X))$ be this optimal solution at the j level. Since $d_{j+1} \neq d_j$, this solution is not a solution at the $j+1$ level. Thus, we have $a_j(X)S(X) = b_j(X) + \delta X^j$ *modulo* X^{j+1} and $\delta_j \neq 0$. We pose $a'_{j+1}(X) = X\delta^{-1}a_j(X)$, $b'_j + 1(X) = X\delta^{-1}b_j(X)$. This pair $(a'_{j+1}(X), b'_{j+1}(X))$ is a solution, since it verifies the polynomial equality, and, moreover, we have $\max(\deg a'(X), 1 + \deg b'(X)) = 1 + \max(\deg a(X), 1 + \deg b(X)) = j + 2 - d_{j+1} = (j+1) + 1 - d_{j+1}$. □

2.6.4.7. *Construction of an optimal solution* $(a_{j+1}(X), b_{j+1}(X))$

This section covers the construction of an optimal solution.

PROPOSITION 2.71. *If an optimal solution $(a_j(X), b_j(X))$ is a solution at the $j+1$ level, then it is optimal at the $j+1$ level, and we have $d_{j+1} = d_j$.*

Proof. This is straightforward, since $d_{j+1} \geq d_j$ (see lemma 2.1). □

Propositions 2.72 and 2.73 give the construction of an optimal solution at the $j+1$ level, in the respective cases where $2d_j > j$ and $2d_j \leq j$.

PROPOSITION 2.72. *If we have an optimal solution $(a_j(X), b_j(X))$ at the j level which is not a solution at the $j+1$ level, if we have a pair $(a'_j(X), b_j(X))$ and if $2d_j > j$, then we can construct an optimal solution $(a_{j+1}(X), b_{j+1}(X))$ at the $j+1$ level, and we have $d_{j+1} = d_j$.*

Proof. Let $(a_j(X), b_j(X))$ be an optimal solution at the j level. Since it is not a solution at the $j+1$ level, we then have: $a_j(X)S(X) = b_j(X) + \delta X^j$ *modulo* X^{j+1}, with $\delta \neq 0$. Let us suppose having a pair $(a'_j(X), b'_j(X))$ such that we have:

$$a'_j(X)S(X) = b'_j(X) + X^j \pmod{X^{j+1}}$$

$$\max(\deg\ a'_j(X), 1 + \deg\ b'_j(X)) = j + 1 - d_j, \quad a'_j(0) = 0$$

Let us pose $a_{j+1}(X) = a_j(X) - \delta_j a'_j(X)$ and $b_{j+1}(X) = b_j(X) - \delta b'_j(X)$. We deduce that the pair $(a_{j+1}(X), b_{j+1}(X))$ is an optimal solution at the $j+1$ level. Indeed, we have $a_j(X)S(X) - b_j(X) = \delta_j X^j = \delta_j(a'_{j+1}(X)S(X) - b'_{j+1}(X))$ *modulo* X_{j+1}. From there: $(a_j(X) - \delta_j a'_j(X))S(X) = b_j(X) - \delta_j b'_j(X)$ *modulo*

X^{j+1}. Let us pose $A = a_j(X) - \delta_j a'_j(X)$ and $B = b_j(X) - \delta_j b'_j(X)$. We can successively deduce:

$$\begin{aligned}
&\max\left(\deg\ (a_j(X) - \delta_j a'_j(X)), 1 + \deg\ (b_j(X) - \delta_j b'_j(X))\right) \\
&\leq \max\left(\max(\deg\ a_j(X), \deg\ a'_j(X)), \max(1 + \deg\ b_j(X), 1 + \deg\ b'_j(X))\right) \\
&= \max\left(\max(\deg\ a_j(X), 1 + \deg\ b_j(X)), \max(\deg\ a'_j(X), 1 + \deg\ b'_j(X))\right) \\
&\leq \max(d_j, j + 1 - d_j)
\end{aligned}$$

By the assumption $2d_j > j$ it thus follows that $\max(d_j, j+1-d_j) = d_j$. Thus, $\max(\deg\ A(X), 1 + \deg\ B(X)) \leq d_j$, but $\max(\deg\ A(X), 1 + \deg\ B(X)) \geq d_{j+1} \geq d_j$. Thus $d_{j+i} = d_j$. □

PROPOSITION 2.73. *If we have an optimal solution* $(a_j(X), b_j(X))$ *at the* j *level, which is not solution at the j+1 level, as well as pair* $(a'_{j+1}(X), b'_{j+1}(X))$ *and if* $2d_j \leq j$*, then we can construct an optimal solution* $(a_{j+1}(X), b_{j+1}(X))$ *at the* $j+1$ *level, and we have* $d_{j+1} = j + 1 - d_j$.

Proof. As in proposition 2.70 we arrive at:

$$\max(\deg a_{j+1}(X), 1 + \deg b_{j+1}(X)) \leq \max\{d_j, j + 1 - d_j\}$$

But since here we have $2d_j \leq j$, then:

$$d_{j+1} \leq \max(\deg\ A(X), 1 + \deg\ B(X)) \leq j + 1 - d_j$$

Since $(a_j(X), b_j(X))$ is an optimal solution at the j level but not a solution at the $j+1$ level, by Proposition 2.69 we also have $d_{j+1} \geq j + 1 - d_j$, from where the required equality $d_{j+1} = j + 1 - d_j$. □

2.6.4.8. *The algorithm*

The basic idea of the algorithm is as follows. We have $j, d_j, (a_j(X), b_j(X)), (a'_j(X), b'_j(X)), \delta_j$:

1) let $(a_j(X), b_j(X))$ be a solution at the $j+1$ level. In this case it is an optimal solution at the $j+1$ level, $d_{j+1} = d_j$, and $(a'_{j+1}(X), b'_{j+1}(X)) = (Xa'_j(X), Xb'_j(X))$;

2) if $(a_j(X), b_j(X))$ is not a solution at the $j+1$ level, and:

– If $2d_j > j$ then $d_{j+1} = d_j$, and:

$$\left(a_{j+1}(X), b_{j+1}(X)\right) = \left(a_j(X) - \delta_j a'_j(X), b_j(X) - \delta b'_j(X)\right)$$

$$\left(a'_{j+1}(X), b'_{j+1}(X)\right) = \left(Xa'_j(X), Xb'_j(X)\right)$$

– if $2d_j \leq j$ then $d_{j+1} = j + 1 - d_j$, and:

$$\left(a_{j+1}(X), b_{j+1}(X)\right) = \left(a_j(X) - \delta_j a'_j(X), b_j(X) - \delta b'_j(X)\right)$$

$$\left(a'_{j+1}(X), b'_{j+1}(X)\right) = \left(X\delta^{-1} a_j(X), X\delta^{-1} b_j(X)\right)$$

Initialization is then performed as follows:

$$\begin{cases} j = 0 \\ (a_0(X), b_0(X)) = (1,0) \\ (a'_0(X), b'_0(X)) = (0,1) \\ d_0 = 0 \end{cases}$$

EXAMPLE 2.42. Let $e(X) = \alpha^2 + \alpha X^4$. We easily find $S(X) = \alpha^4 + \alpha^3 X + X^3$.

We calculate following the algorithm. We obtain:

j	d_j	$(a_j(X), b_j(X))$	$(a_j(X), b_j(X))$	δ_j	$2d_j > j?$
0	0	$(1,0)$	$(0,1)$	α^4	*no*
1	1	$(1,\alpha^4)$	$(\alpha^3 X,0)$	α^3	*yes*
2	1	$(1+\alpha^6 X,\alpha^4)$	$(\alpha^3 X^2,0)$	α^2	*no*
3	2	$(1+\alpha^6 X+\alpha^5 X^2,\alpha^4)$	$(\alpha^5 X+\alpha^4 X^2,0)$	α^3	*yes*
4	2	$(1+\alpha^5 X+\alpha^4 X^2,\alpha^4+\alpha^5 X)$			

We obtain from that $L(X) = 1 + \alpha^5 X + \alpha^4 X^2$, and $LV(X) = \alpha^4 + \alpha^5 X$, which we can easily verify.

2.6.5. *Majority decoding*

Error trapping is a correction technique valid for linear or cyclic, binary or $\mathbb{F}_q$, $q = p^r$, codes. We will only consider cyclic, binary or $\mathbb{F}_q$, $q = 2^r$, codes of length n.

The basic idea of this type of decoding is to choose a position among the n of the received word and to correct an error if it is there. It is clear that there is no statistical reason for an error to occur here rather than there. But by circular *shift* (which leaves the code unchanged) we can bring an error to a selected position. Fundamental questions arise now:

1) what is the mechanism of trapping? What is the link between this mechanism and the code used;

2) how can the trapping be performed efficiently?;

3) what are the bounds of this type of decoding?

2.6.5.1. *The mechanism of decoding, and the associated code*

Any error correction action must have an "empty" result when a word without error is processed, i.e. a codeword. The code must be "transparent". The basic mechanism of error trapping respects this need: we produce scalar products between the received word and the words of the dual of the used code C.

2.6.5.2. *Trapping by words of $C^{\perp}$ incidents between them*

We will say that words are incidents in a set of position S, if all these words have non-zero components in S and if there is no other position where two of these words have a non-zero coefficient.

When S is reduced to a position i, note that we can always make it so that the component of the words in position i is equal to 1 (because the code is a vector subspace in $\mathbb{F}_q$).

2.6.5.2.1. Trapping in one position

In this case the unit S is reduced to one position. Since the code is cyclic, we can suppose that this position is the first (i.e. in X^0). In $C^{\perp}$ we will look for words (i.e. polynomials) with a constant equal to 1, and no other common monomial for any two of these words.

EXAMPLE 2.43. Let us suppose that the code C is binary of length 5. The following trapping polynomials would be appropriate: $1+X+X^5, 1+X^2+X^6, 1+X^3+X^4$. In binary form we have:

$$\begin{array}{c} 1100010 \\ 1010001 \\ 1001100 \end{array}$$

We clearly see the incidence of these three words appearing.

2.6.5.2.2. Trapping in several positions

We can trap the error in several positions, instead of just one. In the following example we show incident words in several positions.

EXAMPLE 2.44. Let us again take the case of a binary code C, but of length 7 this time. Let us give the binary format of the words of $C^{\perp}$ directly:

$$\begin{array}{c} 10110100 \\ 11101000 \\ 10100011 \end{array}$$

These words of $C^{\perp}$ are incidents in positions 1 and 3 (i.e. in X^0, X^2).

2.6.5.3. *Codes decodable in one or two stages*

We will not speak about trappings with more than two stages because the associated decoders are too complex.

2.6.5.3.1. Codes decodable in one stage

This is the case, for example, of a code C whose orthogonal contains word incidents in the first position.

PROPOSITION 2.74. *Let us suppose that $C^{\perp}$ contains δ word incidents in the first position. Then the code C can correct with error trapping up to $[\delta/2]$ errors in the received word. Majority vote gives the value in position* 1. *If there is equality of votes, we take* 0 *as value in position* 1.

Proof. The worst case is if $[\delta/2]$ isolated errors occur.

If an error β is in position 1, then $[\delta/2] - 1$ errors will be distributed in positions different from 1. Thus, at most $[\delta/2]-1$ words can give a vote equal to 0 (in the binary case, these are the ones containing an odd number of errors in positions other than 1). The others $\delta - [\delta/2] + 1$ will give a vote equal to β. It is easy to verify the (strict) inequality $\delta - [\delta/2] + 1 > [\delta/2] - 1$.

If there is no error in position 1, then at most $[\delta/2]$ words will give a non-zero vote. If δ is odd, a majority of words will give a vote equal to 0. If δ is even, in the worst case there can be an equality of votes. By assumption, we then say that the value in position 1 is equal to 0, which is correct. In all cases, the value of the majority vote gives the error value in position 1 (note that the word incidents are selected so that they have a coefficient of 1 in position 1). □

A single stage is then needed to correct an error in position 1.

EXAMPLE 2.45. With $q = 2, n = 7$ let us take the extended Hamming code generated by $g(X)$, with $g(X) = 1 + X + X^2 + X^4$. The generator of its dual is $1 + X^2 + X^3$. The parity check matrix in reduced form is:

$$H = \begin{pmatrix} 1000101 \\ 0100111 \\ 0010110 \\ 0001011 \end{pmatrix}$$

We can find three words that are incidents in position 4 (i.e. in X^3):

$$\begin{matrix} 0001011 \\ 0101100 \\ 1011000 \end{matrix}$$

They correspond to the lines $l_4, l_2 + l_4$ and $l_1 + l_3 + l_4$.

2.6.5.3.2. Codes decodable in two stages

This section covers codes decodable in two stages.

PROPOSITION 2.75. *We have the two following results:*

1) let us suppose that $C^{\perp}$ contains δ' words which are incidents in a set S of positions. Then the code C can trap up to $[\delta'/2]$ errors in S. The majority vote gives the value of the scalar product in S;

2) if we have δ' sets $S_1, S_2, \ldots, S_{\delta'}$ and if these sets are incident in position 1 (for example), then we can correct an error in position 1 by majority vote.

Proof. It is the same we argument as in the proof of proposition 2.74. In this case we have a cascade of two majority votes. □

We can prove, and we will admit it, that we can expect to correct more errors with a code decodable in two stages than with a code decodable in only one.

EXAMPLE 2.46. With $q = 2, n = 7$ let us take the Hamming code generated by $g(X)$, with $g(X) = 1 + X^2 + X^3$. The generator of its dual is $1 + X^2 + X^3 + X^4$. We can find two words that are incidents in positions 1 and 3:

1100101
1110010

as well as two words that are incidents in positions 1 and 2:

1100101
1110010

The two families of cardinal 2 are incidents in position 1. We can trap an error in this position, in two votes:

1) if the error is in position 1, the first vote will give 1 for $\{1, 3\}$ and 1 for $\{1, 2\}$. Thus the majority in position 1 is 1. Therefore, we correct it;

2) if the error is in position 2, the first vote will give 0 for $\{1, 3\}$ and 1 for $\{1, 2\}$. Thus, there is no majority in position for 1 is 1. We thus decide that it is 0, and that is exact;

3) if the error is in position 4, the first vote will give 0 for $\{1, 3\}$ and 0 for $\{1, 2\}$. Thus the majority in position 1 is 0. We do not correct, and that is correct.

2.6.5.4. *How should the digital implementation be prepared?*

Since it does not appear obvious how to carry out the digitalization of this type of decoding, we will develop this point.

PROPOSITION 2.76. *Let there be a code $C = (g(X))$ of length n, and dimension $n-k$. Let us suppose that the parity check matrix H of the code C has the form $(I_{n-k}R)$. Then the product of the received word $(r_0, r_1, \ldots, r_{n-k-1}, c_0, c_1, \ldots, c_{k-1})$ by H (which is the syndrome of the received word) is equal to the remainder of the division of the received word by $g(X)$.*

Proof. The column i of H is the remainder of the division of X^i by $g(X)$ (numbering the columns from 0 to $n-1$, from left to right). The syndrome $(s_0, s_1, \ldots, s_{n-k-1})$ of the received word is thus equal to the remainder of division of the received word by $g(X)$. □

This proposition indicates a way of digital implementation. We look in $C^{\perp}$ for each word used as a voter. Considering one of them r, let us say that it is equal to the sum of certain rows l_i of H, for example, $l_{i_1}, l_{i_2}, \ldots, l_{i_n}$. We note that the scalar product $\langle r, l_{i_j} \rangle$ is equal to s_{i_j}.

PROPOSITION 2.77. *The vote of the word r considered is equal to the scalar product $\sum_j s_{i_j}$.*

Proof. The vote of the word considered is equal to $\langle r, \sum_j l_{i_j} \rangle$, and we have $\langle r, \sum_j l_{i_j} \rangle = \sum_j \langle r, l_{i_j} \rangle = \sum_j s_{i_j}$. □

The remainder of division of the received word is easy to obtain with a register dividing by $g(X)$. To obtain the result of the majority vote, we thus know which sums of oscillations in the dividing register have to be calculated.

We should take into account that the received word is often pre-multiplied while entering the register. If it is pre-multiplied by X^{n-k}, we will take orthogonal voters in position $n-k$ (the position on the left is noted 0).

EXAMPLE 2.47. With the notations of example 2.45, we makes a decision regarding position 1 by making a majority vote between the following syndrome components: $s_1, s_2 + s_4, s_1 + s_3 + s_4$, which is easily done for the register dividing by $g(X)$.

Among the codes that are particularly well adapted to majority decoding there are the codes constructed using projective geometries: PG-codes.

PROPOSITION 2.78. *A PG-code of the order r is decodable in r majority decoding stages.*

Proof. Let there be a distribution of no than $J/2$ errors in the received word. If we take the common projective subspace, the others will be in at most more $J/2-1$ other subspaces. Thus, at most $J/2 - 1$ projective subspaces will be able not to signal an error. There will remain at least $J/2+1$ that will present an error. We can thus perform a majority decoding. □

In practice we fix a trapping position (i.e a point). This point is included in a certain number of level 1 projective spaces, from which we can perform majority decoding. Each projective subspace is included in a certain number of projective subspaces of the order 2, using which we can perform a majority decoding. Thus we correct an error placed at the selected point. To trap other errors we use the cyclic character of the code.

2.6.6. *Hard decoding, soft decoding and chase decoding*

At the receiver level the demodulator is an essential device for the decoding stage.

2.6.6.1. *Hard decoding and soft decoding*

The demodulator is a device able to interpret the signal received during a certain period of time. It can simply estimate that it is a "1", or a "0". It is a *hard* estimate. It can sometimes associate a real numerical value to the estimate that it makes of a received symbol. We can say that it associates a "confidence" to the binary value that it proposes. It is then a *soft* (or "balanced") estimate. If the values of confidence are between 0 and 1, the lowest value of the confidence will be 0 and the highest 1.

EXAMPLE 2.48. We suppose to use a binary code of length 7. For example, the demodulator will give (1011010) as a *hard* estimate of a received word, and (0.2; 0.9; 0.9; 0.7; 0.8; 0.1; 0.6) as the associated *soft* estimate. Using notation in real numbers we have $(0.2; -0.9; 0.9; 0.7; -0.8; 0.1; -0.6)$ The two symbols with the lowest probability are in positions 1 and 6. The two with the highest probability are in positions 2 and 3.

A decoding whose strategy only takes into account the *hard* estimate of the demodulator is known as *hard decoding*. If it takes into account the *soft* estimate, it is known as *soft decoding*. It has been demonstrated by Chase decoding that a *soft* decoding for a $(n, k, 2t+1)$ code makes it possible to correct approximately $2t$ errors per word, as opposed to t for *hard*. *Soft* decoding is thus very powerful. Its disadvantage is the heaviness of associated hardware, and the lengthening of the time to decode each word.

2.6.6.2. *Chase decoding*

Let us suppose using a $(n, k, 2t+1)$ code. Let $R(X)$ be the word received. *Chase decoding* uses the *soft* estimate of the demodulator. The idea of Chase decoding is simple. Since the demodulator is statistically reliable, we should count on the fact that the errors are in positions with the lowest confidence values. Thus we will reverse some of the binary positions of $R(X)$, which will give rise to $R'(X)$, and we will use a *hard* decoder to decode the newly obtained word $R'(X)$. If we have selected r positions with the lowest confidence, we can thus "mask" $0, 1, 2, \ldots, 2^r$ positions. The most powerful strategy is to make 2^r maskings, but often to save time we only make some of the maskings among the 2^r.

The general Chase algorithm is given below:

1) find the t lowest confidence values. n real values should be classified, let us say, in ascending order. It is an expensive operation in terms of silicon surface, and rather long;

2) reverse certain binary values of $R(X)$, in the positions of lowest confidence;

3) for each mask perform a *hard* decoding of the transformed word. If this word can be decoded, we calculate the Euclidean distance between the word proposed by *hard* decoding and the received word;

4) once all maskings are made, as well as all the *hard* decodings, we choose to say that the word suggested by *Chase decoding* is the one with the smallest Euclidean distance from the received word $R(X)$.

EXAMPLE 2.49. We suppose to receive (1011010), with the associated confidence values (0.1; 0.8; 0.9; 0.2; 0.4; 0.8; 0.9). The $(t = 2)$ two lowest confidence values are in positions 1 and 4. We will thus successively add the words (0000000), (1000000), (0001000), (1001000) to the received word. Let us suppose that the real error vector is (0101001). Then the third mask will make it possible for the code to reconstitute the transmitted codeword.

2.7. 2D codes

We calls 2D *binary codes* or *codes with two variables* the codes that are ideal in $A = \mathbb{F}2[X]/(X^n - 1, Y^m - 1)$. It is a generalization of cyclic codes that has been known for a long time.

2.7.1. *Introduction*

Their algebraic structure is much more complex, because we can no longer use the Euclidian or Bezout equalities. The ring A is no longer Euclidean. It is thus difficult, except in particular cases, to know the performances of these codes well.

Any word of a 2D code can be represented as a polynomial with two variables $c(X, Y)$. A 2D code can have a system of generators not reduced to a polynomial, contrary to cyclic codes. The results of Gröbner and B. Buchberger are quite useful to process these codes.

2.7.2. *Product codes*

A very simple case of 2D codes is that of product codes. A product code is an ideal of A generated by the generator $g(X,Y) = g_1(X)g_2(Y)$, where $g_1(X)$ divides $X^n - 1$ and $g_2(Y)$ divides $Y^m - 1$. This code is the set of the multiples of $g(X,Y)$ in A.

PROPOSITION 2.79. *We have the two following results:*

1) Its dimension is k_1k_2, where k_1 is the dimension of the code $(g_1(X))$ in $\mathbb{F}_2[X]/(X^n - 1)$ and k_2 is that of $(g_2(X))$ in $\mathbb{F}_2[Y]/(Y^m - 1)$;

2) Its minimum distance is d_1d_2, where d_1 is the minimum distance of $(g_1(X))$, and d_2 is that of $(g_2(Y))$.

Proof. The proof is outside the scope of this work. □

2.7.3. *Minimum distance of 2D codes*

We do not know much about this minimum distance in the general case. The BCH theorem is no longer valid. A result of J. Jensen gives a lower bound of the minimum distance, but we do not know if it is very good.

2.7.4. *Practical examples of the use of 2D codes*

At present it is primarily the product codes that are being used. In practice we often take $n = m$, $g_1(X) = g_2(X)$, i.e. the same code in rows and columns. The hardware is then simplified compared to the case where the codes would be different.

2.7.5. *Coding*

Any word of a code $g_1(X)g_1(Y)$ is a square table $n \times n$. We can arrange the information in the bottom right square, and then code as in the case of cyclic codes.

EXAMPLE 2.50. Let us pose $n = m = 7, g_1(X) = 1+X+X^3, g_2(Y) = 1+Y+Y^3$. Let us take as information polynomial $1 + Y + X^2Y + XY^2 + X^3Y^2 + X^3Y^3$. It is pre-multiplied by X^3Y^3 in order to place it in the bottom right square of the matrix. We then obtain:

$$\begin{matrix} 0000000 \\ 0000000 \\ 0000000 \\ 0001100 \\ 0000010 \\ 0000100 \\ 0000011 \end{matrix}$$

We perform the coding in columns and obtain:

```
0001011
0001010
0000101
0001100
0000010
0000100
0000011
```

We then code in rows and obtain the codeword:

```
1001011
0011010
1100101
1011100
1110010
0110100
0100011
```

2.7.6. *Decoding*

There does not exist a concept of a polynomial locator, or of a polynomial evaluator. The DFT does exist, but does not allow decoding, to our knowledge. FG-decoding is possible, but is highly complex.

In practice we decode by columns, then by rows and reiterate several times. If for each correction of a word in a row or a column we can create *soft* information, then an iterative decoding is obtained (such as the famous Turbo decoding) which is very powerful.

It should be noticed that rows-columns decoding is not a maximum probability decoding, and that it gives astonishing results.

For example, the code in example 2.50 has a minimum distance of 9, therefore it should correct 4 errors. However, it is unable to correct the error word equal to $1+X+Y+XY$ for example. On the other hand, it can correct $1+XY+X^2Y^2+X^3Y^3+X^4Y^4$, although the weight of this error word is 5.

2.8. Exercises on block codes

2.8.1. *Unstructured codes*

The first series of exercises covers unstructured codes.

EXERCISE 2.1. We pose $x = (101101)$ and $y = (011110)$. Calculate $d_H(x, y)$, $w_H(X)$ and $w_H(y)$.

EXERCISE 2.2. Perform the following operations:

1) build $B_\rho(x)$ with $\rho = 2$ and $x = (10111)$;

2) build $B_1(x) \cap B_2(y)$ with $x = (110111), y = (000110)$;

3) give the parameters of the following binary code:

$$\{(1,0,0,1,1,0,1,1),(0,0,1,0,1,1,1,0),(1,1,0,0,1,0,1,0),$$
$$(0,0,1,0,1,0,0,1),(1,1,1,1,0,0,0,1),(1,1,0,0,1,0,0,1)\}$$

4) give the parameters of the following ternary code:

$$\{(1,2,1,0,1,2,0),(1,0,2,0,2,0,1),(2,1,1,0,1,2,1),$$
$$(0,1,2,0,1,0,2),(1,1,2,1,2,0,0)\}$$

5) construct a binary code of length 5 with the largest possible cardinal that corrects 2 errors per word;

6) construct a binary code of length 7 with the largest possible cardinal that corrects 1 error per word.

EXERCISE 2.3. Let there be a binary code of length n with a cardinal M. What is the volume of memory necessary to make a table decoding using class representatives (in a number of binary positions)?

EXERCISE 2.4. How many elements are there in a sphere of radius r included in $(\mathbb{F}_2)^n$?

EXERCISE 2.5. Let there be a binary code C of length n and cardinal M. Prove that the greatest error correcting capability of the code is the largest integer r such that we have $M \times [\sum_{i=0}^{r} \binom{n}{i}] \leq 2^n$, where $\binom{n}{i}$ is the number of combinations of i objects from n.

2.8.2. *Linear codes*

The second series of exercises covers linear codes.

EXERCISE 2.6. Let there be a binary code C of length 6, whose generator matrix is:

$$G = \begin{pmatrix} 101111 \\ 011101 \\ 111010 \end{pmatrix}$$

Perform the following operations:

1) construct the words of C;

2) give the lateral classes of C, with elements with the smallest weight in their class as representatives;

3) give the parity check matrix H on the basis of G.

EXERCISE 2.7. Let C be a binary linear code of length 7, including a generator matrix:

$$G=\begin{pmatrix}1100101\\0111100\\0101011\\1001100\\1000111\end{pmatrix}$$

Perform the following operations:

1) use the Gauss method to deduce a parity check matrix H;

2) we suppose having received the word $v = (1011011)$. Calculate its syndrome;

3) decode v according to the principle of maximum probability.

EXERCISE 2.8. Let C be the binary code whose generator matrix is:

$$G=\begin{pmatrix}010111\\101101\\100011\end{pmatrix}$$

Perform the following operations:

1) construct the words of C. How many error(s) per word can we correct using this code?;

2) construct the lateral classes of C;

3) construct H, then calculate all the possible syndromes;

4) deduct its possible decoding table. Take a received word and decode it.

EXERCISE 2.9. Demonstrate that the Hsiao code C whose parity check matrix H is provided below:

$$H=\begin{pmatrix}10000111\\01001011\\00101101\\00011110\end{pmatrix}$$

corrects one error per word and detects two of them.

EXERCISE 2.10. Let C be the ternary code whose generator matrix is:

$$G = \begin{pmatrix} 100221 \\ 201010 \\ 001210 \\ 012021 \end{pmatrix}$$

Perform the following operations:

1) construct a parity check matrix H using the Gaussian method;

2) deduce the minimum distance of the code from the study of the columns of H.

EXERCISE 2.11. In $[\mathbb{F}_2]^3$, give all the linear codes of dimension 2. Clarify the formula giving the number of these codes.

EXERCISE 2.12. We take the following parity check matrix H of C:

$$H = \begin{pmatrix} 00001001111 \\ 00010011101 \\ 00100111001 \\ 01000110011 \\ 10000100111 \\ 10101011010 \end{pmatrix}$$

Does C allow the correction of 2 errors per word?

EXERCISE 2.13. Use the Gaussian method to find a parity check matrix of the binary code C generated by the generator matrix:

$$G = \begin{pmatrix} 1011010 \\ 1101100 \\ 0011011 \end{pmatrix}$$

EXERCISE 2.14. Same exercise as above with:

$$G = \begin{pmatrix} 1011011 \\ 0001010 \\ 1101100 \end{pmatrix}$$

EXERCISE 2.15. Let C_1 and C_2 be two codes of length n, in $\mathbb{F}_2$, given by their respective generating matrices G_1 and G_2. Find an algorithm to construct a generator matrix of $C_1 \cap C_2$.

EXERCISE 2.16. Let C be a linear code of length n, in $\mathbb{F}_q$, with a parity check matrix test H. Let $E = \{e_1, \ldots, e_r\}$ be a set of vector-errors (each e_i is thus a n-tuple in $\mathbb{F}_q$):

1) prove that C detects any element e_i of C, if $H[e_i]^t \neq 0$;

2) prove that C corrects any element e_i of C, if for all e_j and e_k of E we have $H[e_j - e_k]^t \neq 0$.

EXERCISE 2.17. Let there be a Hamming code with a parity check matrix:

$$\begin{pmatrix} 0001111 \\ 0110011 \\ 1010101 \end{pmatrix}$$

Write the equations which allow the correction of a codeword $(a_1, a_2, \ldots, a_7)$ using the syndrome.

EXERCISE 2.18. Let there be a binary linear code C whose generator matrix is in the form $G = (I_k R)$. Prove that the number of shortened codes of $C^{\perp}$ that have the same dimension as $C^{\perp}$ is equal to the number of lines $(0 \cdots 0)$ of the submatrix R.

EXERCISE 2.19. Let there be a linear binary code C of length n whose generator matrix G consists of d rows. It is supposed that each row start with 1 and has a weight of d. Finally, we suppose that the minimum distance d of C verifies $d^2 - d + 1 = n$ and that the sum of the rows is the full vector of 1:

1) prove that n and d are odd;

2) prove that in columns $2, 3, \ldots, n$ there is exactly one 1.

2.8.3. *Finite bodies*

The third series of exercises covers finite bodies.

EXERCISE 2.20. Perform the following operations:

1) in $\mathbb{F}_2[X]/(1 + X + X^3)$ compute the (multiplicative) order of $1 + X$, then of $1 + X + X^2$;

2) in $\mathbb{F}_2[X]/(1 + X + X^2 + X^3 + X^4)$ compute the (multiplicative) order of X, of $1 + X$, then of $1 + X^2 + X^3$;

3) $\mathbb{F}_2[X]/(1 + X^3 + X^6)$ compute the (multiplicative) order of X (use the lattice of the divisors of $2^6 - 1$, as well as the binary decomposition of powers). What is the order of $X + X^3$?

EXERCISE 2.21. Perform the following operations:

1) find a primitive element in $\mathbb{F}_2[X]/(1 + X + X^3)$;

2) find a primitive element in $\mathbb{F}_2[X]/(1 + X^3 + X^4)$;

3) let α be a primitive of $\mathbb{F}_{2^4}$. Prove that $\alpha^i = 1$, if i is a multiple of 15, then that α^i is primitive, if $(15, i) = 1$;

4) let there be $\mathbb{F}_2[X]/(1 + X + X^2 + X^3 + X^4)$. Construct all the primitives;

5) let there be a field $\mathbb{F}_q$ and let there be two elements α and β of this field. Let i and j be the respective orders of each of them. Is the order of the product $\alpha\beta$ the LCM of i and j?

EXERCISE 2.22. Perform the following operations:

1) in $\mathbb{F}_2[X]/(1 + X^2 + X^3)$ construct the cyclotomic class of $1 + X$, then that of $X + X2$;

2) in $\mathbb{F}_2[X]/(1+X+X^4)$ construct the cyclotomic class of X, then that of $1+X^2$, then that of $X + X^2 + X^3$;

3) in $\mathbb{F}_2[X]/(1 + X + X^3)$ construct all the cyclotomic classes.

EXERCISE 2.23. Perform the following operations:

1) in $\mathbb{F}_2[X]/(1 + X^2 + X^5)$ construct the minimum polynomial of $1 + X$, then that of $X + X^3 + X^4$;

2) in $\mathbb{F}_2[X]/(1 + X + X^3)$ construct the minimum polynomial of each element of the field, then calculate their product;

3) in $\mathbb{F}_2[X]/(1 + X^3 + X^5)$ construct the minimum polynomial of $1 + X^2 + X^3$, then of $1 + X$.

EXERCISE 2.24. In $[\mathbb{F}_3]^3$ we define a relation of equivalence R, by $R(x, y)$, if $x = \lambda y$, $\lambda \in \mathbb{F}_3$. The class of x will be noted (x):

1) how many such classes are there (they are called *points*)?;

2) let a and b be two representatives of two distinct points. We call *straight* the set of non-zero elements in the form $\alpha a + \beta b$ (α and β in $\mathbb{F}_3$, simultaneously non-zero). How many straights are there?;

3) how many points are there per straight? Provide a straight.

EXERCISE 2.25. Let us consider $\mathbb{F}_2[X]/(P(X)), dgP(X) = 6, P(X)$ palindrome:

1) prove that $X^{-1} = X^5 + u(X)$, where $dgu(X) \leq 4$;

2) prove that the family $\{X, X^2, X^3, (X + X^{-1}), (X^2 + X^{-2}), (X^3 + X^{-3})\}$ is a base.

EXERCISE 2.26. In $\mathbb{F}_{q^n}$ we define a Trace application, noted Tr_q, defined by $Tr_q(\beta) = \beta + \beta^q + \cdots + \beta^{q^{n-1}}$:

1) for $q = 2, n = 4, \mathbb{F}_{2^4} = \mathbb{F}_2[X]/(1 + X + X^4), \beta = 1 + X + X^3$, calculate $Tr_q(\beta)$;

2) for $q = 4, n = 2, \mathbb{F}_{4^2} = \mathbb{F}_4[X]/(1+\alpha X+X^2), \beta = 1+(1+\alpha)X$, where α is a primitive of $\mathbb{F}_4$, calculate $\mathrm{Tr}_q(\beta)$;

3) for $q = 3, n = 2, \mathbb{F}_{3^2} = \mathbb{F}_3[X]/(2+X+X^2), \beta = 1+2X$, calculate $\mathrm{Tr}_q(\beta)$.

EXERCISE 2.27. Let there be the field $\mathbb{F}_{q^n}$ and the Norm function in $\mathbb{F}_q^*$ defined by $N_q(\beta) = \beta\beta^q \cdots \beta^{q^{n-1}}$:

1) for $q = 2, n = 3, \mathbb{F}_8 = \mathbb{F}_2[X]/(1+X^2+X^3), \beta = 1+X$, calculate $N_2(\beta)$,

2) prove that $N_q(\beta)$ is in $\mathbb{F}_q$,

3) prove that N_q is a multiplicative morphism of $\mathbb{F}_{q^n}^*$ in $\mathbb{F}_q^*$.

EXERCISE 2.28. Let $\mathbb{F}_{2^4} = \mathbb{F}_2[X]/(1+X+X^4)$:

1) calculate $\sum_\beta \beta$ for β traversing the field;

2) calculate $\sum_\beta \beta^3$ for β traversing the field;

3) calculate $\sum_\beta \beta^i$ for i non-divisible by 15;

4) calculate $\sum_\beta \beta^{15i}$ for any whole i.

EXERCISE 2.29. Let there be $\mathbb{F}_{2^n}$, and let β be an element of this field:

1) consider the multiplication by 2 in $\mathcal{Z}/(2^n-1)$, and prove that all the cyclotomic classes have a cardinal dividing n;

2) prove that the degree of the minimum polynomial of any element of $\mathbb{F}_{2^n}$ is a divisor of n.

EXERCISE 2.30. Let $\mathbb{F}_{q^{2d}}$ be a finite field. Let there be an a such that $aa^{q^d} = 1$. Prove that there exists a b in $\mathbb{F}_{q^{2d}}$ such that we have $b + ab^{q^d} \neq 0$.

EXERCISE 2.31. Show that there are $(2^3-1)(2^3-2)$ ways of choosing a non-zero vector and another that is not a multiple of the first one. Deduce the number of pairs of vectors that are free. Deduce from this that the number of subspaces of dimension 2 is equal to $(2^3-1)(2^3-2)/(2^2-1)(2^2-2)$. Generalize to the number of subspaces of dimension k in $\mathbb{F}_{2^n}$.

2.8.4. *Cyclic codes*

2.8.4.1. *Theory*

EXERCISE 2.32. Let $g(X) = 1+X^2+X^3+X^4$ be the generator of a cyclic code:

1) what is the length of the code?;

2) construct the generator matrix G of this code whose rows are the (right) shifts of $g(X)$;

3) code the information polynomial $1+X+X^2$ with this matrix;

4) code the same information word using systematic coding;

5) construct the generator matrix G' of the code that corresponds to systematic coding. What is the relationship with G?

EXERCISE 2.33. Construct the binary cyclic code containing the two following elements:

$$x = (1011100)$$
$$y = (0100011)$$

Take the polynomial of the smaller degree, which appears in the code, and calculate the remainder of division of $X^7 - 1$ by this polynomial.

Construct the binary cyclic code that contains the two words:

$$x = (110110110110110)$$
$$y = (111111111111111)$$

Take the polynomial of the smaller degree, which appears in the code, and calculate the remainder of division of $X^{15} - 1$ by this polynomial.

EXERCISE 2.34. Construct the cyclotomic classes of the field $\mathbb{F}_{2^3}$, without using representation. Deduce from that a factorization of $X^8 - X$. Recall that any element of this field different from 1 or 0 has an order equal to 7. Construct the generator of a binary cyclic Hamming code of length 7 correcting 1 error per word.

EXERCISE 2.35. How many binary cyclic codes of length 7 correcting 1 error per word are there (use the previous exercise)?

EXERCISE 2.36. Construct the generator of a cyclic binary BCH code of length 15 correcting 2 errors per word (use the cyclotomic classes in $\mathcal{Z}/(15)$).

EXERCISE 2.37. Construct the generator of a cyclic binary BCH code of length 15 correcting 3 errors per word.

EXERCISE 2.38. Remember that a code RM of length $n = 2^r - 1$ is referred to as of *order* s if its roots α^i are such that the Hamming weight of the binary expression of i is less or equal to $r - s$. Construct the generator of an RM code of length 15, order $s = 2$. According to the BCH theorem, how many errors does it correct per word?

EXERCISE 2.39. Construct the generator of an RS code of binary length 56 correcting 3 binary errors per word. What is its transmission rate?

EXERCISE 2.40. Is there a binary cyclic code of length 31, with a redundancy rate strictly lower than 12% correcting 3 errors per word?

EXERCISE 2.41. Let us consider the following problems:

1) let C be a cyclic code (n, k, d) in $\mathbb{F}_q$ of length n. We suppose having a generator matrix in the systematic form $(I_k R)$. Examine the words in the rows of this matrix and deduce from it the Singleton bound $d \leq n - k + 1$;

2) it is said that a code is MSD (*maximum separable distance*) if it verifies $d = n - k + 1$ (such codes exist, as the RS codes). Prove that there is not an MDS code in $\mathbb{F}_2$ that corrects at least one error.

EXERCISE 2.42. We consider a binary cyclic code of length 7 with the generator $g(X) = 1 + X^2 + X^3$:

1) construct the generator matrix G of the code;

2) using the Gaussian method put it in systematic form $(I_4 R)$;

3) deduce the polynomial $h(X)$ that generates the annihilator of $g(X)$ as follows:

– deduce from $(I_4 R)$ the parity check matrix $(R^t I_{7-4})$. Exchange its i columns 2 by 2 with $n - i + 2$, for $i \neq 1$. Leave column 1 unchanged,

– put the new matrix in the form $(R' I_3)$. Verify that the first row of $(R' I_3)$ is $h(X)$.

EXERCISE 2.43. According to the BCH bound:

1) Is there a BCH code $(15, 7, 7)$?;

2) Is there a BCH code $(31, 12, 5)$?

EXERCISE 2.44. Let $g(X) = (X^{31} - 1)/(1 + X^2 + X^5)$:

1) calculate $g(X)$;

2) calculate a generator matrix H of the code $(g(X^{-1}))$. Put it in systematic form (IR).

EXERCISE 2.45. Let $(g(X))$ be a (n, k, d) binary cyclic code C with $g(X)$ that divides $X^n - 1$. Let $H = (I_k R)$ be a parity check matrix of C. Let $(r_0, \ldots, r_{n-1})$ be the received word. Prove that $H(r_0, r_1, \ldots, r_{n-1})^t$ has as coefficients those of the remainder of the division of $r_0 + r_1 X + \cdots + r_{n-1} X^{n-1}$ by $g(X)$.

EXERCISE 2.46. Let C be a binary cyclic code of length $2r + l$ with a word of odd weight. Prove that it contains the element $j(X) = 1 + X + X^2 + \cdots + X^{2r}$.

EXERCISE 2.47. Let β be a primitive, n^{th} root of 1. Let C be the code having among its roots $\{\beta^{a_0}, \beta^{a_0+a_1}, \beta^{a_0+a_2}, \beta^{a_0+a_1+a_2}\}$, a_1 and a_2 are relatively primes of n. Prove that $d \geq 4$.

EXERCISE 2.48. Let C be a binary cyclic code verifying $C^{\perp} \supseteq C$. Prove that, if C contains a generator of weight $4t$, then all its words have a weight that is a multiple of 4.

EXERCISE 2.49. Let there be a binary code $(g(X))$ of length n, such that $g(X) = g(X^{-1})$ and $g(1) = 0$:

1) prove that this code has a minimum distance at least equal to 6;

2) find such a cyclic binary code (33, 22, 6) (note that $1 + X^3 + X^{10}$ is primitive).

2.8.4.2. *Applications*

The following exercises relate to the applications of cyclic codes.

EXERCISE 2.50. Plot the curve giving the output of a Hamming code (15, 11, 3). Plot the curve without coding. What benefit do we have in decibels with 10^{-5}?

EXERCISE 2.51. We have a Hamming code decoder (7, 4, 3). We want to protect the information passing through a channel with *bursts* of length 3. Propose a coding and decoding system.

EXERCISE 2.52. We wish to code information blocks correcting two errors per word. Is there a binary cyclic code of length 21 and dimension 11 that answers the question?

EXERCISE 2.53. We wish to transmit strings of length 75. The input (i.e. the channel) error rate is 10^{-3}. The redundancy rate r must be no larger than 0.2. Is there a code which lowers the error rate?

EXERCISE 2.54. An industry specialist has a list of specifications to satisfy. It must transmit strings of information of length 43307. The channel by the transmission is carried has an error rate equal to 2×10^{-3}. For the application concerned the residual rate must be with not more than 10^{-6}. For reasons of transmission costs the redundancy rate must be no more than 0.24:

1) give the possible lengths for the information blocks;

2) for each choice of k give the possible values of n;

3) does there exist a cyclic code which solves this problem?

EXERCISE 2.55. We wish to transmit strings of information of length 258. The transmission line is an SBC (symmetrical binary channel), the errors are isolated with rate (p_c) equal to 10^{-3}. We wish to bring this rate to (p_r) equal to 10^{-4} at the output, but we can accept only accept a redundancy rate of no more than 7%. See, if it is possible to satisfy the customer who has required these specifications.

EXERCISE 2.56. We wish to code information words of length 23. Is it possible to find a corrector code enabling to obtain:

– a binary error correction of 1 symbol per codeword?;

– a redundancy rate strictly lower than 10%?

EXERCISE 2.57. Let us consider the following situations:

1) We have strings to be transmitted by satellite, which have an imposed length equal here to 93. The error rate of the atmospheric channel equals 10^{-2}. We wish to obtain after decoding a residual error rate equal to 10^{-3}. Find the generator of a binary cyclic code (if it exists) making it possible to solve this problem.

2) We have strings for transmission via satellite of length equal to 217 in the application concerned. The error rate of the channel is in this case equal to 3×10^{-2}. We wish to obtain after decoding a residual error rate equal to 7×10^{-3}. Find the generator of a binary cyclic code (if it exists) providing a solution.

3) In the following specifications there are the constraints. Study if there is a solution. If solutions exist, provide the shortest code;

– length of string: 126,

– channel rate: 10^{-2},

– requested residual rate: no more than 6×10^{-4},

– redundancy rate no more than 0.1.

EXERCISE 2.58. We wish to transmit images through a disturbed channel (Gaussian white noise inducing an error rate of 10^{-3}). These images have 69 rows and 69 columns. We neglect the case where the error word is in the code. We wish that no more than one residual error per image remains.

1) what are the two cases where decoding introduces errors?;

2) which cutting of each row must we perform to satisfy the specifications?;

3) find the generator of a cyclic code which solves the presented problem.

2.8.5. *Exercises on circuits*

The fifth series of exercises covers circuits.

We say that a register is associated to a polynomial, if it is a register with shifts whose *feedback* strands represent the coefficients of the polynomial (there is a strand, if the corresponding coefficient is 1).

EXERCISE 2.59. Let there be the register associated to $1 + X + X^5$, initialized at (10000) (see Figure 2.14):

1) show that $p(X) = 1 + X + X^5$ is not irreducible;

2) the output sequence of the register is periodic. Write a period;

3) does the sequence 11111 appear at output?;

4) does the content (11111) appear?

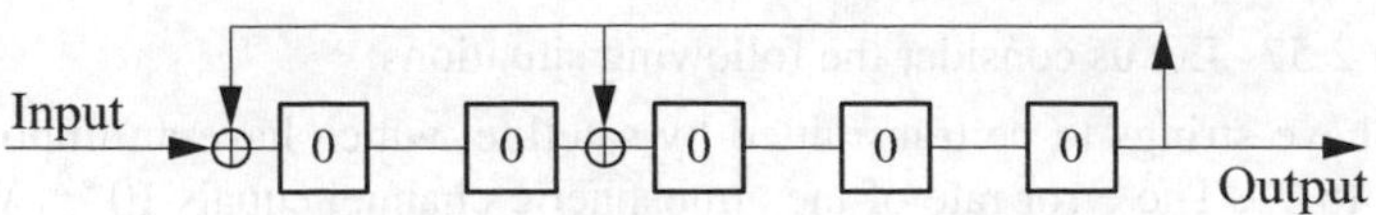

Figure 2.14. *Illustration of exercise 2.59*

EXERCISE 2.60. Let there be the following circuit in Figure 2.15:

1) for an input representing $1+X+X^3+X^5+X^8$ calculate the output, as well as the final content of the register;

2) write the Euclidean equality between the input and the polynomial giving the *feedback*. What is the relationship with question 1?

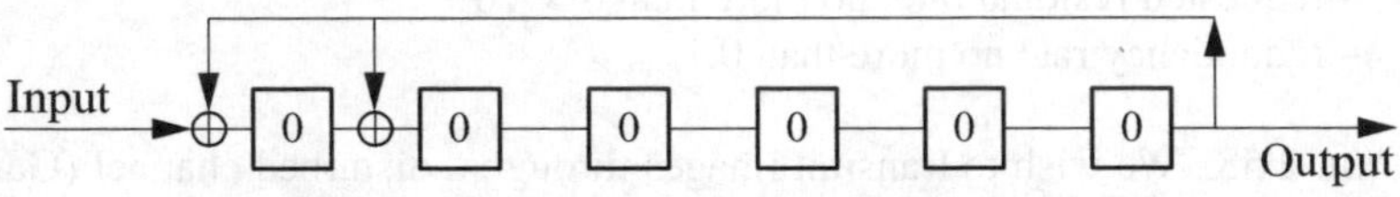

Figure 2.15. *Illustration of exercise 2.60*

EXERCISE 2.61. We initialize the following circuit in Figure 2.16 with 111000:

1) how many clock signals are necessary to find the same contents?;

2) find the theoretical justification proving this result.

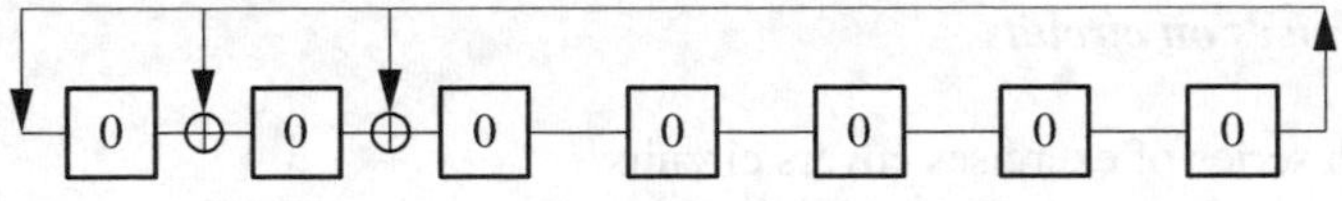

Figure 2.16. *Illustration of exercise 2.61*

EXERCISE 2.62. Consider the following circuit in Figure 2.17, initialized with (000):

1) input $X^7(1+X+X^2)$, and take at output the last 7 binary symbols (the 3 first correspond to the emptying of the register and thus do not mean anything). What do we obtain?;

2) verify that the last 7 symbols $s_0, s_1, \ldots, s_6$ (first to come out on the right) correspond to a polynomial $s_0+s_1X+\cdots+s_6X^6$, which is divisible by $1+X+X^2+X^4$.

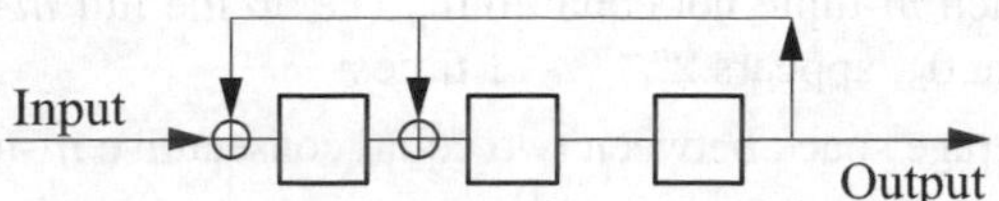

Figure 2.17. *Illustration of exercise 2.62*

EXERCISE 2.63. Consider the following circuit in Figure 2.18, initialized with (0000):

1) input $X^{15}(1+X+X^3)$. What does the output represent?;

2) calculate this output in the form of a polynomial;

3) demonstrate that the set of these polynomials is the binary cyclic code of length 15 generated by $1+X+X^2+X^3+X^5+X^7+X^8+X^{11}$.

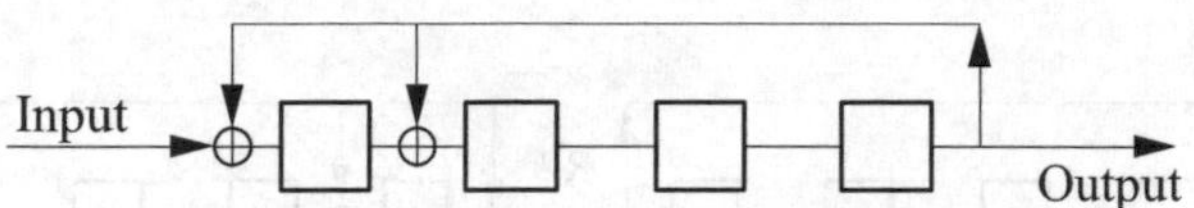

Figure 2.18. *Illustration of exercise 2.63*

EXERCISE 2.64. Supplement the Hsiao decoder given in the course representing the outputs indicating whether there is an error, if an incorrigible error has been detected.

EXERCISE 2.65. Let $p(X)$ be a binary primitive irreducible polynomial with $p(X)=p_0+\cdots+p_{n-1}X^{n-1}+p_nX^n$. Let us consider its associated register R:

1) prove that $p_0=1$, and that $p(1)=1$;

2) prove that any n-tuple binary appears exactly once in R;

3) let us note the contents of R by the power of the associated α : $(10\cdots0)=\alpha^0, (010\cdots0)=\alpha^1$, etc. Calculate k, such that $\alpha^k=(p_0, p_0+p_1, p_0+p_1+p_2, \ldots, p_0+p_1+\cdots+p_{n-1})$.

EXERCISE 2.66. Let $p(X)$ be an irreducible binary polynomial, $p(X)=p_0+\cdots+p_nX^n$, with $p_0=p_n=1$, and the associated register with shifts. Prove that at the register output we find the n-tuple $(1\ldots1)$, if and only if the register contains at a given time the sequence $(p_0, p_0+p_1, \ldots, p_0+p_1+\cdots+p_{n-1})$.

EXERCISE 2.67. We consider the register with shifts whose connections are the coefficients of the minimum polynomial $\mathcal{M}_\alpha(X)$ of a primitive element a of $\mathbb{F}_{2^n}$ with $2^n - 1$ prime. We considers the m positions on the left, $m \leq n$:

1) prove that each m-tuple not equal to 0_m (i.e. to the full m-tuple of 0) appears 2^{n-m} times and that 0_m appears $2^{n-m} - 1$ times;

2) calculate average space between two equal consecutive m-tuples;

3) take the example of $1 + X + X^4, m = 2$. Calculate the successive spaces and compare them with the theoretical result.

EXERCISE 2.68. We have a cyclic binary code C of length 15. Its generator $g(X)$ equals $1 + X^3 + X^4 + X^5 + X^6$:

1) using the circuit of division by $g(X)$, examine whether $1{+}X{+}X^7{+}X^9$ is in C;

2) draw the circuit giving the redundancy for the systematic coding of the information word associated with $1 + X + X^3 + X^4 + X^8$. Give the codeword.

EXERCISE 2.69. Let there be the following double register in Figure 2.19:

1) Turn the double register until obtaining (10000) in R_1,

2) Interpret the contents of R_2 at that point.

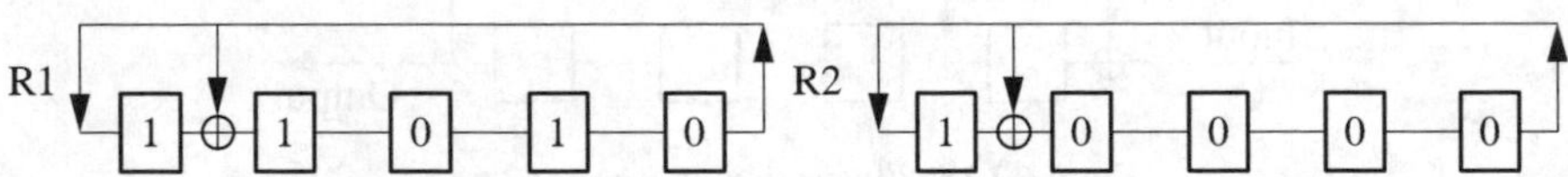

Figure 2.19. *Illustration of exercise 2.69*

EXERCISE 2.70. Let there be the following set-up in Figure 2.20 (Meggitt decoder) where we will admit that emptying of R2 into R′2 followed by a RTZ (reset to zero) of R2 can be carried out without spending time (for example, on a raising clock face). The dotted arrows descending towards R′2 indicate a parallel emptying of all the 7 signals (i.e. when the remainder of division by $1 + X + X^3$ of the input word is calculated in R2). Moreover, as soon as R2 is emptied, an RTZ is immediately performed. Simulate the circuit with the sequence 0110100110111111100111 at input (first input on the right). What do we obtain?

EXERCISE 2.71. Let $g(X) = g_0 + g_1X + \cdots + g_4X^4 + X^5$:

1) prove that at the register output associated with $g(X)$ we find a sequence 11111, whatever the initialization of the register;

2) what is the content of the register when we read the last 1 of the sequence 11111, expressed using the coefficients $g_0, g_1, \ldots, g_4$?;

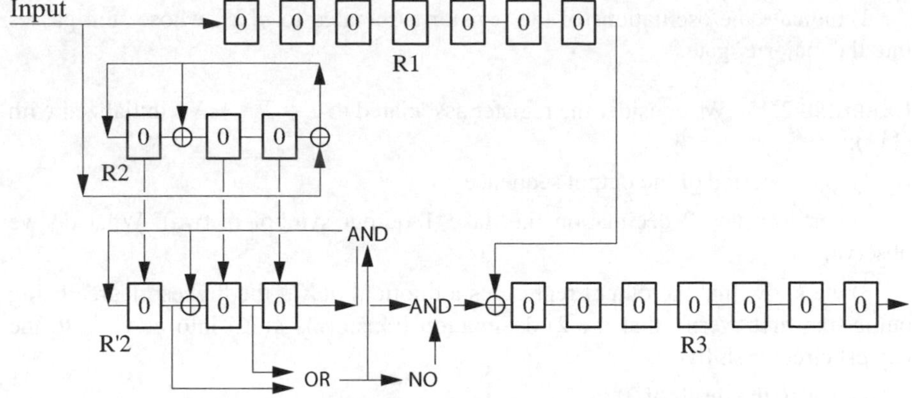

Figure 2.20. *Illustration of exercise 2.70*

3) deduce from it a hardware set-up giving the coefficients of $g(X)$ regardless of the initialization.

EXERCISE 2.72. Let there be a register associated to $1 + X + X^4 = p(X)$, initialized with (1101). We consider the sequence $s_0, s_1, \ldots, s_{14}$, (s_{14} output first) of 15 consecutive binary values, which we obtain at the register output. This sequence can be regarded as a polynomial $s_0 + s_1 X + \cdots + s_{14} X^{14}$, an element $\mathbb{F}_2[X]/(X^{15} - 1)$:

1) prove that this polynomial belongs to the cyclic code generated by $1+X+X^2+X^3 + X^5 + X^7 + X^8 + X^{11}$ in $\mathbb{F}_2[X]/(X^{15} - 1)$;

2) prove the result in the general case, when $p(X)$ is an unspecified irreducible of degree n and exhibitor m.

EXERCISE 2.73. Let there be a register associated with a primitive irreducible polynomial $p(X)$ of degree n. We initialize this register with arbitrary non-zero contents. Prove that the output of each oscillation yields the same sequence to the nearest circular permutation.

EXERCISE 2.74. Let there be the PG-code $(m = 4, s = 2, r = 1)$ with generator $g(X)$ equal to $1+X^3+X^5+X^6+X^9+X^{10}+X^{11}+X^{12}+X^{13}+X^{17}+X^{18}+X^{20}+X^{21}+X^{22} + X^{24} + X^{26}$ that divides $X^{31} - 1$ (see section 2.4.3.7). We wish to produce a decoder by majority decoding using the R2 register of the Meggitt set-up (see exercise 2.70), but replacing R′2 by a majority gate.

1) verify that $\{1, \alpha, \alpha^{18}\}$, $\{1, \alpha^2, \alpha^5\}$, $\{1, \alpha^3, \alpha^{29}\}$, $\{1, \alpha^4, \alpha^{10}\}$, $\{1, \alpha^6, \alpha^{27}\}$, $\{1, \alpha^7, \alpha^{22}\}$, $\{1, \alpha^8, \alpha^{20}\}$, $\{1, \alpha^9, \alpha^{16}\}$, $\{1, \alpha^{11}, \alpha^{19}\}$, $\{1, \alpha^{12}, \alpha^{23}\}$, $\{1, \alpha^{13}, \alpha^{14}\}$, $\{1, \alpha^{15}, \alpha^{24}\}$, $\{1, \alpha^{17}, \alpha^{30}\}$, $\{1, \alpha^{21}, \alpha^{25}\}$, $\{1, \alpha^{26}, \alpha^{28}\}$ are the projective rows (or 1-*flats*) orthogonal in 1 (primitive α, such that $1 + \alpha^2 + \alpha^5 = 0$);

2) indicate the oscillations of the register associated to $g(X)$ whose sum passes into the majority gate.

EXERCISE 2.75. We consider the register associated to $1 + X^2 + X^3$ initialized with (111):

1) give a period of the output sequence;

2) perform step 2 decimation (i.e. take 1 output symbol in two). What do we observe;

3) we know that the output represents a quotient $q(X)$, the highest degree being output the first. Verify that the 2) decimation transforms $q(X)$ into $q(X^{-4})$ to the nearest circular shift;

4) justify the result of 3).

Chapter 3

Convolutional Codes

3.1. Introduction

Convolutional codes, invented in 1954 by P. Elias[1], constitute a family of error correcting codes whose decoding simplicity and good performances, in particular for the Gaussian channel, are, without doubt, very much at the origin of their success.

For a convolutional code at every moment k the encoder delivers a block of N binary symbols[2] $c_k = (c_{k,1}, c_{k,2}, \ldots, c_{k,N})$, a function of the block of K information symbols $d_k = (d_{k,1}, d_{k,2}, \ldots d_{k,K})$ present at its input along with m preceding blocks. convolutional codes consequently introduce an memory effect of the order m.

The quantity $\nu = m + 1$ is called the *constraint length of the code* and the ratio $R = K/N$ is called the *code rate*. If K information symbols at the encoder input are found explicitly in the coded block c_k, that is:

$$c_k = (d_{k,1}, \ldots, d_{k,K}, c_{k,K+1}, \ldots, c_{k,N}) \qquad [3.1]$$

then the code is known as *systematic*. In the contrary case it is known as *non-systematic*. The general diagram of an encoder with output K/N and memory m is represented in Figure 3.1.

1. ELIAS P., "Error-free coding", *IEEE Transactions on Information Theory*, p. 29–37, September 1954.

2. The notations used in this chapter are represented in Table 3.1.

Chapter written by Alian GLAVIEUX and Sandrine VATON.

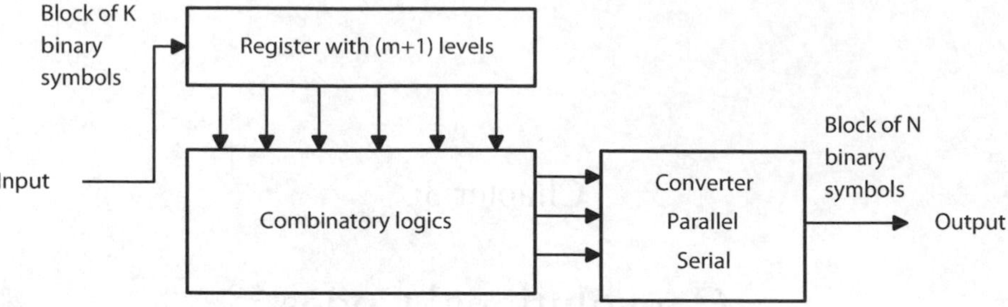

Figure 3.1. *General diagram of a convolution encoder with an output of K/N and memory m*

$d_k \triangleq (d_{k,1}, \ldots, d_{k,K}), 1 \leq k \leq M$	Encoder input at the indexed moment k
$\boldsymbol{d} \triangleq (d_1, \ldots, d_M)$	Encoder input in the interval $1 \leq k \leq M$
$c_k \triangleq (c_{k,1}, \ldots, c_{k,N}), 1 \leq k \leq M$	Encoder output at the indexed moment k
$\mathbf{c} \triangleq (c_1, \ldots, c_M)$	Encoder output in the interval $1 \leq k \leq M$
$S_k \triangleq (S_{k,1}, \ldots, S_{k,m})$	State of the encoder at the indexed moment k
$\boldsymbol{S} \triangleq (S_0, \ldots, S_M)$	Sequence of encoder states in the interval $0 \leq k \leq M$
$y_k \triangleq (y_{k,1}, \ldots, y_{k,N}), 1 \leq k \leq M$	Channel output at the indexed moment k
$\boldsymbol{y} \triangleq (y_1, \ldots, y_M)$	Channel output in the interval $1 \leq k \leq M$
$\hat{d}_k \triangleq (\hat{d}_{k,1}, \ldots, \hat{d}_{k,K}), 1 \leq k \leq M$	Estimate of d_k calculated by the decoder
$\hat{\boldsymbol{d}} \triangleq (\hat{d}_1, \ldots, \hat{d}_M)$	Estimate of d calculated by the decoder

Table 3.1. *Notations*

At every moment k, the encoder, has m blocks of K information symbols in memory. These m K binary symbols define the S_k state of the encoder:

$$S_k = (d_k, d_{k-1}, \ldots, d_{k-m+1}) \quad [3.2]$$

If the input of the encoder is permanently fed by blocks of K information symbols, then the encoder output consists of N infinite sequences of coded symbols, which, for the output i, have the form:

$$(c_{1,i}, c_{2,i}, \ldots, c_{k,i}, \ldots) \quad i = 1, \ldots, N \quad [3.3]$$

Let us note that convolutional codes are well adapted to code transmissions with continuous flow of data. Indeed, the sequences of data to be coded can have any length.

To each coded sequence $i = 1, \ldots, N$ we can associate its transform in D defined by:

$$C_i(D) = \sum_k c_{k,i} D^k \quad [3.4]$$

where D (*delay*) is the operator delay, equivalent to the variable z^{-1} of the z-transform.

Each coded symbol $c_{k,i}$ may be expressed as a linear combination of K information symbols present at the encoder input and $K\ m$ symbols contained in its memory:

$$c_{k,i} = \sum_{l=1}^{K} \sum_{j=0}^{m} g_{l,j}^{i} d_{k-j,l} \qquad [3.5]$$

where the $g_{l,j}^{i}$ coefficients take the values 0 or 1 and where the sum operations are made *modulo* 2. Relation [3.5] is a convolutional product between the sequence of symbols to be coded and the impulse response of the encoder defined by the $g_{l,j}^{i}$ coefficients.

Taking into account expression [3.5], $C_i(D)$ may also be expressed as:

$$C_i(D) = \sum_{l=1}^{K} G_l^i(D) d_l(D) \qquad [3.6]$$

where $G_l^i(D)$ and $d_l(D)$ are, respectively, the transforms in D of the encoder response and the sequence to be coded for the input $l, l = 1, \ldots, K$:

$$G_l^i(D) = \sum_{j=0}^{m} g_{l,j}^{i} D^j$$
$$d_l(D) = \sum_{p} d_{p,l} D^p \qquad [3.7]$$

The quantities $G_l^i(D), 1 \leq l \leq K, 1 \leq i \leq N$ are called the *generator polynomials of the code.*

Introducing matrix notations:

$$d(D) = (d_1(D), \ldots, d_K(D))$$
$$C(D) = (C_1(D), \ldots, C_N(D)) \qquad [3.8]$$

encoder output can be expressed as a function of its input by the following matrix relation:

$$C(D) = d(D)G(D) \qquad [3.9]$$

where $G(D)$ is a matrix with K rows and N columns and the generator matrix of the code:

$$G(D) = \begin{bmatrix} G_1^1(D) & \cdots & \cdots & G_1^i(D) & \cdots & \cdots & G_1^N(D) \\ \vdots & \ddots & \ddots & \vdots & \ddots & \ddots & \vdots \\ G_l^1(D) & \cdots & \cdots & G_l^i(D) & \cdots & \cdots & G_l^N(D) \\ \vdots & \ddots & \ddots & \vdots & \ddots & \ddots & \vdots \\ G_K^1(D) & \cdots & \cdots & G_K^i(D) & \cdots & \cdots & G_K^N(D) \end{bmatrix} \quad [3.10]$$

Let us consider two examples of convolutional codes and determine their respective generating matrices.

EXAMPLE 3.1. Let there be a non-systematic code with memory 2 (constraint length $\nu = 3$) and output $1/2$ ($K = 1, N = 2$) whose encoder is represented in Figure 3.2. The generator matrix of this code is of the form:

$$G(D) = (G^1(D), G^2(D)) \quad [3.11]$$

We have omitted to index the generator polynomials in relation [3.11] since $K = 1$ for this code. Examining the diagram in Figure 3.2 it is easy to verify that:

$$\begin{aligned} g_{1,0}^1 &= g_{1,1}^1 = g_{1,2}^1 = 1 \\ g_{1,0}^2 &= g_{1,2}^2 = 1 \quad g_{1,1}^2 = 0 \end{aligned} \quad [3.12]$$

and, thus, the generator polynomials have the respective expressions:

$$\begin{aligned} G^1(D) &= 1 + D + D^2 \\ G^2(D) &= 1 + D^2 \end{aligned} \quad [3.13]$$

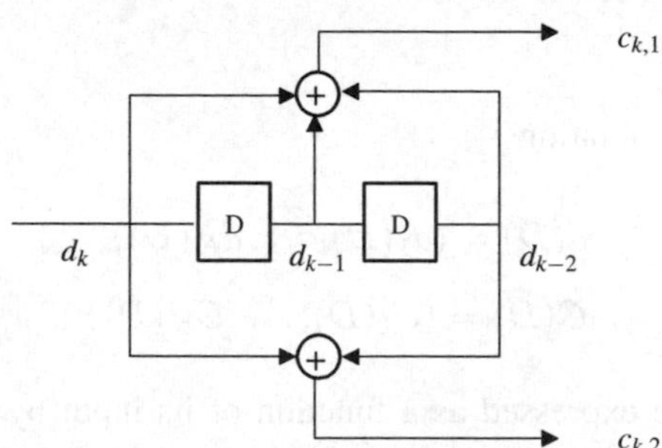

Figure 3.2. *General diagram of a convolution encoder with parameters $m = 2, R = 1/2$*

We can also associate a binary representation to the generator polynomials. For example 3.1, we have:

$$\begin{aligned} G^1(D) &\rightarrow g^1 = (1,1,1) \\ G^2(D) &\rightarrow g^2 = (1,0,1) \end{aligned} \qquad [3.14]$$

Using an octal representation, the relations [3.14] can be also written as:

$$g^1 = 7_{\text{octal}} \qquad g^2 = 5_{\text{octal}} \qquad [3.15]$$

EXAMPLE 3.2. Let there be a systematic code with memory $m = 1$ (constraint length $\nu = 2$) and output 2/3 ($K = 2, N = 3$) whose encoder is represented in Figure 3.3. Using the diagram in Figure 3.3 and taking into account the relations [3.5], [3.7] and [3.10], the generator matrix of this code equals:

$$G(D) = \begin{bmatrix} 1 & 0 & 1+D \\ 0 & 1 & D \end{bmatrix} \qquad [3.16]$$

This code has six generator polynomials. Two are zero, two are equal to 1 and the two others equal to $(1 + D)$ and D respectively.

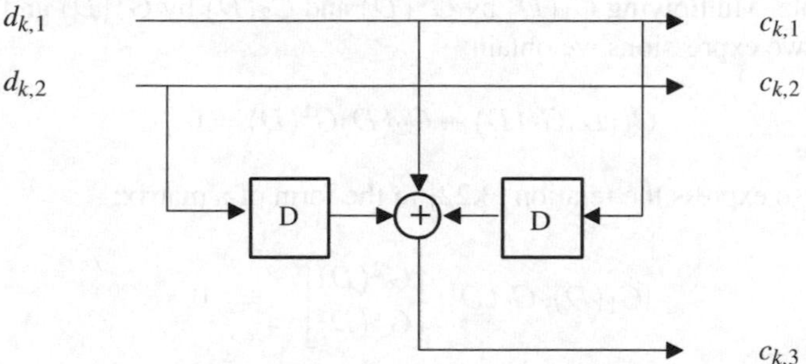

Figure 3.3. *General diagram of a convolution encoder with parameters $m = 1, R = 2/3$*

In Chapter 2, it has been demonstrated that a code was entirely defined by its generator matrix G and its parity check matrix H with:

$$GH^t = 0 \qquad [3.17]$$

where t indicates the transposition.

For a convolutional code we can also define a parity check matrix verifying the relation [3.17]. If the generator matrix $G(D)$ is expressed in the reduced form (case of systematic codes):

$$G(D) = [I_K, P(D)] \qquad [3.18]$$

where I_K indicates the identity matrix of dimension K, then the matrix $H(D)$ has the form:

$$H(D) = [P^t(D), I_{N-K}] \tag{3.19}$$

If the matrix G is not in reduced form, the matrix H is obtained by solving equation [3.17] or, similarly, solving:

$$C(D)H^t(D) = 0 \tag{3.20}$$

where the vector $C(D)$ is defined by relations [3.8] and [3.9].

For the code in example 3.1, the matrix $H(D)$ is obtained by solving relation [3.20]. According to relation [3.6] we can write:

$$\begin{aligned} C_1(D) &= d(D)G^1(D) \\ C_2(D) &= d(D)G^2(D) \end{aligned} \tag{3.21}$$

In relation [3.21] we have omitted the indices noted l in relation [3.6], since $K = 1$ in this example. Multiplying $C_1(D)$ by $G^2(D)$ and $C_2(D)$ by $G^1(D)$ and then summing up the two expressions we obtain:

$$C_1(D)G^2(D) + C_2(D)G^1(D) = 0 \tag{3.22}$$

We can also express the relation [3.22] in the form of a matrix:

$$[C_1(D), C_2(D)] \begin{bmatrix} G^2(D) \\ G^1(D) \end{bmatrix} = 0 \tag{3.23}$$

Taking into account relations [3.20] and [3.23], the parity check matrix $H(D)$ equals:

$$H(D) = [G^2(D), G^1(D)] \tag{3.24}$$

Replacing the generator polynomials $G^1(D)$ and $G^2(D)$ by their expressions [3.13] and using expression [3.20], the coded symbols verify the following parity relation:

$$c_{k,1} + c_{k-2,1} + c_{k,2} + c_{k-1,2} + c_{k-2,2} = 0 \quad \forall k \tag{3.25}$$

This relation is sometimes used to synchronize the decoder, i.e. to locate in the sequence of the symbols received by the decoder the beginning of each block of N coded symbols transmitted by the encoder at every moment k.

3.2. State transition diagram, trellis, tree

The operation of a convolution encoder may be represented by a graph called a *state transition diagram*. This graph reveals the various states of the encoder and the possible transitions between states. Let us recall that for an encoder with memory m and output K/N, the number of states is equal to 2^{Km}. We can show that the succession of the states of the encoder is a Markov chain. The state transition diagram of the convolutional code from example 3.1 is represented in Figure 3.4. A binary pair whose first symbol is $c_{k,1}$ and the second is $c_{k,2}$ is associated with each branch of the diagram.

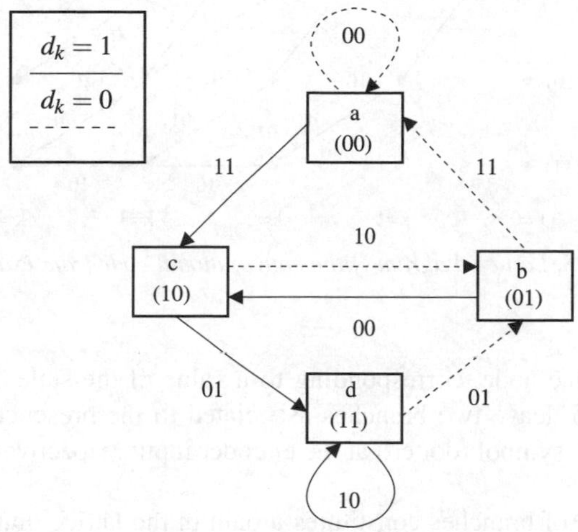

Figure 3.4. *State transition diagram of the convolutional code from example 3.1*

The state of the encoder at the moment k noted $S_k = (d_k, d_{k-1})$ can take four values regardless of k. We will note them here:

$$a = (00) \qquad b = (01) \qquad c = (10) \qquad d = (11) \qquad [3.26]$$

Two transitions are possible for each state according to the value of the symbol presented at the encoder input.

Binary pairs carried by the transitions between states, or buckling on the same state, correspond to the blocks transmitted by the encoder at every moment k. The transitions in solid (respectively dotted) lines correspond to the presence of one "1" (respectively "0") at the encoder input.

Another representation of the encoder operation, revealing the evolution of its S_k state through time, is possible. It is the trellis diagram. The interest of this representation compared to the state transition diagram will appear more clearly in the discussion on the decoding of convolutional codes (see section 3.6).

The lattice diagram of the convolutional code from example 3.1 is plotted in Figure 3.5 with the assumption that $S_0(k = 0)$ is $a = (00)$ generally referred to as the "all at zero" state. A binary pair whose first symbol is $c_{k,1}$ and the second is $c_{k,2}$ is associated to each branch of the diagram.

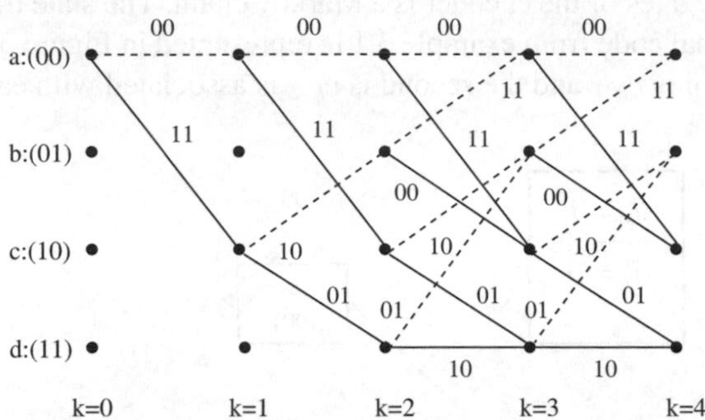

Figure 3.5. *Lattice diagram of the convolutional code from example 3.1*

From each lattice node, corresponding to a value of the state S_k and noted by a point in Figure 3.5, leave two branches associated to the presence of a "1" symbol (solid) and of a "0" symbol (dotted) at the encoder input respectively.

The succession of branches constitutes a path in the lattice diagram; each path is associated to particular sequence transmitted by the encoder (here, sequence indicates the *sequence* of coded symbols transmitted by the encoder, placed in a series).

Examining Figure 3.5 we can notice that the encoder produces particular sequences known by the decoder. If the sequence received by the decoder does not correspond to a path of the lattice diagram, the decoder will be able to detect the presence of transmission error(s). Moreover, as we will see later on, it will be able to correct the error(s) by finding in the lattice diagram the sequence "nearest" to the received sequence.

The complexity of the lattice diagram depends on the number of states of the encoder (2^{Km}) and the number of branches that leave (or merge towards) each node (2^K). Its complexity thus grows exponentially with the memory of the encoder but also with the length K of the coded blocks.

For packet transmissions, the state of the encoder is generally initialized at zero at the beginning of each coded block. That can be achieved by adding $m\ K$ zero symbols at the end of each coded block. All the paths of the lattice diagram then merge towards the zero state. These $m\ K$ symbols are called *tail-biting symbols* of the lattice diagram.

The convolution encoder then associates a unique coded sequence with each information sequence. Convolutional codes are in this case perfectly similar to block codes and the coded sequences are codewords.

The lattice diagram is mainly used for the decoding of convolutional codes.

A last graph, called a *tree diagram*, also makes it possible to represent the operation of a convolution encoder. This diagram is represented in Figure 3.6 for the convolution encoder from example 3.1 making the assumption that its initial state S_0 is equal to $a = (00)$. Two branches associated to the presence of the symbol "1" (respectively "0") at the encoder input leave from each state S_k of the tree diagram, just as for the lattice diagram. This diagram is also used to decode convolutional codes, in particular, when the encoder has a large memory m (typically more than 10).

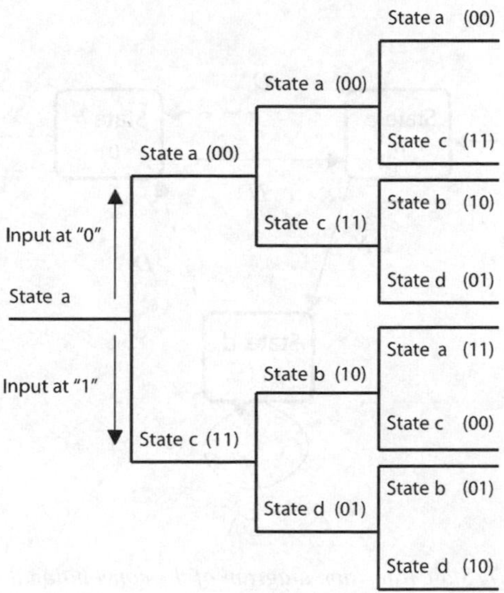

Figure 3.6. *Tree diagram of the convolutional code from example 3.1*

3.3. Transfer function and distance spectrum

The performances of a convolutional code are functions of the Hamming distance between the coded sequences associated to the paths of the lattice diagram that diverge and then merge again. For a given coding output, the correction capacity of the code is higher the greater these distances are.

The Hamming distance between two coded sequences is also equal to the Hamming weight of their sum. Convolutional codes being linear, the sum of two coded sequences is a coded sequence and, thus, the evaluation of distances between coded sequences can be reduced to determining the weight of the non-zero coded sequences (we exclude the "all at zero" paths).

These weights can be obtained on the basis of the transfer function of the code. To illustrate the calculation of the transfer function let us again consider the convolutional code from example 3.1.

The state $a = (00)$ of the state transition diagram is, first of all, divided into two states; the input state $a_e = (00)$ and the output state $a_s = (00)$. This operation is carried out to stop the transition from state a to itself and thus to avoid taking into account the "all at zero" path (corresponding to the emission of a succession of zero symbols by the encoder). This new state transition diagram is represented in Figure 3.7.

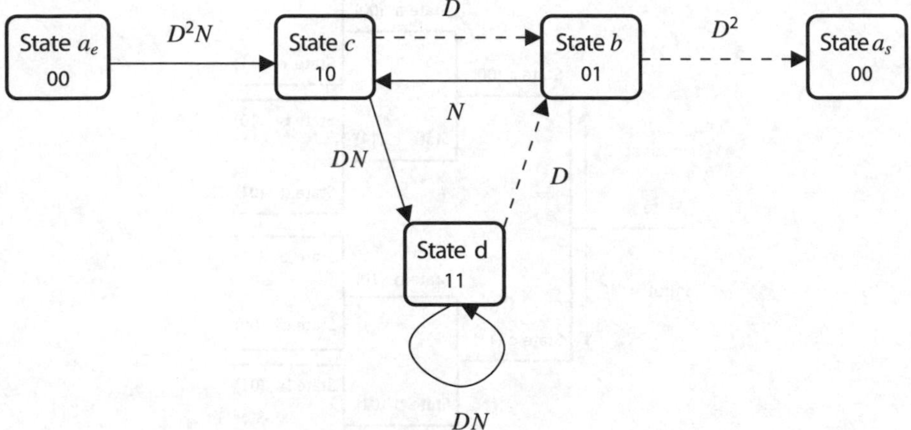

Figure 3.7. *State transition diagram of the convolutional code from example 3.1. The state has been divided into two states: a_e and a_s*

A label $D^j N^i$ is attached to each branch (transition between states) where i and j are respectively the Hamming weight of the block of K symbols at the encoder input and the Hamming weight of the block of N coded symbols, corresponding to this branch. For example, to pass from the state $a_e = (00)$ to the state $c = (10)$, it is necessary to place a symbol at 1 ($i = 1$) at the encoder, which will produce the coded pair (11), that is, $j = 2$. The branch joining a_e to c will thus receive the label D^2N.

The transfer function $T(D, N)$ of the code from example 3.1 is defined by:

$$T(D, N) = a_s/a_e \tag{3.27}$$

For each state we can write:

$$\begin{aligned} c &= D^2 N a_e + N b \\ b &= Dc + Dd \\ d &= DNc + DNd \\ a_s &= D^2 b \end{aligned} \qquad [3.28]$$

By solving this system of four equations, we obtain:

$$T(D, N) = D^5 N/(1 - 2DN) \qquad [3.29]$$

The transfer function can develop in series:

$$T(D, N) = D^5 N(1 + 2DN + 4D^2N^2 + 8D^3N^2 + \cdots) \qquad [3.30]$$

From relation [3.30], it is easy to see that the transfer function can be still put in the form:

$$T(D, N) = \sum_{d=5}^{+\infty} 2^{d-5} D^d N^{d-4} \qquad [3.31]$$

The encoder can transmit a coded sequence of weight 5 generated by an information sequence of weight 1, two sequences of weight 6 generated by two information sequences of weight 2, etc.

More generally, the encoder can transmit $n(d) = 2^{d-5}$ coded sequences of weight $d(d \geq 5)$ generated by information sequences of weight $d - 4$.

The minimum weight of the non-zero coded sequences (in our case 5), also equal to the minimum distance between the coded sequences, is called *the free distance of the code*. It is generally noted d_f ("f" stands for *free*).

Examining the lattice diagram in Figure 3.5 we can verify that an information sequence of weight 1 of the form:

$$(0 \cdots\cdots 0 \quad 1 \quad 0 \cdots\cdots 0)$$

generates a coded sequence of weight 5 equal to:

$$(0 \quad 0 \quad \cdots \quad \cdots 0 \quad 0 \quad 1 \quad 1 \quad 1 \quad 0 \quad 1 \quad 1 \quad 0 \quad 0 \quad \cdots \quad \cdots 0 \quad 0)$$

On the basis of relation [3.31] we can also write:

$$T(D, 1) = \sum_{d=d_f}^{+\infty} n(d) D^d \qquad [3.32]$$

Finally, deriving [3.31] with respect to N and taking $N = 1$ we obtain:

$$\left.\frac{\partial T(D,N)}{\partial N}\right|_{N=1} = \sum_{d=5}^{+\infty} w(d)D^d \qquad [3.33]$$

with $w(d) = (d-4)2^{d-5}$. The $w(d)$ coefficients constitute the distance spectrum of the code involved in the calculation of the code performances.

The transfer function is calculated easily if the number of encoder states is not high. Otherwise the calculations quickly become very tiresome and we limit ourselves to evaluating the first terms of the serial development of the transfer function on the basis of the lattice diagram using an adapted algorithm. Later we will see that these first terms are enough to evaluate the performances of convolutional codes.

3.4. Perforated convolutional codes

For a convolutional code with output K/N there are 2^K transitions starting at each node. Thus, for a high-rate code, i.e. for which the coefficient K is large, the complexity of the lattice diagram can quickly become rather great, and the code, consequently, become rather complex to decode.

It is possible to build higher rate convolutional codes on the basis of codes with 1/2 output ($K = 1$). To obtain such codes we use the technique of perforation.

Perforation consists of not transmitting all the coded symbols delivered by the encoder with 1/2 output; certain symbols are removed or punched out. Let us consider an example illustrated in Figure 3.8. This figure represents the output of an encoder with 1/2 output consisting of a succession of blocks of two coded symbols.

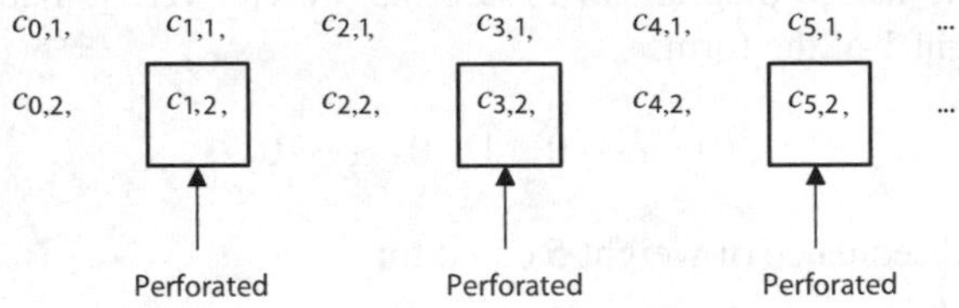

Figure 3.8. *Coded symbols at the output of a convolution encoder with 1/2 output*

If we punch out one coded symbol in four, for example, the framed symbols in Figure 3.8, then two information symbols of the encoder are associated to three coded symbols, i.e. a rate of 2/3.

The rule of perforation is defined on the basis of a mask M which, for our example, can be represented by a matrix with two rows and two columns:

$$M = \begin{bmatrix} 1 & 1 \\ 1 & 0 \end{bmatrix} \quad [3.34]$$

The element of the matrix M equal to 0 indicates the coded symbol to punch out. Here it is one in every two symbols delivered to the output number 2 of the encoder.

The lattice diagram of a perforated code with 2/3 output built on the basis of the convolution encoder from example 3.1 and mask M defined by relation [3.34] is represented in Figure 3.9.

We can verify on the diagram in Figure 3.9 that the free distance of the perforated code is 3 whereas it was 5 for the non-perforated code. Perforation increases the output of the code but reduces its free distance.

More generally, for a perforated code of output $p/(p+1), p > 1$, built on the basis of a code with 1/2 output, the perforation mask is represented by a matrix with 2 rows and p columns with $(p-1)$ elements equal to 0.

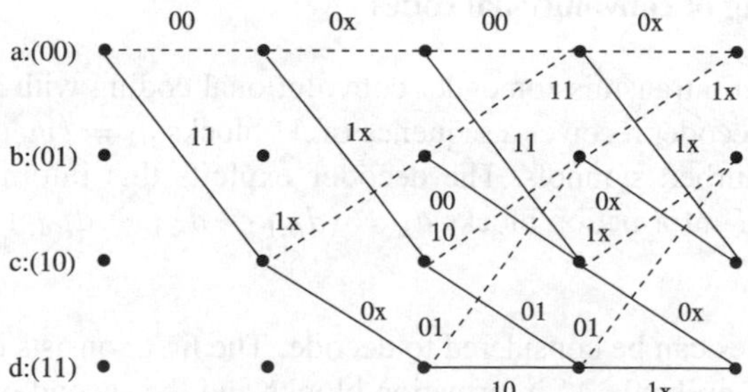

Figure 3.9. *Lattice diagram of a perforated code of 2/3 output built on the basis of the convolutional code from example 3.1. x indicates the punched out symbol*

Of course, the choice of the perforation mask is not indifferent. It must be selected so as to optimize the properties of code distance and, in particular, to maximize its free distance. The search for the best perforation mask is carried out using data-processing programs that evaluate, for a given mask, the free distance of the perforated code on the basis of its lattice diagram.

The performances of perforated codes are generally a little lower than those of non-punctured codes of the same rate and of the same constraint length. It is the price paid to have a simplified lattice diagram.

3.5. Catastrophic codes

There are convolutional codes called *catastrophic codes*, for which a finite number of errors at the decoder input can generate an infinite number of errors at the decoder output.

For these codes there exists at least one information sequence of infinite weight, which at the encoder output produces a coded sequence of finite weight. It follows from this comment that systematic codes are never catastrophic.

We can show that a convolutional code of output $R = 1/N$ is catastrophic if the highest common factor (HCF) of its generator polynomials is different from one.

EXAMPLE 3.3. The code with rate $R = 1/2$ with generator polynomials $G^1(D) = 1 + D$ and $G^2(D) = 1 + D^2$ is catastrophic since the HCF of $1 + D$ and $1 + D^2$ is equal to $1 + D$. Similarly, the code with generator polynomials $G^1(D) = 1 + D + D^2$ and $G^2(D) = 1 + D^3$ is catastrophic, since the HCF of $G^1(D)$ and of $G^2(D)$ is equal to $1 + D + D^2$. On the other hand, the convolutional code with generator polynomials $G^1(D) = 1 + D + D^2$ and $G^2(D) = 1 + D^2$ is not catastrophic because the HCF of $G^1(D)$ and $G^2(D)$ is equal to 1.

3.6. The decoding of convolutional codes

Let us consider a transmission under convolutional coding with K/N output, and suppose that the decoder receives a sequence of M blocks $y_k = (y_{k,1} \cdots y_{k,i} \cdots y_{k,N})$ composed of disturbed symbols. The decoder exploits this information to make a decision on the M information blocks $d_k = (d_{k,1} \cdots d_{k,i} \cdots d_{k,K})$ introduced at the encoder input.

Two approaches can be considered to decode. The first consists of looking for the sequence of more probable M information blocks and the second of determining the most probable information block at every moment k.

The first approach can be implemented rather simply using the Viterbi algorithm. This approach makes it possible to guarantee a minimum error probability for the decoded sequences and not a minimum error probability for the information blocks. However, as soon as the signal to noise ratio exceeds a few decibels, the Viterbi algorithm also leads, with good approximation, to a minimum error probability for the information blocks.

The second approach that works at the level of information blocks is known as the MAP (Maximum *a posteriori*) criterion, sometimes also referred to as the *BCJR algorithm* (Bahl, Cock, Jelinek, Raviv; 1974). This approach guarantees a minimum probability of error for the information blocks. Its disadvantage is its complexity compared to the Viterbi algorithm, while its advantage is that it makes it possible to associate reliability information to each decoded information block.

In the rest of this section we will note:

– $d = (d_1 \cdots d_k \cdots d_M)$ the information sequence;
– $c = (c_1 \cdots c_k \cdots c_M)$ the coded sequence;
– $y = (y_1 \cdots y_k \cdots y_M)$ the noisy sequence ("observation") used by the decoder.

3.6.1. *Viterbi algorithm*

This algorithm is used to look for the sequence $\hat{\boldsymbol{d}}$, such that:

$$p(\hat{\boldsymbol{d}}|\boldsymbol{y}) \geq p(\boldsymbol{d}|\boldsymbol{y}) \quad \forall \boldsymbol{d} \qquad [3.35]$$

where $p(\boldsymbol{d}|\boldsymbol{y})$ is the probability of the information sequence d conditionally to observation $\boldsymbol{y}$. In an equivalent manner the criterion [3.34] can also be stated:

$$\hat{\boldsymbol{d}} = \text{Arg}\max_{\boldsymbol{d}} p(\boldsymbol{d}/\boldsymbol{y}) \qquad [3.36]$$

where Arg indicates the argument $\boldsymbol{d}$ of $p(\boldsymbol{d}|\boldsymbol{y})$.

By definition of conditional probability we have:

$$p(\boldsymbol{d} \mid \boldsymbol{y}) = p(\boldsymbol{d}, \boldsymbol{y})/p(\boldsymbol{y}) \qquad [3.37]$$

The standardization term, $p(y)$ does not depend on d and the criterion [3.36] can also be expressed in the form:

$$\hat{\boldsymbol{d}} = \text{Arg}\max_{\boldsymbol{d}} p(\boldsymbol{d}, \boldsymbol{y}) \qquad [3.38]$$

Let us introduce the sequence $\boldsymbol{S} = (S_0 \cdots S_k \cdots S_M)$ where $S_k = (d_k, d_{k-1}, \ldots, d_{k-\nu+2})$ is the state of the encoder at the moment k and ν is its constraint length. Knowing the sequence S makes it possible to find the information sequence d. Indeed, each transition $S_{k-1} \rightarrow S_k$ depends only on one block d_k. Thus, decoding the sequence of blocks d_k is equivalent to estimating the succession of S_k encoder states.

Reasoning on the decoder states find the sequence $\hat{\boldsymbol{S}}$, such that:

$$\hat{\boldsymbol{S}} = \text{Arg}\max_{\boldsymbol{S}} p(\boldsymbol{S}, \boldsymbol{y}) \qquad [3.39]$$

The complete probability $p(\boldsymbol{S}, \boldsymbol{y})$ can be factorized into a product of M conditional probabilities. Indeed, by definition of conditional probability, we can write:

$$\begin{aligned} p(\boldsymbol{S}, \boldsymbol{y}) = {} & p(y_M, S_M \mid y_{M-1}, \ldots, y_1, S_{M-1}, \ldots, S_0) \\ & p(y_{M-1}, \ldots, y_1, S_{M-1}, \ldots, S_0) \end{aligned} \qquad [3.40]$$

We make the assumption of a source with mutually independent symbols and a channel without memory. The first term of relation [3.40] can then be simplified:

$$p(y_M, S_M \mid y_{M-1}, \ldots, y_1, S_{M-1}, \ldots, S_0) = p(y_M, S_M \mid S_{M-1}) \qquad [3.41]$$

From equations [3.40] and [3.41] we deduce a factorization of the complete probability as a product of two terms:

$$p(\boldsymbol{S}, \boldsymbol{y}) = p(y_M, S_M \mid S_{M-1})\, p(y_{M-1}, \ldots, y_1, S_{M-1}, \ldots, S_0) \qquad [3.42]$$

Reiterating the reasoning M times, we obtain a factorization of complete probability as a product of M terms:

$$p(\boldsymbol{S}, \boldsymbol{y}) = p(S_0) \prod_{k=1,M} p(y_k, S_k \mid S_{k-1}) \qquad [3.43]$$

where $p(S_0)$ represents the distribution of the initial encoder state S_0.

Taking the logarithm of relation [3.43] we obtain an additive form for the complete log-probability:

$$\log p(\boldsymbol{S}, \boldsymbol{y}) = \log p(S_0) + \sum_{k=1,M} \log p(y_k, S_k \mid S_{k-1}) \qquad [3.44]$$

that is, also:

$$\log p(\mathbf{S}, \mathbf{y}) = \log p(S_0) + \sum_{k=1,M} (\log p(y_k \mid S_{k-1}, S_k) + \log p(S_k \mid S_{k-1})) \quad [3.45]$$

since, by definition of conditional probability, we have:

$$p(y_k, S_k \mid S_{k-1}) = p(y_k \mid S_{k-1}, S_k)\, p(S_k \mid S_{k-1}) \qquad [3.46]$$

Let us analyze the terms that appear in equation [3.45] and give their respective expressions under certain assumptions.

3.6.1.1. *The term* $\log p(S_0)$

If we suppose that the initial state of the encoder is the "all at zero" state, then:

$$p(S_0) = \begin{cases} 1 & \text{for } S_0 = (0 \cdots 0 \cdots 0) \\ 0 & \text{otherwise} \end{cases} \qquad [3.47]$$

and thus:

$$\log p(S_0) = \begin{cases} 0 & \text{for } S_0 = (0 \cdots 0 \cdots 0) \\ -\infty & \text{otherwise} \end{cases} \qquad [3.48]$$

If the value of the initial encoder state is unknown, and if all the values of S_0 *a priori* have equal probability, we have:

$$p(S_0) = 1/2^{(\nu-1)K} \quad \forall S_0 \qquad [3.49]$$

where $2^{(\nu-1)K}$ is the number of possible values for S_0.

3.6.1.2. *The term* $\log p(S_k|S_{k-1})$

The transition between states $S_{k-1} \rightarrow S_k$ depends only on the block d_k placed at the encoder input. If the transition $S_{k-1} = s \rightarrow S_k = s'$ is possible, then a single value of d_k corresponds to this transition, which we note $d(s, s')$.

$$p(S_k = s' \mid S_{k-1} = s) = p(d_k = d(s, s')) \qquad [3.50]$$

The transition probability $p(S_k|S_{k-1})$ thus translates *a priori* that we have on the source of information. If the binary information symbols are mutually independent and take values 0 and 1 with the same probability, we have:

$$p(S_k = s' \mid S_{k-1} = s) = 1/2^K \qquad [3.51]$$

3.6.1.3. *The term* $\log p(y_k|S_k, S_{k-1})$

The observation y_k is a random function of the transition $S_{k-1} \rightarrow S_k$ and only of this transition, if the channel is without memory. The observation y_k randomly depends only on the coded block c_k, which is a deterministic function of the transition $S_{k-1} \rightarrow S_k$ between the encoder states and, consequently:

$$\log p(y_k \mid S_{k-1} = s, S_k = s') = \log p\,(y_k \mid c_k = c(s, s')) \qquad [3.52]$$

where $c(s, s')$ represents the coded block at the encoder output when it passes from state s to state s'.

3.6.1.3.1. The term $\log p(y_k|S_k, S_{k-1})$ in the case of the symmetric binary channel

For this channel with binary input and output the demodulator makes hard decisions that can be affected with errors. The observation y_k comprises N binary symbols; we speak of *hard decoding*. We will note the bit error rate of this channel p, i.e. the probability of inversion of a binary symbol. This channel having no memory, the errors or binary inversions are mutually independent.

Conditionally to a transition $S_{k-1} = s \rightarrow S_k = s'$ the law of y_k has the probability:

$$p(y_k \mid S_{k-1} = s, S_k = s') = p^{d_H^k(s,s')}(1-p)^{N-d_H^k(s,s')} \qquad [3.53]$$

where $d_H^k(s, s') = d_H(y_k, c(s, s'))$ is the Hamming distance between the observation y_k and the coded block $c(s, s')$. Taking the logarithm of relation [3.53] we obtain:

$$\log p(y_k \mid S_{k-1} = s, S_k = s') = d_H^k(s, s') \log \frac{p}{1-p} + N \log(1-p) \qquad [3.54]$$

The second term of the right hand side of equation [3.54] is the same for all the pairs (s, s'). Consequently, in the complete expression of log-probability $\log p(\boldsymbol{S}, \boldsymbol{y})$ this term will appear in the form of an additive constant that does not depend on $\boldsymbol{S}$. The problem of optimization of relation [3.39] will thus have the same solution whether we take this term into account or not. This is why we neglect it in the calculation of conditional log-probability [3.54]; we can thus write that:

$$\log p(y_k \mid S_{k-1} = s, S_k = s') = d_H^k(s, s') \log \frac{p}{1-p} \qquad [3.55]$$

except for an additive constant.

3.6.1.3.2. The term $\log p(y_k|S_k, S_{k-1})$ in the case of the Gaussian channel with binary input

For a Gaussian channel with binary input the observation y_k has the form:

$$y_k \quad = \quad c'_k + n_k \qquad [3.56]$$

where c'_k results from the coded block c_k through a modulation operation, which we suppose to have 2 or 4 phase states. The block c'_k thus contains N binary components taking their values in $\{+1, -1\}$ ($c'_k = 2c_k - 1_N$ noting 1_N the vector line consisting in N components at 1). The term n_k is an N-dimensional Gaussian noise whose N components are centered, not correlated and have the same variance σ_n^2. Moreover, the successive noise samples $n_0, n_1, n_2, \ldots$ are independent. Let us recall that for this channel the demodulator did not make hard decisions and that the N components of y_k are the sampled outputs of an adapted filter; the observation thus takes its values in $\mathbb{R}^N$. We then speak of *soft decoding*.

The probability of y_k conditionally to the transition $S_{k-1} = s \rightarrow S_k = s'$ is equal to:

$$\begin{aligned} & p(y_k \mid S_{k-1} = s, S_k = s') \\ & = 1/\left(\sqrt{2\pi\sigma_n^2}\right)^N \exp\left(-\frac{1}{2\sigma_n^2}||y_k - c'(s, s')||^2\right) \end{aligned} \qquad [3.57]$$

where $||y_k - c'(s, s')||^2$ indicates the square of the Euclidean distance between y_k and $c'(s, s')$, and where $c'(s, s') = 2c(s, s') - 1_N$ is the coded and modulated block associated to the transition $S_{k-1} = s \rightarrow S_k = s'$.

Conditional log-probability easily results from expression [3.57]:

$$\log p(y_k \mid S_{k-1} = s, S_k = s') = -\frac{N}{2}\log(2\pi\sigma_n^2) - \frac{1}{2\sigma_n^2}||y_k - c'(s,s')||^2 \quad [3.58]$$

The first term of the right member of expression [3.58] can be neglected. Indeed, it does not depend on the pair (s, s') and the result of optimization [3.39] will be the same, whether we take it into account in [3.58] or not. We can thus write that:

$$\log p(y_k \mid S_{k-1} = s, S_k = s') = -\frac{1}{2\sigma_n^2}||y_k - c'(s,s')||^2 \quad [3.59]$$

to the nearest additive constant.

3.6.1.3.3. The term $\log p(y_k|S_k, S_{k-1})$ in the case of the Rayleigh channel with binary input

For a Rayleigh channel with binary input, the observation y_k has the form:

$$y_k = \rho_k c'_k + n_k \quad [3.60]$$

In equation [3.60] c'_k and n_k have the same definitions as in the case of the Gaussian channel (see section 3.6.1.3.2). n_k represents an additive noise, whose successive samples are independent, Gaussian, have zero average and a variance σ_n^2. We suppose that ρ_k is an attenuation whose successive manifestations are constant over the duration of the coded block c'_k, independent among themselves and with respect to the additive noise n_k and distributed according to a Rayleigh law with $\mathbb{E}(\rho_k^2) = 2\sigma_\rho^2$.

The attenuation ρ_k has as a probability density:

$$p(\rho_k) = \frac{1}{\sigma_\rho^2}\rho_k \exp\left(-\frac{\rho_k^2}{2\sigma_\rho^2}\right) 1\!\!1_{\rho_k} \geq 0 \quad [3.61]$$

where the term $1\!\!1_{\rho_k} \geq 0$ represents the indicating function of the unit $\{\rho_k \geq 0\}$ that equals 1 if $\rho_k \geq 0$ and 0 if not.

In the case of the Rayleigh channel with binary input, the term $p(y_k|S_{k-1} = s, S_k = s')$ equals:

$$\begin{aligned} p(y_k \mid S_{k-1} = s, S_k = s') &= p(y_k \mid c'_k = c'(s,s')) \\ &= \textstyle\int_0^\infty p(\rho_k)\, p(y_k \mid \rho_k, c'_k = c'(s,s'))\, d\rho_k \end{aligned} \quad [3.62]$$

The density of the observation y_k conditionally to the attenuation ρ_k and the coded and modulated symbol $c'k = c'(s, s')$ is Gaussian N-dimensional:

$$p(y_k \mid c'_k = c'(s,s')) = \frac{1}{\sqrt{2\pi}\sigma_n}\exp\left(-\frac{1}{2\sigma_n^2}\|y_k - \rho_k c'(s,s')\|^2\right) \quad [3.63]$$

Integrating [3.61] and [3.63] in [3.62] we obtain:

$$p(y_k \mid S_{k-1} = s, S_k = s') \tag{3.64}$$

$$= \int_0^\infty 1/(\sigma_\rho^2 \sqrt{2\pi}\sigma_n)\, \rho_k\, \exp(-\rho_k^2/2\sigma_\rho^2)\, \exp(-\|y_k - \rho_k c'(s,s')\|^2/2\sigma_n^2)\, d\rho_k$$

Relation [3.64] does not lead to a simple expression of conditional log-probability $\log p(y_k|S_{k-1}, S_k)$.

The approach adopted in practice does not consist of using expression [3.64]. It is preferred to estimate the attenuation ρ_k for each coded block c'_k. Thus, this attenuation being known through its estimate, the Rayleigh channel can be treated as a Gaussian channel.

Conditional log-probability is then, according to [3.59], equal, to the nearest additive factor, to:

$$\log p(y_k \mid S_{k-1} = s, S_k = s') = -\frac{1}{2\sigma^2}\|y_k - \hat{\rho}_k c'(s,s')\|^2 \tag{3.65}$$

where $\hat{\rho}_k$ is the estimate of the attenuation ρ_k.

We can verify that if the estimate of the ρ_k is found satisfactorily, the performances of the decoder for a Rayleigh channel are better when using [3.65] than [3.64].

Let us reconsider the decoding rule [3.39] and take the particular, but very frequent, case where the initial state of the encoder is the "all at zero" state ($S_0 = (0 \cdots 0 \cdots 0)$) and where the binary information symbols take values 0 and 1 with an equal probability.

Under these assumptions, the most probable sequence $\boldsymbol{S}$ has the initial value $S_0 = (0, 0, \ldots, 0)$ and the transition probability $S_{k-1} = s \to S_k = s'$ is independent of the pair (s, s'); it equals:

$$p(S_k = s' \mid S_{k-1} = s) = 1/2^K \tag{3.66}$$

provided that the transition $S_{k-1} = s \to S_k = s'$ is possible, which corresponds to the presence of a branch of the state s reaching towards the state s' in the lattice of the code.

The complete log-probability $\log p(\boldsymbol{S}, \boldsymbol{y})$ is then reduced, to the nearest additive constant that we can neglect, to the sum of M conditional log-probabilities:

$$\log p(\boldsymbol{S}, \boldsymbol{y}) = \sum_{k=1}^{M} \log p(y_k \mid S_{k-1} = s, S_k = s') \tag{3.67}$$

EXAMPLE 3.4 (SYMMETRIC BINARY CHANNEL). We suppose that the probability of binary inversion of the symmetric binary channel is strictly less than 1/2: $p < 1/2$. In this case $\log(p/(l-p)) < 0$; according to relation [3.55] the most probable sequence $\hat{S}$ is that which minimizes the Hamming distance between the sequence y at the decoder input and the sequence $\boldsymbol{c} = (c_1, \ldots, c_M)$ of the coded words:

$$\begin{aligned} \hat{\mathbf{S}} &= \operatorname{Arg}\min_{\mathbf{S}} \sum_{k=1,M} d_H\left(y_k, c(S(k-1), S(k))\right) \\ &= \operatorname{Arg}\min_{\mathbf{S}} d_H(\boldsymbol{y}, \boldsymbol{c}(\boldsymbol{S})) \end{aligned} \qquad [3.68]$$

noting $\boldsymbol{c}(\boldsymbol{S})$ the sequence of coded and modulated words corresponding to the sequence $\boldsymbol{S}$.

EXAMPLE 3.5 (GAUSSIAN CHANNEL WITH BINARY INPUT). According to relation [3.59] the most probable sequence $\boldsymbol{S}$ is that minimizing the Euclidean distance between $\boldsymbol{y}$ and $c'(\boldsymbol{S})$:

$$\begin{aligned} \hat{\mathbf{S}} &= \operatorname{Arg}\min_{\mathbf{S}} \sum_{k=1}^{M} ||y_k - c'(S_{k-1}, S_k)||^2 \\ &= \operatorname{Arg}\min_{\mathbf{S}} ||\mathbf{y} - \mathbf{c}'(\mathbf{S})||^2 \end{aligned} \qquad [3.69]$$

In the case of the Gaussian channel with binary input, decoding according to the most probable sequence $\boldsymbol{S}$ leads to choosing the sequence $\hat{\boldsymbol{S}}$ that minimizes the Euclidean distance between the decoder input $\boldsymbol{y}$ and the sequence of coded and modulated blocks $c'(\boldsymbol{S})$.

An exhaustive search in the set of 2^{KM} values that $\boldsymbol{d}$ can take would have a numerical complexity exponentially increasing with M and K, which is not possible in practice as soon as K or M exceed a few units.

Several algorithms exist to circumvent this difficulty. Among these algorithms we can cite the sequential Fano algorithm which uses the tree diagram to find the most probable sequence $\boldsymbol{S}$. This algorithm is generally reserved for the decoding of convolutional codes with large constraint lengths (typically $\nu \geq 10$).

The Viterbi algorithm uses the lattice diagram to find the most probable sequence $\boldsymbol{S}$. The numerical complexity of this algorithm being proportional to $2^{\nu K} \times M$, it is well adapted for the decoding of convolutional codes with constraint length lower than $\nu = 10$.

3.6.1.4. *Viterbi algorithm*

Let us introduce the quantity $\gamma_k(s, s')$ defined by:

$$\gamma_k(s, s') \quad = \quad -\log p(y_k, S_k = s' \mid S_{k-1} = s) \qquad [3.70]$$

With this notation the decoding rule [3.39], taking into account relation [3.44], is stated:

$$\hat{\mathbf{S}} = \operatorname{Arg}\min_{S} - \log p(S_0) + \sum_{k=1}^{M} \gamma_k(S_{k-1}, S_k) \qquad [3.71]$$

Rule [3.71] is interpreted as the search for a shorter path in the lattice code diagram. Let us assign to the branch stemming from the node $S_{k-1} = s$ to the node $S_k = s'$ a "length" $\gamma_k(s, s')$ called *branch metric.*

To the nearest term – $\log p(S_0)$ the quantity minimized by the rule of decoding [3.71] is the sum of branch metrics along the path corresponding to the sequence $\boldsymbol{S}$ of encoder states. This sum can thus be interpreted as the length of the path $\boldsymbol{S}$ in the lattice and the decoding according to [3.71] is thus reduced to finding the shortest path in the lattice.

Before presenting the Viterbi algorithm, let us define two quantities:

1) $M_k^s(s')$ is equal to the sum of the branch metric of the shortest path from the node S_0 ending at the node $S_k = s'$. This path is called *surviving* in the terminology of the Viterbi algorithm, from where the exhibitor s of $M_k^s(s')$; $M_k^s(s')$ is called the *cumulated metric* of the surviving path;

2) $A_k(s')$ is called the *previous better node* for the node $S_k = s'$; $A_k(s')$ is the node by which passes at the S_{k-1} level the shortest (surviving) path among all the paths starting at S_0 and ending at the $S_k = s'$ node.

3.6.1.4.1. Initialization ($k = 0$)

Calculation of the cumulated metrics $M_0^s(s')$:

$$M_0^s(s') = -\log p(S_0 = s') \qquad [3.72]$$

For the calculation of this metric we generally make the assumption that the encoder is in the "all at zero" state and thus:

$$M_0^s(s') = \begin{cases} 0 & \text{for } s' = (0 \cdots 0 \cdots 0) \\ -\infty & \text{otherwise} \end{cases} \qquad [3.73]$$

3.6.1.4.2. Calculation of the branch metrics ($k = 1, 2, \ldots, M$)

For any moment k and any pair of nodes $(S_{k-1} = s, S_k = s')$ that communicate with each other, we calculate the metric $\gamma_k(s, s')$, that is $2^{\nu\ K}$ branch metrics to evaluate at every moment k.

3.6.1.4.3. Calculation of the $M_k^s(s')$ metrics of the surviving paths ($k = 1, 2, \ldots, M$)

The $M_k^s(s')$ metric at the level of the node $S_k = s'$ is equal to the minimum value, the minimum being taken among all the possible previous nodes $S_{k-1} = s$ of the node $S_k = s'$, the sum of the branch metric $\gamma_k(s, s')$ and of the surviving metric $M_{k-1}^s(s)$ at the node $S_{k-1} = s$:

$$M_k^s(s') = \min_s (M_{k-1}^s(s) + \gamma_k(s, s')) \qquad [3.74]$$

3.6.1.4.4. Determination of the best previous node ($k = 1, 2, \ldots, M$)

The best previous node of the node $S_k = s'$ is the S_{k-1} node by which passes the shortest path arriving at $S_k = s'$, i.e. the path with a cumulated metric $M_k^s(s')$:

$$A_k(s') = \text{Arg}\min_s (M_{k-1}^s(s) + \gamma_k(s, s')) \qquad [3.75]$$

At the level of the node $S_k = s'$, $A_k(s')$ points to the previous node S_{k-1}, by which passes the shortest path among all the path arriving at the node $S_k = s'$.

The metrics of the surviving paths $M_k^s(s')$ and the best previous nodes $A_k(s')$ must be memorized at each moment k and for all the nodes of the lattice diagram.

3.6.1.4.5. Determination of the shortest path by back tracing the lattice diagram

After the calculation of all the surviving metrics $M_k^s(s')$ and all the best previous nodes $A_k(s')$, it is enough to choose the shortest path in $k = M$, that is the one whose cumulated metric is the weakest. Let $\hat{S}_M$ be the arrival node of this path:

$$\hat{S}_M = \text{Arg}\min_s M_M^s(s) \qquad [3.76]$$

This path comes from the node $\hat{S}_{M-1}$, which is the best previous node of $\hat{S}_M$, and thus, gradually, we reassemble the lattice up to the S_0:

$$\hat{S}_{k-1} = A_k(\hat{S}_k) \quad k = M, \ldots, 1 \qquad [3.77]$$

Once the most probable path $\hat{S} = (\hat{S}_0, \hat{S}_1, \ldots, \hat{S}_M)$ has been determined, we decode the information blocks d_k.

To complete the presentation of the Viterbi algorithm, let us provide the expressions of branch metrics in the case of the symmetric binary channel and in the case of the Gaussian channel with binary input.

Let us recall that, by definition, the branch metric is equal to:

$$\gamma_k(s, s') = -\log p(y_k, S_k = s' \mid S_{k-1} = s) \qquad [3.78]$$

that can also be written:

$$\begin{aligned}\gamma_k(s,s') = &-\log p(y_k \mid S_{k-1}=s, S_k=s') \\ &-\log p(S_k=s' \mid S_{k-1}=s)\end{aligned} \quad [3.79]$$

Symmetric binary channel
According to relations [3.50] and [3.55] the branch metric has the expression:

$$\gamma_k(s,s') = -d_H^k(s,s')\log\left(\frac{p}{1-p}\right) - \log p(d_k = d(s,s')) \quad [3.80]$$

Gaussian channel with binary input
According to relations [3.50] and [3.59], the branch metric has the expression:

$$\gamma_k(s,s') = \frac{1}{2\sigma^2}||y_k - c'(s,s')||^2 - \log p(d_k = d(s,s')) \quad [3.81]$$

For these two channels the calculation of branch metrics requires knowing the probability of bit inversion at output of the demodulator p (symmetric binary channel) or the variance σ^2 of the noise (Gaussian channel with binary input).

If we suppose that the source of information delivers binary symbols taking values "0" or "1" with equal probability, then, for these two channels, the branch metric can be simplified. Indeed, in this case, the term $p(d_k = d(s,s'))$ equals 1/2 when the branch (s,s') belongs to the lattice; this term does not depend on the couple (s,s') and it is not necessary to take it into account in the calculation of the branch metric; thus we obtain:

1) symmetric binary channel:

$$\gamma_k(s,s') = d_H^k(s,s') \quad [3.82]$$

2) Gaussian channel with binary input:

$$\gamma_k(s,s') \quad = \quad ||y_k - c'(s,s')||^2 \quad [3.83]$$

where $c'(s,s')$ is the coded and modulated block associated to the transition $S_{k-1} = s \rightarrow S_k = s'$. We can thus again simplify the branch metric:

$$\gamma_k(s,s') \quad = \quad \langle y_k, c'(s,s')\rangle \quad [3.84]$$

where $\langle\cdot,\cdot\rangle$ indicates the scalar product. This quantity is the scalar product of the observation y_k and the coded and modulated block $c'(s,s')$. Expression [3.84] of the metric was obtained using expression [3.83], omitting the terms independent of the pair (s,s') and multiplying the metric by -1. Consequently, with expression [3.84] of the metric, minimization is transformed into maximization and it is necessary to replace min by max in relations [3.74] and [3.75];

3) Rayleigh channel with binary input: adopting the assumption that we have an estimate $\hat{\rho}_k$ of the attenuations ρ_k, the branch metric is equal to:

$$\gamma_k(s, s') = \|y_k - \hat{\rho}_k\, c'(s, s')\|^2 \qquad [3.85]$$

This metric can be further simplified taking into account the observations made previously (case of the Gaussian channel with binary input):

$$\gamma_k(s, s') = \hat{\rho}_k\, \langle y_k, c'(s, s')\rangle \qquad [3.86]$$

To illustrate the Viterbi algorithm, let us consider an example. Let us suppose that the sequence to be coded is 1001 and that the convolution encoder is the one represented in Figure 3.2. By making the assumption that the encoder is in the "all at zero" state at the initial moment ($k = 0$) the coded sequence is 11 10 11 11. Let us consider, for example, a symmetric binary channel introducing an error at position 3 so that the observation y is 11 00 11 11. The Viterbi algorithm develops as follows; the lattice diagram of the convolutional code is repeated in Figure 3.10:

1) $k = 1$:

– Branch metrics:

$$\gamma_1(a, a) = 2$$
$$\gamma_1(a, c) = 0$$

– Cumulated surviving metrics and the best previous nodes:

$$M_1^s(a) = 2 \qquad A_1(a) = a$$
$$M_1^s(c) = 0 \qquad A_1(c) = a$$

2) $k = 2$:

– Branch metrics:

$$\gamma_2(a, a) = 0 \quad \gamma_2(c, b) = 1$$
$$\gamma_2(a, c) = 2 \quad \gamma_2(c, d) = 1$$

– Cumulated surviving metrics and the best previous nodes:

$$M_2^s(a) = 2 \qquad A_2(a) = a$$
$$M_2^s(b) = 1 \qquad A_2(b) = c$$
$$M_2^s(c) = 4 \qquad A_2(c) = a$$
$$M_2^s(d) = 1 \qquad A_2(d) = c$$

3) $k = 3$:
– Branch metrics:

$$\gamma_3(a,a) = 2 \quad \gamma_3(b,a) = 0 \quad \gamma_3(c,b) = 1 \quad \gamma_3(d,b) = 1$$

$$\gamma_3(a,c) = 0 \quad \gamma_3(b,c) = 2 \quad \gamma_3(c,d) = 1 \quad \gamma_3(d,d) = 1$$

– Cumulated surviving metrics and the best previous nodes:

$$M_3^s(a) = 1 \qquad A_3(a) = b$$

$$M_3^s(b) = 2 \qquad A_3(b) = d$$

$$M_3^s(c) = 2 \qquad A_3(c) = a$$

$$M_3^s(d) = 2 \qquad A_3(d) = d$$

4) $k = 4$:
– Branch metrics:

$$\gamma_4(a,a) = 2 \quad \gamma_4(b,a) = 0 \quad \gamma_4(c,b) = 1 \quad \gamma_4(d,b) = 1$$

$$\gamma_4(a,c) = 0 \quad \gamma_4(b,c) = 2 \quad \gamma_4(c,d) = 1 \quad \gamma_4(d,d) = 1$$

– Cumulated surviving metrics and the best previous nodes:

$$M_4^s(a) = 2 \qquad A_4(a) = b$$

$$M_4^s(b) = 3 \qquad A_4(b) = c$$

$$M_4^s(c) = 1 \qquad A_4(c) = a$$

$$M_4^s(d) = 3 \qquad A_4(d) = d$$

The weakest cumulated metric is $M_4^s(c) = 1$. The most probable sequence $\hat{\boldsymbol{S}}$ thus arrives at the node $\hat{S}_4 = c$; the node $S_4 = c$ has the node $A_4(c) = a$ as best previous and thus $\hat{S}_3 = a$. Similarly, the node $S_3 = a$ has the node $A_3(a) = b$ as best previous and, thus, $\hat{S}_2 = b$. By back-tracking up the lattice diagram in this manner, from right to left, we obtain the most probable sequence $\hat{\boldsymbol{S}}$:

$$c, a, b, c, a \qquad \hat{\boldsymbol{S}} = (a, c, b, a, c)$$

which corresponds to the information sequence:

$$\boldsymbol{d} = (1001)$$

The single error introduced by the channel has been corrected.

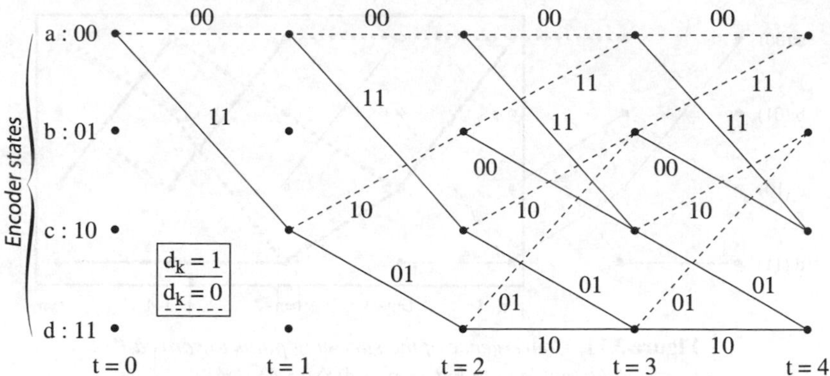

Figure 3.10. *Lattice diagram of the convolutional code from example 3.1*

3.6.1.5. *Viterbi algorithm for transmissions with continuous data flow*

To finish this section on Viterbi decoding, let us say that for transmissions with continuous data flow, it is generally difficult to wait until the entire coded sequence has been received to begin decoding. Indeed, that would introduce a too large a delay into the decoding, and would require to have a prohibitively large memory to memorize all the surviving paths of the lattice.

Back-tracking the surviving paths arriving at $t = n$ at each node $S_n = s'$ we realize that these paths almost always merge towards the same node, in $t = n - \Delta$ for sufficiently large Δ. This phenomenon is illustrated in Figure 3.11 where the four paths surviving in $t = n$ merge towards the node $a = (00)$ in $t = n - 4(\Delta = 4)$. If we retrace starting from the node $S_{n-4} = a = (00)$, along the single surviving path we follow the branch $S_{n-5} = (00) \rightarrow S_{n-4} = (00)$; binary data of information equal to $d_{n-4} = 0$ corresponds to this branch, represented by a dotted line in the lattice in Figure 3.11; the Viterbi algorithm with sliding window thus leads to the decision $\hat{d}_{n-4} = 0$.

To decode the information block $d_{n-\Delta}$ it is not necessary to observe the received sequence beyond $t = n$. In practice, memorizing the surviving paths can be limited to a temporal window of depth Δ. The decoding delay thus remains finite for an infinite or at least very large sequence to be decoded.

We can demonstrate by simulation that the window Δ must be all the larger as the code rate and its code constraint length are higher. Thus, for a code with an output, for example, $R = 1/2$, we can take Δ to be equal to 5 or 6 times the length of the constraint ν.

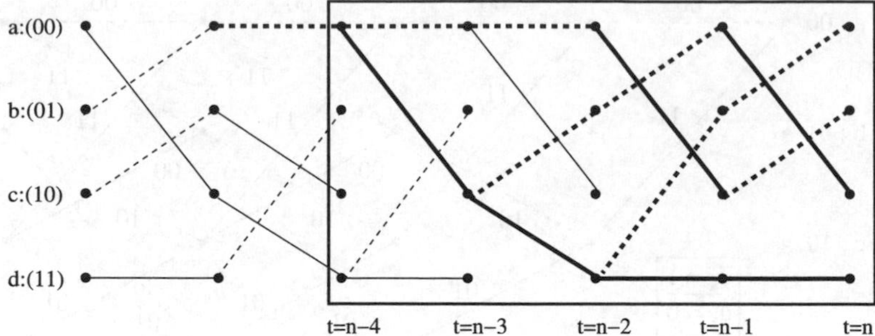

Figure 3.11. *Convergence of the surviving paths towards a single path in* $t = n - 4(\Delta = 4)$

3.6.2. *MAP criterion or BCJR algorithm*

For this decoding criterion, the decoder looks for the information given $\hat{d}_k$ for any $k = 1, \ldots, M$, such that:

$$p(\hat{d}_k \mid \mathbf{y}) \quad \geq \quad p(d_k \mid \mathbf{y}) \quad \forall d_k \qquad [3.87]$$

where $p(\hat{d}_k|\boldsymbol{y})$ is the probability of the information given $\hat{d}_k$ conditionally to the observation $\boldsymbol{y}$. In an equivalent fashion the criterion is also stated:

$$\hat{d}_k \quad = \quad \text{Arg}\max_{d_k} p(d_k \mid \mathbf{y}) \qquad [3.88]$$

where Arg indicates the argument d_k of $p(d_k|\boldsymbol{y})$.

Before presenting the BCJR algorithm let us introduce three quantities, which make it possible to implement decoding using the MAP criterion:

– $\alpha_k k(s)$ represents the joint probability of the observations $y_1, \ldots, y_k$ and of the encoder state $S_k = s$:

$$\alpha_k(s) \quad = \quad p(y_1, y_2, \ldots, y_k, S_k = s) \qquad [3.89]$$

The quantity $\alpha_k(s)$ is called the *Front filter*;

– $\beta_k(s)$ represents the joint probability of the observations $y_{k+1}, \ldots, y_M$ conditionally to the $S_k = s$ encoder state:

$$\beta_k(s) \quad = \quad p(y_{k+1}, \ldots, y_M \mid S_k = s) \qquad [3.90]$$

The quantity $\beta_k(s)$ is called the *Back filter*;

– $\psi_k(s, s')$ represents the joint probability of $y = (y_1, y_2, \ldots, y_M)$, of $S_{k-1} = s$ and $S_k = s'$:

$$\psi_k(s, s') = p(S_{k-1} = s, S_k = s', \mathbf{y}) \qquad [3.91]$$

The quantity $\psi_k(s, s')$ is proportional to the *a posteriori* probability of the branch $S_{k-1} = s \rightarrow S_k = s'$:

$$\psi_k(s, s') \propto p(S_{k-1} = s, S_k = s' \mid \mathbf{y}) \qquad [3.92]$$

where the $\propto$ sign represents equality to the nearest proportionality factor.

3.6.2.1. *BCJR algorithm (Bahl, Cock, Jelinek, Raviv; 1974)*

The BCJR algorithm calculates the Front filter $\alpha_k(s)$ and the Back filter $\beta_k(s)$ iteratively:

$$\begin{aligned} \alpha_{k+1}(\cdot) &\leftarrow \alpha_k(\cdot), y_{k+1} \\ \beta_k(\cdot) &\leftarrow \beta_{k+1}(\cdot), y_{k+1} \end{aligned} \qquad [3.93]$$

In equation [3.93] the "$\cdot$" in $\alpha_k(\cdot)$ means that we consider the quantities $\alpha_k(s)$ for the $2^{(\nu-1)K}$ possible values of s. Similarly, the "$\cdot$" in $\beta_k(\cdot)$ means that we consider the quantities $\beta_k(s)$ for the $2^{(\nu-1)K}$ possible values of s, where s represents an encoder state.

Relation [3.93] shows symbolically that the quantity $\alpha_{k+1}(\cdot)$ is obtained from the observation y_{k+1} and from the quantity $\alpha_k(\cdot)$; similarly, the quantity $\beta_k(\cdot)$ is obtained from the observation y_{k+1} and from the quantity $\beta_{k+1}(\cdot)$.

This makes it possible to represent the Front filter and the Back filter as a lattice. We initialize $\alpha_0(\cdot)$, then calculate $\alpha_1(\cdot)$ using y_1 and $\alpha_0(\cdot)$, the we calculate $\alpha_2(\cdot)$ using y_2 and $\alpha_1(\cdot)$ and so on until $\alpha_M(\cdot)$.

Similarly, we initialize $\beta_M(\cdot)$, then we calculate $\beta_{M-1}(\cdot)$ using y_M and of $\beta_M(\cdot)$; we then calculate $\beta_{M-2}(\cdot)$ using y_{M-1} and of $\beta_{M-1}(\cdot)$, and so on until $\beta_0(\cdot)$.

3.6.2.1.1. Initialization of the Front filter ($k = 0$)

We pose:

$$\alpha_0(s) = p(S_0 = s) \qquad [3.94]$$

For example, if the encoder is initialized in the "all at zero" state, we have:

$$\alpha_0(s) = \begin{cases} 1 & \text{for } s = (0, 0, \ldots, 0) \\ 0 & \text{if not} \end{cases} \qquad [3.95]$$

If we have no *a priori* information on the initial state of the encoder, we decide that all the states have equal probability at the start:

$$\alpha_0(s) = 1/2^{(\nu-1)K}, \quad \forall s \qquad [3.96]$$

3.6.2.1.2. Propagation of the Front filter ($k = 1, 2, \ldots, M$)

The event $y_1, y_2, \ldots, y_k$ and $S_k = s$ is also equal to the sum of the events $y_1, y_2, \ldots, y_k, S_{k-1} = s'$ and $S_t = s$, this sum being taken for all of the states s', such that the branch $S_{k-1} = s' \rightarrow S_k = s$ exists in the lattice. Moreover, this sum is discrete.

From this reasoning and from definition [3.89] we get:

$$\alpha_k(s) = \sum_{s'/s' \rightarrow s} p(y_1, \ldots, y_k, S_{k-1} = s', S_k = s) \qquad [3.97]$$

$s'/s' \rightarrow s$ means that the sum [3.97] relates to all of the states s', such that the node $S_{k-1} = s'$ communicates with the node $S_k = s$.

By definition of conditional probability we have:

$$\begin{aligned} & p(y_1, \ldots, y_k, S_{k-1} = s', S_k = s) \\ & = p(y_k, S_k = s \mid y_1, \ldots, y_{k-1}, S_{k-1} = s')\, p(y_1, \ldots, y_{k-1}, S_{k-1} = s') \end{aligned} \qquad [3.98]$$

We recognize in the second term of the right hand side of equation [3.98] the quantity:

$$\alpha_{k-1}(s') = p(y_1, \ldots, y_{k-1}, S_{k-1} = s') \qquad [3.99]$$

Conditionally to S_{k-1}, the quantities y_k and S_k do not depend on $y_{k-1}, y_{k-2}, \ldots, y_1$. Consequently:

$$\begin{aligned} & p(y_k, S_k = s \mid y_1, \ldots, y_{k-1}, S_{k-1} = s') \\ & = p(y_k, S_k = s \mid S_{k-1} = s') \end{aligned} \qquad [3.100]$$

From [3.100] and definition [3.70] of the branch metric it follows that:

$$p(y_k, S_k = s \mid y_1, \ldots, y_{k-1}, S_{k-1} = s') = \exp(-\gamma_k(s', s)) \qquad [3.101]$$

Using expressions [3.99] and [3.101] in [3.98] we obtain:

$$p(y_1, \ldots, y_k, S_{k-1} = s', S_k = s) = \alpha_{k-1}(s') \exp(-\gamma_k(s', s)) \qquad [3.102]$$

and using [3.102] in equation [3.97] we obtain the Front filter reinitialization equation:

$$\alpha_k(s) = \sum_{s'/s' \to s} \alpha_{k-1}(s') \exp\left(-\gamma_k(s', s)\right) \qquad [3.103]$$

Let us pose $\boldsymbol{A}_k$ as the vector line of the components $\alpha_k(s)$ and $\boldsymbol{G}_k$ as the matrix whose element on in row s' and column s is $g_k(s', s) = \exp(-\gamma k(s', s))$:

$$\begin{aligned} \boldsymbol{G}_k &= [g_k(s', s)]_{s,s'} \\ g_k(s', s) &= p(y_k, S_k = s \mid S_{k-1} = s') \end{aligned} \qquad [3.104]$$

Equation [3.103] is also expressed in matrix form:

$$\boldsymbol{A}_k = \boldsymbol{A}_{k-1} \boldsymbol{G}_k \qquad [3.105]$$

3.6.2.1.3. Initialization of the Back filter ($k = M$)

We pose:

$$\beta_M(s) = 1 \forall s \qquad [3.106]$$

3.6.2.1.4. Propagation of the filter Back ($k = M - 1, \ldots, 0$)

The event $y_{k+1}, \ldots, y_M$ equals the sum of the events $y_{k+1}, \ldots, y_M$ and $S_{k+1} = s'$, the sum being calculated for all encoder states, such that the branch $S_k = s \to S_{k+1} = s'$ is a branch of the lattice:

$$\begin{aligned} \beta_k(s) &= p(y_{k+1}, \ldots, y_M \mid S_k = s) \\ &= \sum_{s'/s \to s'} p(y_{k+1}, \ldots, y_M, S_{k+1} = s' \mid S_k = s) \end{aligned} \qquad [3.107]$$

By definition of conditional density we have:

$$p(y_{k+1}, \ldots, y_M, S_{k+1} = s' \mid S_k = s) \qquad [3.108]$$

$$= p(y_{k+1}, S_{k+1} = s' \mid S_k = s) p(y_{k+2}, \ldots, y_M \mid y_{k+1}, S_{k+1} = s', S_k = s)$$

Knowing that $S_{k+1} = s'$ the variables $y_{k+2}, \cdots, y_M$ are independent of y_{k+1} and $S_k = s$. Consequently:

$$\begin{aligned} &p(y_{k+2}, \ldots, y_M \mid y_{k+1}, S_{k+1} = s', S_k = s) \\ &= p(y_{k+2}, \ldots, y_M \mid S_{k+1} = s') \end{aligned} \qquad [3.109]$$

Substituting [3.109] in [3.108] and using definition [3.70] of the branch metric we obtain:

$$p(y_{k+1},\dots,y_M,S_{k+1}=s' \mid S_k=s)=\exp(-\gamma_{k+1}(s,s'))\beta_{k+1}(s') \qquad [3.110]$$

By substituting [3.110] in [3.107] we obtain the reinitialization equation of the Back filter:

$$\beta_k(s) \;=\; \sum_{s'/s\to s'} \exp(-\gamma_{k+1}(s,s'))\beta_{k+1}(s') \qquad [3.111]$$

Let us pose $\boldsymbol{B}_k$ as the vector column of the $\beta_k(s)$ components, relation [3.111] is written in algebraic form:

$$\boldsymbol{B}_k \;=\; \boldsymbol{G}_{k+1}\boldsymbol{B}_{k+1} \qquad [3.112]$$

where $\boldsymbol{G}_{k+1}$ is the matrix with $2^{(\nu-1)K}$ rows and $2^{(\nu-1)K}$ columns defined in [3.104].

OBSERVATION (ON THE INITIALIZATION OF THE BACK FILTER). The quantity $\beta_k(s)$ is defined as:

$$\beta_k(s) \;=\; p(y_{k+1},\dots,y_M \mid S_k=s) \qquad [3.113]$$

The notation $y_{k+1},\dots,y_M$ only makes sense for $k \le M-1$. If we apply equation [3.111] of Back filter propagation, posing $k=M-1$ and initializing $\beta_M(s)$ according to equation [3.106] we obtain:

$$\begin{aligned}\beta_{M-1}(s) &= \sum_{s'/s\to s'} \exp(-\gamma_M(s,s'))\\ &= \sum_{s'/s\to s'} p(y_M,S_M=s' \mid S_{M-1}=s)\\ &= p(y_M \mid S_{M-1}=s)\end{aligned} \qquad [3.114]$$

This result is coherent with definition [3.90] of the Back filter; this validates the initialization [3.106] of the Back filter.

3.6.2.1.5. A *posteriori* calculation of branch probability

We will express the quantity $\psi_k(s,s')$ according to $\beta_{k-1}(s)$, of $\beta_k(s')$ and of $\gamma_k(s,s')$.

By definition:

$$\psi_k(s,s') \;=\; p(y_1,\dots,y_{k-1},S_{k-1}=s,y_k,S_k=s',y_{k+1},\dots,y_M) \qquad [3.115]$$

According to the definition of conditional probability we have:

$$\begin{aligned}\psi_k(s,s') &= p(y_1,\ldots,y_{k-1},S_{k-1}=s)\\ &\quad p(y_k,S_k=s',y_{k+1},\ldots,y_M \mid y_1,\ldots,y_{k-1},S_{k-1}=s)\end{aligned} \qquad [3.116]$$

and by applying the definition of conditional probability a second time:

$$\begin{aligned}\psi_k(s,s') &= p(y_1,\ldots,y_{k-1},S_{k-1}=s)\\ &\quad p(y_{k+1},\ldots,y_M \mid y_k,S_k=s',y_1,\ldots,y_{k-1},S_{k-1}=s)\\ &\quad p(y_k,S_k=s' \mid y_1,\ldots,y_{k-1},S_{k-1}=s)\end{aligned} \qquad [3.117]$$

The first term of the right hand side of equation [3.117] is equal to $\alpha_{k-1}(s)$:

$$p(y_1,\ldots,y_{k-1},S_{k-1}=s) = \alpha_{k-1}(s) \qquad [3.118]$$

Knowing that $S_k = s'$, the quantities $y_{k+1},\ldots,y_M$ are independent of $S_{k-1}=s$ and $y_1,\ldots,y_k$; from that we deduce that the second term of equation [3.117] is equal to $\beta_k(s')$:

$$\begin{aligned}&p(y_{k+1},\ldots,y_M \mid y_k,S_k=s',y_1,\ldots,y_{k-1},S_{k-1}=s)\\ &\quad = p(y_{k+1},\ldots,y_M \mid S_k=s')\\ &\quad = \beta_k(s')\end{aligned} \qquad [3.119]$$

Knowing that $S_{k-1} = s, S_k = s'$ and y_k are independent of $y_1,y_2,\ldots,y_{k-1}$; consequently, the third term of the right hand side of equation [3.117] is equal to:

$$\begin{aligned}&p(y_k,S_k=s' \mid S_{k-1}=s,y_1,\ldots,y_{k-1})\\ &\quad = p(y_k,S_k=s' \mid S_{k-1}=s)\\ &\quad = \exp(-\gamma_k(s,s'))\end{aligned} \qquad [3.120]$$

By integrating expressions [3.118], [3.119] and [3.120] in equation [3.117] we obtain:

$$\psi_k(s,s') = \alpha_{k-1}(s)\exp\left(-\gamma_k(s,s')\right)\beta_k(s') \qquad [3.121]$$

3.6.2.1.6. Decision taken by the decoder according to the MAP criterion

The MAP criterion leads to the decision:

$$\hat{d}_k = \operatorname*{Arg\,max}_{d_k} p(d_k \mid \mathbf{y}) \qquad [3.122]$$

The probability of the information block d_k conditionally to the observations y provided by the demodulator has the expression:

$$p(d_k \mid \boldsymbol{y}) = p(d_k, \boldsymbol{y}) \;/\; p(\boldsymbol{y}) \tag{3.123}$$

Since the term $p(y)$ does not depend on d_k, criterion [3.122] is also written:

$$\hat{d}_k = \text{Arg}\max_{d_k} p(d_k, \mathbf{y}) \tag{3.124}$$

Taking account the definition [3.91] of $\psi_k(s, s')$, the joint probability $p(d_k, y)$ is written as the sum:

$$p(d_k, \mathbf{y}) = \sum_{(s,s')/d_k = d(s,s')} \psi_k(s, s') \tag{3.125}$$

The notation $(s, s')/d_k = d(s, s')$ means that the sum relates to all of the branches $S_{k-1} = s \rightarrow S_k = s'$ that correspond to an information block equal to d_k.

From [3.124] and [3.125] it results that the decision taken according to the MAP criterion is:

$$\hat{d}_k = \text{Arg}\max_{d_k} \sum_{(s,s')/d_k = d(s,s')} \psi_k(s, s') \tag{3.126}$$

3.6.2.1.7. Probability of the decision taken according to the MAP criterion

The probability of the decision $\hat{d}_k$ conditionally to the data $\boldsymbol{y}$ provided by the demodulator is expressed as:

$$p(\hat{d}_k \mid \boldsymbol{y}) = p(\hat{d}_k, \boldsymbol{y}) \;/\; p(\boldsymbol{y}) \tag{3.127}$$

It results from [3.125] that the probability of observations $\boldsymbol{y}$ has the expression:

$$\begin{aligned} p(\mathbf{y}) &= \sum_{d_k} p(d_k, \mathbf{y}) \\ &= \sum_{d_k} \sum_{(s,s')/d_k = d(s,s')} \psi_k(s, s') \\ &= \sum_{(s,s')} \psi_k(s, s') \end{aligned} \tag{3.128}$$

and we deduce from equations [3.125], [3.127] and [3.128] the expression of the probability of the decision $\hat{d}_k$:

$$p(\hat{d}_k \mid \boldsymbol{y}) = \frac{\sum_{(s,s')/d_k = d(s,s')} \psi_k(s, s')}{\sum_{(s,s')} \psi_k(s, s')} \tag{3.129}$$

In conclusion, decoding using the MAP criterion at first recursively evaluates the quantities $\alpha_k(s)$ of the Front filter using relation [3.103] and the quantities $\beta_k(s)$ of the Back filter using relation [3.111], then calculates the quantities $\psi_k(s, s')$ representing the *a posteriori* probabilities of branches according to relation [3.121]. The joint probability of d_k and y is calculated using relation [3.125] for each of the 2^K possible values of the information block d_k, and, finally, we look for the value of d_k that maximizes this joint probability according to relation [3.126].

The major interest of decoding using the MAP criterion lies in the fact that it makes it possible to associate reliability information in the form of probability of $\hat{d}_k$ to each decoded value $\hat{d}_k$ conditionally to the sequence of observations $\boldsymbol{y}$ presented at the input of the decoder; this *soft decision* is calculated according to relation [3.129]. This information on reliability can then be used for decoding with soft input of external code in the case of a serial concatenation of two codes as illustrated in Figure 3.12, or for iterative decoding of a turbocode. On the basis of this reliability information we also work out the extrinsic information (see Chapter 5) used in iterative turbocode decoding.

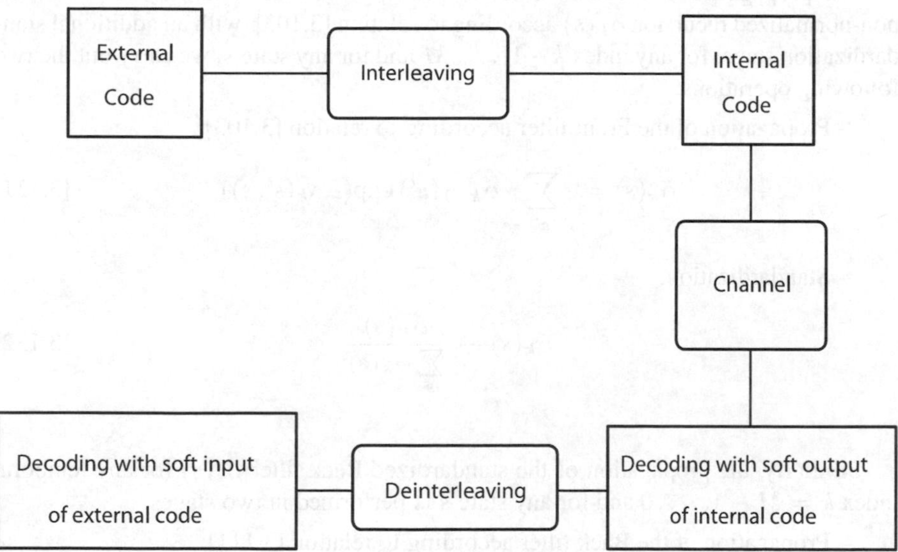

Figure 3.12. *Use of a decoder with soft output for the decoding of internal code in a serial concatenation of two codes*

3.6.2.1.8. Normalization of the Front and Back filters

Definitions [3.89] and [3.90] of the Front and Back filters lead to the equations of propagation [3.103] and [3.111], or, equivalently, to the equations of propagation

[3.105] and [3.112]. Given the multiplicative form of these equations, it is clear that the order of magnitude of the filters $\alpha_k(\cdot)$ and $\beta_k(\cdot)$ decreases exponentially, along the course of the iterations.

This would quickly lead $\alpha_k(\cdot)$ and $\beta_k(\cdot)$ below the precision threshold of calculators. To circumvent this difficulty in practice we work not with $\alpha_k(\cdot)$ and $\beta_k(\cdot)$ as defined by relations [3.89] and [3.90], but with these same quantities standardized:

$$\begin{aligned} \tilde{\alpha}_k(s) &= \alpha_k(s)/\sum_{s'}\alpha_k(s') \\ \tilde{\beta}_k(s) &= \beta_k(s)/\sum_{s'}\beta_k(s') \end{aligned} \qquad [3.130]$$

The interest to work with standardized filters $\tilde{\alpha}_k(s)$ and $\tilde{\beta}_k(s)$ is that the order of magnitude of these filters remains the same for any value of k; this makes it possible to overcome the problem of overflow, which any programmer comes across when trying to work with the recursions $\alpha_k(s)$ and $\beta_k(s)$ without standardization.

The propagation of the standardized filter $\tilde{\alpha}_k(s)$ is made similarly to that of the non-normalized recursion $\alpha_k(s)$ according to relation [3.103], with an additional standardization stage; for any index $k-1,\ldots,M$ and for any state s, we carry out the two following operations:

– Propagation of the Front filter according to relation [3.103]:

$$\tilde{\alpha}_k(s) = \sum_{s'/s'\to s} \tilde{\alpha}_{k-1}(s')\exp(-\gamma_k(s',s)) \qquad [3.131]$$

– Standardization:

$$\tilde{\alpha}_k(s) \leftarrow \frac{\tilde{\alpha}_k(s)}{\sum_s \tilde{\alpha}_k(s)} \qquad [3.132]$$

Similarly, the propagation of the standardized Back filter $\tilde{\beta}_k(s)$ for any temporal index $k = M-1,\ldots,0$ and for any state s is performed in two stages:

– Propagation of the Back filter according to relation [3.111]:

$$\tilde{\beta}_k(s) = \sum_{s'/s\to s'} \exp(-\gamma_{k+1}(s,s'))\tilde{\beta}_{k+1}(s') \qquad [3.133]$$

– Standardization:

$$\tilde{\beta}_k(s) \leftarrow \frac{\tilde{\beta}_k(s)}{\sum_s \tilde{\beta}_k(s)} \qquad [3.134]$$

Once the propagation of the standardized Front and Back filters has been carried out, the decoding algorithm develops further without modifications, simply replacing the quantities $\alpha_k(s)$ and $\beta_k(s)$ by their standardized counterparts $\tilde{\alpha}_k(s)$ and $\tilde{\beta}_k(s)$.

EXAMPLE 3.6 (SYMMETRIC BINARY CHANNEL). By definition, the quantity $g_k(s, s')$ has the expression:

$$g_k(s, s') \;\; = \;\; p(y_k, S_k = s' \mid S_{k-1} = s) \qquad [3.135]$$

From the definition of conditional probability it results that:

$$\begin{aligned} g_k(s, s') &= p(y_k \mid S_{k-1} = s, S_k = s')p(S_k = s' \mid S_{k-1} = s) \\ &= p(y_k \mid S_{k-1} = s, S_k = s')p(d_k = d(s, s')) \end{aligned} \qquad [3.136]$$

In the case of the symmetric binary channel, the quantity $p(y_k|S_{k-1} = s, S_k = s')$ according to relation [3.53] has the expression:

$$p(y_k \mid S_{k-1} = s, S_k = s') = (\frac{p}{1-p})^{d_H^k(s,s')}(1-p)^N \qquad [3.137]$$

noting by $d_H^k(s, s') = d_H(y_k, c(s, s'))$ the Hamming distance between y_k and the coded block $c(s, s')$.

In [3.137] the term $(1 - p)^N$ is a multiplicative constant, which does not depend on s or s'; we can therefore neglect it:

$$p(y_k \mid S_{k-1} = s, S_k = s') = \left(\frac{p}{1-p}\right)^{d_H^k(s,s')} \qquad [3.138]$$

In the case of the symmetric binary channel we thus have:

$$g_k(s, s') = \left(\frac{p}{1-p}\right)^{d_H^k(s,s')} p\,(d_k = d(s, s')) \qquad [3.139]$$

Moreover, if the given units of information d_k have equal probability:

$$p\,(d_k = d(s, s')) = 1/2^K \qquad [3.140]$$

This term does not depend on the pair (s, s'); it can therefore be neglected in [3.139]. Consequently, in the case of the symmetric binary channel, when the information given is distributed with equal probability, we have:

$$g_k(s, s') = \left(\frac{p}{1-p}\right)^{d_H^k(s,s')} \qquad [3.141]$$

EXAMPLE 3.7 (GAUSSIAN CHANNEL WITH BINARY INPUTS). In the case of the Gaussian channel with binary inputs, the quantity $p(y_k|S_{k-1} = s, S_k = s')$ has the expression:

$$p(y_k \mid S_{k-1} = s, S_k = s') = 1 / \sqrt{2\pi\sigma^2}^N \exp\left(-\frac{1}{2\sigma^2}||y_k - c'(s,s')||^2\right) \quad [3.142]$$

Since the term $1/\sqrt{2\pi\sigma^2}^N$ does not depend on (s, s'), it can be neglected:

$$p(y_k \mid S_{k-1} = s, S_k = s') = \exp\left(-\frac{1}{2\sigma^2}||y_k - c'(s,s')||^2\right) \quad [3.143]$$

Using expression [3.143] in [3.136] we have:

$$g_k(s,s') = \exp\left(-\frac{1}{2\sigma^2}||y_k - c'(s,s')||^2\right) p\left(d_k = d(s,s')\right) \quad [3.144]$$

If, moreover, the binary information units have equal probability, the term $p(d_k = d(s, s'))$ does not depend on the pair (s, s') and can be neglected; in [3.144]; in this case we obtain:

$$g_k(s,s') = \exp\left(-\frac{1}{2\sigma^2}||y_k - c'(s,s')||^2\right) \quad [3.145]$$

The square of the Euclidean distance between the observation y_k and the coded and modulated block $c'(s, s')$ breaks up into a sum of three terms:

$$||y_k - c'(s,s')||^2 = ||y_k||^2 + ||c'(s,s')||^2 - 2\langle y_k, c'(s,s')\rangle \quad [3.146]$$

$||y_k||^2$ does not depend on the pair (s, s'); thus, it can be neglected in [3.146]. For a modulation with binary symbols ($+1$ or -1) (case of the MDP-2 and MDP-4) the quantity $||c'(s, s')||^2$ does not depend on (s, s') either and can therefore also be neglected in [3.146]:

$$||y_k - c'(s,s')||^2 = -2\langle y_k, c'(s,s')\rangle \quad [3.147]$$

to the nearest additive constant. Using [3.147] in [3.145] we obtain:

$$g_k(s,s') = \exp\left(\frac{1}{\sigma^2}\langle y_k, c'(s,s')\rangle\right) \quad [3.148]$$

3.6.2.2. *Example*

To illustrate decoding using the MAP criterion we consider the same example as in section 3.6.1. We suppose that the sequence to be coded is 1001 and that the convolution encoder is the one represented in Figure 3.2. Making the assumption that the encoder is in the "all at zero" state at the initial moment ($k = 0$), the coded sequence is 11 10 11 11. We consider a symmetric binary channel introducing an error in position 3 so that the observation there is 11 00 11 11. We suppose that the d_k units are equal to 0 or 1 with equal probability.

Decoding using the MAP criterion requires that the probability of error p be known, which was not the case for the Viterbi algorithm. We suppose that p verifies:

$$p/(1-p) = 1/10 \qquad [3.149]$$

According to [3.141] we have:

$$g_k(s,s') = \left(\frac{p}{1-p}\right)^{d_H^k(s,s')} = 1/10^{d_H^k(s,s')} \qquad [3.150]$$

For this example, we work with the non-standardized Front and Back filters $\alpha_k(s)$ and $\beta_k(s)$. Indeed, M being small ($M = 4$), the problems of calculation's precision mentioned above do not appear in this academic case.

Decoding using the MAP criterion proceeds as follows:

3.6.2.2.1. Calculation of the $g_k(s,s')$ quantities

For $k = 1$:

$$g_1(a,a) = 1/10^2$$
$$g_1(a,c) = 1$$

For $k = 2$:

$$g_2(a,a) = 1 \quad g_2(c,b) = 1/10$$
$$g_2(a,c) = 1/10^2 \quad g_2(c,d) = 1/10$$

For $k = 3$:

$$g_3(a,a) = 1/10^2 \quad g_3(b,a) = 1 \quad g_3(c,b) = 1/10 \quad g_3(d,b) = 1/10$$
$$g_3(a,c) = 1 \quad g_3(b,c) = 1/10^2 \quad g_3(c,d) = 1/10 \quad g_3(d,d) = 1/10$$

For $k = 4$:

$$g_4(a,a) = 1/10^2 \quad g_4(b,a) = 1 \quad g_4(c,b) = 1/10 \quad g_4(d,b) = 1/10$$
$$g_4(a,c) = 1 \quad g_4(b,c) = 1/10^2 \quad g_4(c,d) = 1/10 \quad g_4(d,d) = 1/10$$

3.6.2.2.2. Front filter

The encoder is initialized in the "all at zero" state:

$$\alpha_0(a) = 1 \quad \alpha_0(b) = 0 \quad \alpha_0(c) = 0 \quad \alpha_0(d) = 0$$

We then apply the propagation equation [3.103], or as an equivalent [3.105]:

$\alpha_1(a) = 0.01$ $\alpha_1(b) = 0$ $\alpha_1(c) = 1$ $\alpha_1(d) = 0$

$\alpha_2(a) = 0.01$ $\alpha_2(b) = 0.1$ $\alpha_2(c) = 0.0001$ $\alpha_2(d) = 0.1$

$\alpha_3(a) = 0.1001$ $\alpha_3(b) = 0.01$ $\alpha_3(c) = 0.011$ $\alpha_3(d) = 0.01001$

$\alpha_4(a) = 0.011001$ $\alpha_4(b) = 0.002101$ $\alpha_4(c) = 0.1002$ $\alpha_4(d) = 0.002101$

3.6.2.2.3. Back filter

Initialization:

$$\beta_4(a) = 1 \quad \beta_4(b) = 1 \quad \beta_4(c) = 1 \quad \beta_4(d) = 1$$

We then apply the propagation equation [3.111], or equivalently [3.112]:

$\beta_3(a) = 1.01$ $\beta_3(b) = 1.01$ $\beta_3(c) = 0.2$ $\beta_3(d) = 0.2$

$\beta_2(a) = 0.2101$ $\beta_2(b) = 1.0120$ $\beta_2(c) = 0.1210$ $\beta_2(d) = 0.1210$

$\beta_1(a) = 0.21131$ $\beta_1(b) = 0$ $\beta_1(c) = 0.1133$ $\beta_1(d) = 0$

$\beta_0(a) = 0.1154131$ $\beta_0(b) = 0$ $\beta_0(c) = 0$ $\beta_0(d) = 0$

3.6.2.2.4. A *posteriori* probabilities of branches

We apply relation [3.121]. For $k = 1$:

$$\psi_1(a, a) = 0.0021131$$
$$\psi_1(a, c) = 0.1133$$

For $k = 2$:

$$\psi_2(a, a) = 0.0021131 \quad \psi_2(c, b) = 0.1012$$
$$\psi_2(a, c) = 1.21 \cdot 10^{-5} \quad \psi_2(c, d) = 0.0121$$

For $k = 3$:

$$\psi_3(a, a) = 0.0001 \quad \psi_3(b, a) = 0.1010$$
$$\psi_3(c, b) = 1.01 \cdot 10^{-5} \quad \psi_3(d, b) = 0.0101$$
$$\psi_3(a, c) = 0.002 \quad \psi_3(b, c) = 0.0002$$
$$\psi_3(c, d) = 2 \cdot 10^{-6} \quad \psi_3(d, d) = 0.002$$

For $k = 4$:

$$\psi_4(a, a) = 0.001 \quad \psi_4(b, a) = 0.01$$
$$\psi_4(c, b) = 0.0011 \quad \psi_4(d, b) = 0.001$$
$$\psi_4(a, c) = 0.1001 \quad \psi_3(b, c) = 0.0001$$
$$\psi_4(c, d) = 0.0011 \quad \psi_4(d, d) = 0.001$$

3.6.2.2.5. Hard decision and soft decision

We apply relation [3.125] to calculate the joint probability of $\boldsymbol{y}$ and of $d_k = d$, then by using the decision rule [3.126] we decode $\hat{d}_k$ and evaluate the reliability of the decision according to relation [3.129].

For $k = 1$:

$$p(d_1 = 0, \mathbf{y}) = 0.0021 \quad p(d_1 = 1, \mathbf{y}) = 0.1133$$

Consequently:

$$\hat{d}_1 = 1 \quad \text{and} \quad p(d_1 = 1 \mid \mathbf{y}) = 0.9817$$

For $k = 2$:

$$p(d_2 = 0, \mathbf{y}) = 0.1033 \quad p(d_2 = 1, \mathbf{y}) = 0.0121$$

Consequently,

$$\hat{d}_2 = 0 \quad \text{and} \quad p(d_2 = 0 \mid \mathbf{y}) = 0.8951$$

For $k = 3$:

$$p(d_3 = 0, \boldsymbol{y}) = 0.1112 \quad p(d_3 = 1, \boldsymbol{y}) = 0.0042$$

Consequently:

$$\hat{d}_3 = 0 \quad \text{and} \quad p(d_3 = 0 \mid \boldsymbol{y}) = 0.9636$$

For $k = 4$:

$$p(d_4 = 0, \boldsymbol{y}) = 0.0131 \quad p(d_4 = 1, \boldsymbol{y}) = 0.1023$$

Consequently:

$$\hat{d}_4 = 1 \quad \text{and} \quad p(d_4 = 1 \mid \boldsymbol{y}) = 0.8864$$

The sequence of binary information units estimated by the MAP is thus:

$$\hat{d} = (1001)$$

The single error introduced by the channel has been corrected.

3.6.3. *SubMAP algorithm*

The MAP algorithm is relatively complex to implement in a circuit and, therefore, we try to define the sub-optimal algorithms derived from MAP, whose implementation in a circuit would be simpler. One of the most powerful among these algorithms is known by the name of subMAP and is presented immediately hereafter.

Before presenting the subMAP algorithm we will define the following quantities:

– $\bar{\alpha}_k(s)$ is equal to the logarithm of $\alpha_k(s)$:

$$\bar{\alpha}_k(s) = \log \alpha_k(s) \tag{3.151}$$

Hereinafter we will call the quantity $\bar{\alpha}_k(s)$ the *Front filter*;

– $\bar{\beta}_k(s)$ is equal to the logarithm of $\bar{\beta}_k(s)$:

$$\bar{\beta}_k(s) = \log \beta_k(s) \tag{3.152}$$

Hereinafter we will call the quantity $\bar{\beta}_k(s)$ the *Back filter*;

– $\bar{\psi}_k(s, s')$ is equal to the logarithm of $\psi_k(s, s')$:

$$\bar{\psi}_k(s, s') = \log \psi_k(s, s') \tag{3.153}$$

3.6.3.1. *Propagation of the Front filter*

From relation [3.103] it results that:

$$\begin{aligned} \bar{\alpha}_k(s) &= \log \sum_{s'/s' \to s} \alpha_{k-1}(s') \exp\left(-\gamma_k(s', s)\right) \\ &= \log \sum_{s'/s' \to s} \exp\left(\bar{\alpha}_{k-1}(s')\right) \exp\left(-\gamma_k(s', s)\right) \\ &= \log \sum_{s'/s' \to s} \exp\left(\bar{\alpha}_{k-1}(s') - \gamma_k(s', s)\right) \end{aligned} \tag{3.154}$$

We can note that:

$$\begin{aligned} \log(e^x + e^y) &= \log\left(e^{\max(x,y)}(1 + e^{-|y-x|})\right) \\ &= \max(x, y) + \log(1 + e^{-|y-x|}) \end{aligned} \tag{3.155}$$

If one of variables x or y is larger than the other one, we can neglect the second term in the right hand side of equation [3.155]:

$$\log(e^x + e^y) \simeq \max(x, y) \quad \text{when} \quad |y - x| \gg 0 \tag{3.156}$$

Equation [3.155] can be generalized to a sum of n terms:

$$\begin{aligned} &\log(e^{x_1} + e^{x_2} + \cdots + e^{x_n}) \\ &= \log(e^{\max(x_1,\ldots,x_n)}(e^{-|x_1-\max(x_1,\ldots,x_n)|} + \cdots + e^{-|x_n-\max(x_1,\ldots,x_n)|})) \\ &= \max(x_1, \ldots, x_n) \\ &\quad + \log(e^{-|x_1-\max(x_1,\ldots,x_n)|} + \cdots + e^{-|x_n-\max(x_1,\ldots,x_n)|}) \end{aligned} \tag{3.157}$$

If one of the variables $x_1, \ldots, x_n$ is large compared to the others, we can simplify equation [3.157]:

$$\log(e^{x_1} + e^{x_2} + \cdots + e^{x_n}) \simeq \max(x_1, \ldots, x_n)$$
$$\text{when } \exists\, i \,/\, x_i \gg x_j,\ \forall\, j \neq i \qquad [3.158]$$

and using [3.158] in [3.154] we obtain:

$$\bar{\alpha}_k(s) \quad = \quad \max_{s'/s' \to s} \left(\bar{\alpha}_{k-1}(s') - \gamma_k(s', s)\right) \qquad [3.159]$$

Relation [3.159] is valid when the signal to noise ratio of the transmission is sufficiently high, i.e., typically, higher than a few decibels.

3.6.3.2. *Propagation of the Back filter*

From relation [3.111] it stems that:

$$\bar{\beta}_k(s) = \log \sum_{s'/s \to s'} \exp(\bar{\beta}_{k+1}(s') - \gamma_{k+1}(s, s')) \qquad [3.160]$$

and using [3.158] in [3.160] we obtain:

$$\bar{\beta}_k(s) = \max_{s'/s \to s'} (\bar{\beta}_{k+1}(s') - \gamma_{k+1}(s, s')) \qquad [3.161]$$

The approximation [3.161] is valid under the same conditions as [3.159].

3.6.3.3. *Calculation of the $\bar{\psi}_k(s, s')$ quantities*

From relation [3.121] it stems that:

$$\bar{\psi}_k(s, s') = \bar{\alpha}_{k-1}(s) - \gamma_k(s, s') + \bar{\beta}_k(s') \qquad [3.162]$$

3.6.3.4. *Calculation of the joint probability of d_k and $\boldsymbol{y}$*

Using relation [3.125] and definition [3.153] we obtain:

$$\log p(d_k, \mathbf{y}) \quad = \quad \log \sum_{(s,s')/d_k = d(s,s')} \exp(\bar{\psi}_k(s, s')) \qquad [3.163]$$

Then using approximation [3.158] we obtain:

$$\log p(d_k, \mathbf{y}) \quad = \quad \max_{(s,s')/d_k = d(s,s')} \bar{\psi}_k(s, s') \qquad [3.164]$$

Let us note $\bar{\phi}_k(d_k)$ the joint log-probability of d_k and $\boldsymbol{y}$:

$$\bar{\phi}_k(d_k) \quad = \quad \log p(d_k, \boldsymbol{y}) = \max_{(s,s')/d_k = d(s,s')} \bar{\psi}_k(s, s') \qquad [3.165]$$

The subMAP algorithm leads to the following decision:

$$\hat{d}_k = \operatorname{Arg}\max_{d_k} \bar{\phi}_k(d_k) \qquad [3.166]$$

3.7. Performance of convolutional codes

The performance of convolutional codes is generally evaluated by calculating the probability of error for the decoded information symbols.

Since convolutional codes are linear, we can demonstrate that the calculation of the probability of error can be performed by arbitrarily choosing the transmitted coded sequence. Thus, to simplify calculations, let us suppose that the encoder delivers the sequence comprised of a sequence of zero symbols. This sequence, called the *zero sequence*, is represented on the lattice diagram by the path noted p_0.

Considering a decoding according to the most likely sequence (Viterbi algorithm), a first error event will occur at the moment $t = k$, if the surviving path is not the p_0 path but a p_d path with the distance $d \geq d_f$ from p_0. This situation occurs, if:

$$M_k(p_d) < M_k(p_0) \quad [3.167]$$

where $M_k(p_i)$ is the cumulated metric at the moment $t = k$ of the path $p_i, i = 0$ or d.

The probability of this first error event is thus equal to:

$$P_k(p_d) = \mathbb{P}(M_k(p_d) < M_k(p_0)) \quad [3.168]$$

where $\mathbb{P}(A)$ indicates the probability of event A.

We will show hereafter that this probability does not in fact depend on the path p_d, but only on the Hamming weight d of this path. Therefore, this probability is noted hereafter $P_k(d)$ (it is the same one for all the paths with a Hamming weight d that diverge from the path p_0, then merge again in $t = k$).

If $n(d, i)$ is the number of paths of weight d generated by an information sequence of weight i, then the average number of erroneous information symbols is upper bounded by:

$$\bar{n_k} \leq \sum_{d=d_f}^{\infty} \sum_{i=1}^{\infty} i\, n(d,i) P_k(d) \quad [3.169]$$

It clearly is an upper bound, since the cumulated metrics appearing in expression [3.169] relate to paths with common branches.

Posing:

$$w(d) = \sum_{i=1}^{\infty} i\, n(d,i) \quad [3.170]$$

expression [3.169] can also be written:

$$\bar{n_k} \le \sum_{d=d_f}^{\infty} w(d) P_k(d) \quad [3.171]$$

The set of terms $w(d)$ is called the *distance spectrum* of the code. Its terms can be determined on the basis of the transfer function $T(D, N)$. Indeed, we can write:

$$T(D, N) = \sum_{d=d_f}^{\infty} \sum_{i=1}^{\infty} n(d, i) D^d N^i \quad [3.172]$$

Deriving the transfer function with respect to N and then making $N = 1$ we obtain:

$$\left. \frac{\partial T(D, N)}{\partial N} \right|_{N=1} = \sum_{d=d_f}^{\infty} w(d) D^d \quad [3.173]$$

The coefficients of the monomials D^d are precisely the $w(d)$ terms.

If the average number of error symbols gives a first outline of the code's performance, the probability of error for the decoded symbols is generally a more relevant indicator.

Considering a code with output K/N and a transmission of L blocks of K information symbols, the symbol error probability can be set an upper bound by:

$$P_e \le \frac{1}{LK} \sum_{k=1}^{L} \bar{n_k} \quad [3.174]$$

If the probability $P_k(d)$ is independent of the moment k, which is roughly true for large L, then the probability of error can also be bounded by:

$$P_e \le \frac{1}{K} \sum_{d=d_f}^{\infty} w(d) P(d) \quad [3.175]$$

We will now evaluate the probability $P(d)$ considering a channel with binary input and continuous output and a channel with binary input and output.

3.7.1. *Channel with binary input and continuous output*

Let us consider for this channel that the coded symbols are transmitted using phase modulation with 2 or 4 states. Thus, a block of N modulation symbols of the following form is associated to a block of N coded symbols:

$$a_{k,i} = 2c_{k,i} - 1 \quad i = 1, 2, \ldots, N \quad c_{k,i} \in \{0, 1\} \quad [3.176]$$

Considering a coherent demodulation and a transmission disturbed by zero-mean white Gaussian noise with double sided power spectral density equal to $N0/2$, the decoder receives samples with the expression:

$$y_{k,i} = \rho_{k,i} A a_{k,i} + n_{k,i} \qquad [3.177]$$

where A is a constant amplitude, $\rho_{k,i}$ is an attenuation introduced by the transmission medium and $n_{k,i}$ is a sample of zero-mean white Gaussian noise with a variance $\sigma_n^2 = N0/2$.

For this channel the metrics $M_k(p_d)$ and $M_k(p_O)$ are respectively equal to:

$$\begin{aligned} M_k(p_d) &= \sum_{j=1}^{k} \sum_{i=1}^{N} [y_{j,i} - A\rho_{j,i} a_{j,i}(p_d)]^2 \\ M_k(p_O) &= \sum_{j=1}^{k} \sum_{i=1}^{N} [y_{j,i} + A\rho_{j,i}]^2 \end{aligned} \qquad [3.178]$$

where the symbols $a_{j,i}(p_d)$ correspond to the coded sequence associated with the path p_d.

Replacing the metrics with their expression [3.178], the probability $P_k(d)$ is equal to:

$$P_k(d) = \mathbb{P}\left(\sum_{j=1}^{k} \sum_{i=1}^{N} y_{j,i}\rho_{j,i}\left(1 + a_{j,i}(p_d)\right) > 0 \right) \qquad [3.179]$$

The path p_d being at a Hamming distance d from the path p_O, the sequence $\{a_{j,i}(p_d)\}$ has d symbols at $+1$, the others being -1. Expression [3.179] is thus reduced to a sum of d terms.

To continue the calculation let us distinguish between two situations. The first corresponds to a channel where the attenuation $\rho_{j,i}$ is constant ($\rho_{j,i} = \rho$); it is a Gaussian channel. For the second the $\rho_{j,i}$ are random independent Rayleigh variables; it is a Rayleigh channel without memory.

3.7.1.1. *Gaussian channel*

For this channel the samples received by the decoder have the expression:

$$y_{j,i} = -A\rho + n_{j,i} \qquad [3.180]$$

since it is the zero sequence that has been transmitted so that $a_{j,i}(p_O) = -1, \forall i, \forall j$.

After replacing $y_{j,i}$ by its expression [3.180] in relation [3.179] the probability $P_k(d)$ is still equal to:

$$P_k(d) = \mathbb{P}(Z > d\rho A) \qquad [3.181]$$

where Z is a sum of d random independent variables, distributed identically according to law $\mathcal{N}(0, \sigma_n^2)$ so that Z follows a law $\mathcal{N}(0, d\sigma_n^2)$. We may note that the probability $P_k(d)$ does not depend on k; it will, therefore, be hereafter noted $P(d)$.

Introducing the additional error function,

$$\text{erfc}(x) = \frac{2}{\sqrt{\pi}} \int_x^{\infty} \exp(-u^2)\, du \qquad [3.182]$$

we can finally express the probability $P(d)$ as:

$$P(d) = \frac{1}{2}\text{erfc}\sqrt{\frac{d\rho^2 A^2}{N0}} \qquad [3.183]$$

Let E_b be the average energy received by an information symbol and R be the code output. The probability $P(d)$ can then be written in the form:

$$P(d) = \frac{1}{2}\,\text{erfc}\sqrt{\frac{d\,R\,E_b}{N0}} \qquad [3.184]$$

Finally, for a convolutional code with output K/N, the probability of error on a Gaussian channel is greatly bounded by:

$$P_e \leq \frac{1}{2K} \sum_{d=d_f}^{\infty} w(d)\,\text{erfc}\sqrt{\frac{d\,R\,E_b}{N0}} \qquad [3.185]$$

The terms $w(d)$ generally increase with d, whereas the complementary error function decreases with d all the faster the larger the $E_b/N0$ ratio is.

For low values of $E_b/N0$ (typically lower than 3 dB) the bound [3.185] is rather rough and its use is not very precise. It is then preferable to evaluate code performance by determining the error rate by simulation.

For an average $E_b/N0$ ratio, the complementary error function decreases sufficiently quickly with d so that we can limit ourselves to the calculation of the first terms (4 to 5 terms) of the bound [3.185].

With a strong $E_b/N0$ ratio the probability of error is approximated well by the first term of the bound [3.185]:

$$P_e \simeq \frac{1}{2K} w(d_f)\text{erfc}\sqrt{Rd_f\frac{E_b}{N0}} \quad E_b/N0 \gg 1 \qquad [3.186]$$

For a Gaussian channel without coding and phase modulation with 2 or 4 states with coherent demodulation, the probability of error is equal to:

$$P_e = \frac{1}{2}\,\mathrm{erfc}\sqrt{\frac{E_b}{N0}} \qquad [3.187]$$

Neglecting at first approximation the term $w(d_f)/K$ (which amounts to considering that it is close to 1) we see that expressions [3.186] and [3.187] lead to the same probability of error if the $E_b/N0$ ratio without coding (expressed in dB) is greater than $10\log_{10}(Rd_f)$ as the same ratio with coding.

The difference between these two signal to noise ratios that makes it possible to obtain the same probability of error with and without coding is called *asymptotic coding gain* (because we take a strong $E_b/N0$):

$$G_{dB} = 10\log_{10}(Rd_f) \qquad [3.188]$$

For the convolutional code of example 3.1 ($R = 1/2, d_f = 5, w(d_f) = 1$) the asymptotic coding gain is 4 dB.

Real coding gain is, of course, lower than asymptotic gain, since we need it to take into account the term $w(d_f)/K$ in the comparison of performances. In addition, if the ratio $E_b/N0$ is not very large, we cannot limit ourselves to the first term of relation [3.185] to evaluate the probability of error.

The product Rd_f that fixes the asymptotic coding gain provides information on the potentialities of a code in terms of error correction.

There exist expressions other than relation [3.185] to upper bound the probability of error. Indeed, by taking into account the fact that:

$$\mathrm{erfc}\sqrt{R\,d\,\frac{E_b}{N0}} < \exp\left(-Rd\frac{E_b}{N0}\right) \qquad [3.189]$$

and the relations [3.173] and [3.175], the probability of error can also be expressed in the form:

$$P_e < \frac{1}{2K}\left.\frac{\partial T(D,N)}{\partial N}\right|_{N=1,D=\exp(-R\,E_b\,/\,N0)} \qquad [3.190]$$

This new bound, which reveals the partial derivative of the transfer function is obviously less fine (see relation [3.189]) than bound [3.185]. However, for a strong signal to noise ratio these two bounds lead to equivalent results.

Another bound, finer than [3.190], can be obtained by replacing the complementary error function by:

$$\mathrm{erfc}\sqrt{x+y} \leq \mathrm{erfc}\sqrt{x}\ \exp(-y) \qquad [3.191]$$

The approximation used in [3.191] is better than that used in [3.189].

Posing:

$$d = d_f + d - d_f \qquad [3.192]$$

and taking into account relation [3.191], the probability $P(d)$ (see relation [3.184]) can be bounded by:

$$P(d) < \frac{1}{2}\,\text{erfc}\sqrt{Rd_f\frac{E_b}{N0}}\ \exp\left(Rd_f\frac{E_b}{N0}\right)\ \exp\left(-Rd\frac{E_b}{N0}\right) \qquad [3.193]$$

Thus, using expressions [3.173] and [3.175] and relation [3.193], the probability of error is bounded by:

$$P_e < \frac{1}{2K}\text{erfc}\sqrt{Rd_f\frac{E_b}{N0}}\exp\left(Rd_f\frac{E_b}{N0}\right)\left.\frac{\partial T(D,N)}{\partial N}\right|_{N=1,D=\exp(-R\,E_b/N0)} \qquad [3.194]$$

3.7.1.2. *Rayleigh channel*

For this channel the $\rho_{j,i}$ attenuations are random independent Rayleigh variables (channel without memory) of probability density:

$$p(\rho_{j,i}) = \frac{1}{\sigma_\rho^2}\rho_{j,i}\exp(-\rho_{j,i}^2/(2\sigma_\rho^2))\mathbb{1}_{\rho_{j,i}\geq 0} \qquad [3.195]$$

where $\mathbb{E}(\rho_{j,i}^2) = 2\sigma_\rho^2$ and where $\mathbb{1}_{\rho_{j,i}} \geq 0$ is the indicator of the set $\{\rho_{j,i} \geq 0\}$, which equals 1 if $\rho_{j,i} \geq 0$, and 0 if not.

Let us recall that it has been established (see relation [3.179]) that the probability $P_k(d)$ equals:

$$P_k(d) = \mathbb{P}\left(\sum_{j=1}^{k}\sum_{i=1}^{N} y_{j,i}\rho_{j,i}(1 + a_{j,i}(p_d)) > 0\right) \qquad [3.196]$$

Let us note by $\mathcal{I}$ the set of pairs (i, j), for which $a_{j,i}(p_d) = +1$; these pairs number d, because the path p_d has a Hamming weight equal to d:

$$\mathcal{I} = \{(i, j); a_{j,i}(p_d) = +1\} \quad \text{card}\,(\mathcal{I}) = d$$

Using the notation $\mathcal{I}$ the expression of $P_k(d)$ is reduced to:

$$P_k(d) = \mathbb{P}\left(\sum_{(i,j)\in\mathcal{I}} y_{j,i}\rho_{j,i} > 0\right) \qquad [3.197]$$

The sequence transmitted by the encoder corresponds to the path p_0, so that:

$$y_{j,i} = -A\rho_{j,i} + n_{j,i} \quad [3.198]$$

Replacing $y_{j,i}$ with its expression [3.198] in [3.197] it follows:

$$P_k(d) = \mathbb{P}\left(\sum_{(i,j)\in\mathcal{I}} n_{j,i}\rho_{j,i} > A\sum_{(i,j)\in\mathcal{I}} \rho_{j,i}^2\right) \quad [3.199]$$

Let us suppose initially that the series of the attenuations $\rho_{j,i}$ is known; let us note by α_d the sum of the squares of the attenuations for the all the indices (i,j) in $\mathcal{I}$:

$$\alpha_d = \sum_{(i,j)\in\mathcal{I}} \rho_{j,i}^2 \quad [3.200]$$

The random variable α_d follows a Chi-square law with $2d$ degrees of freedom, a law whose density is given by:

$$p(\alpha_d) = \frac{1}{(2\sigma_\rho)^d(d-1)!}\alpha^{d-1}\exp\left(-\frac{\alpha_d}{2\sigma_\rho}\right)\mathbb{1}_{\alpha_d\geq 0} \quad [3.201]$$

where $\mathbb{1}_{\alpha d} \geq 0$ is the indicator of the set $\{\alpha_d \geq 0\}$.

Conditionally to the attenuations $\rho_{j,i}$ the random variable is a sum of:

$$W = \sum_{(i,j)\in\mathcal{I}} n_{j,i}\rho_{j,i} \quad [3.202]$$

independent Gaussian variables; thus it is also a random Gaussian variable, with a zero mean and a variance $\sigma_n^2\alpha_d$:

$$[W \mid (\rho_{j,i})_{(i,j)\in\mathcal{I}}] \quad \sim \quad \mathcal{N}(0, \sigma_n^2\alpha_d) \quad [3.203]$$

Conditionally to $\rho_{j,i}$ the probability that the Viterbi decoder chooses the path p_d rather than the path p_0 is thus equal to:

$$\mathbb{P}(W > A\alpha_d) = \mathbb{P}(Z > A\sqrt{\alpha_d} \,/\, \sigma_n) = \frac{1}{2}\text{erfc}\sqrt{A^2\alpha_d \,/\, (2\sigma_n^2)} \quad [3.204]$$

noting Z the random variable $Z = W/(\sigma_n\sqrt{\alpha_d})$ which follows a law $\mathcal{N}(0,1)$.

This conditional probability can be expressed as a function of the output R of the code, of the ratio $E_b/N0$ and of α_d:

$$\mathbb{P}(W > A\alpha_d) = \frac{1}{2}\text{erfc}\sqrt{R\frac{E_b}{N0}\alpha_d} \quad [3.205]$$

Focusing on the various manifestations of the attenuation $\rho_{j,i}$ we obtain the probability that the path p_d be preferred to the path p_0:

$$P(d) = \int_0^{+\infty} \frac{1}{2} \operatorname{erfc}\sqrt{R\frac{E_b}{N0}\alpha_d} \; p(\alpha_d) \; d\alpha_d \qquad [3.206]$$

Taking into account [3.201] and after integration, the probability $P(d)$ is, finally, equal to:

$$P(d) = \left(\frac{1-\lambda}{2}\right)^d \sum_{i=0}^{d-1} C_{d-1+i}^i \left(\frac{1+\lambda}{2}\right)^i \qquad [3.207]$$

with:

$$\lambda = [(R\bar{E}_b/N0)/(1+R\bar{E}_b N0)]^{1/2} \quad C_{d-1+i}^i = \frac{(d-1+i)!}{i!(d-1)!}$$

$$\bar{E}_b = \mathbb{E}(\rho_{j,i}^2)E_b = 2\sigma_\rho^2 E_b$$

The quantity $\bar{E}_b$ represents the average energy received per symbol of transmitted information.

For a strong signal to noise ratio ($\bar{E}_b/N0 \gg 1$) expression [3.207] is well approximated by:

$$P(d) \simeq C_{2d-1}^d \left(\frac{1}{4R\bar{E}b/N0}\right)^d \quad \bar{E}b/N0 \gg 1 \qquad [3.208]$$

and thus the probability of error is bounded by:

$$P_e \leq \frac{w(d_f)}{K} C_{2d_f-1}^{d_f} \left(\frac{1}{4R\bar{E}b/N0}\right)^{d_f} \qquad [3.209]$$

For a modulation with 2 or 4 states of phase and without coding and a coherent demodulation, the probability of error for the Rayleigh channel is equal to:

$$P_e = \frac{1}{2}\left[1 - \sqrt{\frac{\bar{E}b/N0}{1+\bar{E}b/N0}}\right] \qquad [3.210]$$

Still taking a strong signal to noise ratio ($\bar{E}_b/N0 \gg 1$) expression [3.210] is well approximated by:

$$P_e \simeq \frac{1}{4\bar{E}b/N0} \qquad [3.211]$$

For the Rayleigh channel the asymptotic coding gin is no longer expressed as a function of the product Rd_f. Besides, there is no analytical expression of this gain.

To illustrate the performance of convolutional codes for the Rayleigh channel let us consider the code from example 3.1 ($R = 1/2, K = 1, d_f = 5, w(d_f) = 1$):

– With coding and a strong $\bar{E}_b/N0$:

$$P_e \leq 126 \left(\frac{1}{2\bar{E}b/N0}\right)^5$$

– Without coding:

$$P_e \simeq \frac{1}{4\bar{E}b/N0}$$

For example, for a probability of error of 10^{-5} the coding gain is 33 dB.

3.7.2. *Channel with binary input and output*

Now let us consider the performances of the Viterbi decoder in the case of the symmetric binary channel (SBC). With this channel model, the demodulator makes hard decisions; the decoder is thus supplied with binary symbols.

The metrics used by the Viterbi decoder are the Hamming distances between the received sequence and the surviving sequences in each node of the lattice. In the same way as before, we suppose that the sequence p_0 has been transmitted. Moreover, we suppose that the concurrent path of the p_0 path in a certain node B – corresponding to the moment $t = k$ – is a path p_d whose Hamming weight out in B equals d.

If d is odd, the path p_0 will be selected correctly, if the error count in the received sequence is strictly lower than $(d+1)/2$; if not, it is the p_d path that is selected. The probability of the former error event thus equals:

$$P(d) = \sum_{k=(d+1)/2}^{d} C_d^k \, p^k \, (1-p)^{d-k} \quad \text{with} \quad C_d^k = \frac{d!}{k!(d-k)!} \qquad [3.212]$$

If d is even, the path p_d is selected, if the error count exceeds $d/2$; if this error count is exactly equal to $d/2$, the paths p_0 and p_d have an equal probability – the choice can thus be arbitrary. The probability of the first error event thus has the expression:

$$P(d) = \sum_{k=d/2+1}^{d} C_d^k \, p^k \, (1-p)^{d-k} + \frac{1}{2} C_d^{d/2} \, p^{d/2} \, (1-p)^{d/2} \qquad [3.213]$$

The probability of error after decoding is still bounded by expression [3.175] with the term $P(d)$ substituted by one of the two expressions above (odd or even d).

Instead of using one of the expressions [3.212] or [3.213] we can use the following upper bound:

$$P(d) < [4p(1-p)]^{d/2} \tag{3.214}$$

The use of this upper bound in [3.175] makes it possible to establish another upper bound of the probability of error for the symmetric binary channel, less fine than that obtained by combining the bound [3.175] with expressions [3.212] or [3.213]:

$$P_e < \left.\frac{\partial T(D,N)}{\partial N}\right|_{N=1, D=\sqrt{4p(1-p)}} \tag{3.215}$$

To illustrate the preceding calculations for the bounds of the probability of error, let us again consider the code from example 3.1 and take, for example, a strong signal to noise ratio. Under this assumption, every bound is as fine as another. We will thus use the bound [3.190] for the Gaussian channel and the bound [3.215] for the symmetric binary channel.

The transfer function of the code from example 3.1 is given by relation [3.29], and its partial derivative with respect to N is equal to:

$$\left.\frac{\partial T(D,N)}{\partial N}\right|_{N=1} = D^5 \;/\; (1-2D)^2 \tag{3.216}$$

We then easily obtain:

$$P_e < \frac{1}{2}\frac{\exp\left(-\frac{5}{2}\frac{E_b}{N0}\right)}{\left(1-2\exp\left(-\frac{E_b}{2\,N0}\right)\right)^2} \quad \text{(Gaussian channel)} \tag{3.217a}$$

$$P_e < \frac{(2p(1-p))^{5/2}}{(1-2\sqrt{2p(1-p)})^2} \quad \text{(symmetric binary channel)} \tag{3.217b}$$

With a strong signal to noise ratio ($E_b/N0 \gg 1$ for the Gaussian channel, $p \ll 1$ for the symmetric binary channel) the expressions above can be simplified:

$$P_e < \frac{1}{2}\exp\left(-\frac{5}{2}\frac{E_b}{N0}\right) \quad \text{(Gaussian channel)} \tag{3.218a}$$

$$P_e < 2^{5/2}p^{5/2} \quad \text{(symmetric binary channel)} \tag{3.218b}$$

Let us suppose that for the symmetric binary channel we use phase modulation with 2 or 4 states with coherent demodulation:

$$p = \frac{1}{2}\,\text{erfc}\sqrt{\frac{Eb}{2\,N0}} < \frac{1}{2}\exp\left(-\frac{Eb}{2\,N0}\right) \tag{3.219}$$

Under this assumption the [3.218b] bounds become:

$$P_e < \exp(-5\ Eb\ /\ (4\ N0)) \quad \text{(symmetric binary channel)} \qquad [3.220]$$

Comparing relations [3.220] (symmetric binary channel) and [3.218a] (Gaussian channel) we realize that the coding gain for the Gaussian channel is 3 dB higher than the coding gain for the symmetric binary channel. This result clearly shows the interest of *soft input decoding* compared to *hard input decoding*. This conclusion, which we have reached for a particular code, remains true for any code considered.

3.8. Distance spectrum of convolutional codes

We saw that the distance spectrum of convolutional codes was defined by the sequence of terms $w(d)$, $d \geq d_f$. These terms can be evaluated using the transfer function of the code.

For codes whose constraint length is greater than a few units (typically, $\nu \geq 5$), the calculation of the transfer function can prove to be complex. We then prefer to determine the spectrum of the code, or at least the first terms of this spectrum, using an algorithm that explores the various paths of the lattice diagram.

Tables 3.2 and 3.3 contain the first terms of the distance spectrum of some systematic and non-systematic convolutional codes with output 1/2, for lengths of constraint ranging from 3 to 9. While examining Tables 3.2 and 3.3 we can see that the coefficients $w(d)$ increases with d, apart from the fact that for certain codes, these coefficients are periodically cancelled. We can also note that for a given constraint length the free distance of systematic codes is always lower than that of non-systematic codes.

ν	G^2 in octal	d_f	$w(d) \quad d = d_f, d_f+1, d_f+2, \ldots$
3	(7)	4	3, 0, 15, 0, 58, 0, 201, 0, 655, 0, 2,052, 0, 6,255, 0, 18,687, ...
4	(15)	4	1, 0, 16, 0, 62, 0, 360, 0, 1,502, 0, 6,870, 0, 28,555, 0, 120,347, ...
5	(35)	5	4, 4, 6, 46, 79, 138, 488, 1,044, 2,016, 5,292, 12,053, 24,824, 130,206, 278,834, ...
6	(73)	5	6, 0, 44, 0, 245, 0, 1,661, 0, 9, 508, 0, 53,394, 0, 302,811, 0, 1,642,869, ...
7	(153)	6	4, 0, 28, 0, 158, 0, 1,311, 0, 7,433, 0, 48,102, 0, 282,803, 0, 1,675,543, ...
8	(247)	6	1, 0, 33, 0, 133, 0, 1,165, 0, 7,110, 0, 44,344, 0, 273,751, 0, 1,628,601, ...
9	(715)	7	4, 6, 17, 68, 110, 318, 917, 2,256, 6,276, 15,124, 36,890, 96,972, 240,104, 591,988, 1,478,753, ...

Table 3.2. *Spectrum of some systematic convolutional codes with output* $R = 1/2$

The performance of systematic codes is thus, for medium and strong signal to noise ratio, lower than that of non-systematic codes (weaker asymptotic coding gain). For values of d large when compared with d_f the coefficients $w(d)$ are lower for systematic codes than for non-systematic codes. Thus, for weak signal to noise ratios systematic codes make it possible to achieve better performance that non-systematic codes. In practice non-systematic codes are used, because they are more effective for bit error rates corresponding to applications (bit error rates lower than 10^{-3}).

To illustrate the performance evaluation of convolutional codes we traced the probability of error of the code with generator polynomials 133–171 and with output 1/2 using its spectrum $w(d)$ provided in Table 3.3, and the bounds [3.185]. In Figure 3.13, the performance of this code is traced using 1 coefficient then 15 coefficients $w(d)$ at the bound [3.185]. In this figure we have also traced the probability of error for the uncoded case, as well as the bit error rate obtained on the basis of a simulation.

ν	G^1, G^2 in octal	d_f	$w(d) \quad d = d_f, d_f+1, d_f+2, \ldots$
3	(5), (7)	5	1, 4, 12, 32, 80, 192, 448, 1,024, 2,304, 5,120, 11,264, 24,576, 53,248, 114,688, 245,760, ...
4	(15), (17)	6	2, 12, 20, 48, 126, 302, 724, 1,732, 4,112, 9,714, 22,850, 53,538, 125,008, 393,635, 925,334, ...
5	(23), (35)	7	4, 12, 20, 72, 225, 500, 1,324, 3,680, 8,967, 22,270, 57,403, 142,234, 348,830, 867,106, 2,134,239, ...
6	(53), (75)	8	2, 36, 32, 62, 332, 701, 2,342, 5,503, 12,506, 36,234, 88,576, 225,685, 574,994, 1,400,192, 3,554,210, ...
7	(133), (171)	10	36, 0, 211, 0, 1,404, 0, 11,633, 0, 77,433, 0, 502,690, 0, 3,322,763, 0, 21,292,910, ...
8	(247), (371)	10	2, 22, 60, 148, 340, 1,008, 2,642, 6,748, 18,312, 48,478, 126,364, 320,062, 821,350, 2,102,864, 5,335,734, ...
9	(561), (753)	12	33, 0, 281, 0, 2,179, 0, 15,035, 0, 105,166, 0, 692,330, 0, 4,138,761, 0, ...

Table 3.3. *Spectrum of some non-systematic convolutional codes of output* $R = 1/2$

For a strong signal to noise ratio, the various curves merge and the coding gain, with a 10^{-10} probability of error, is approximately 6.2 dB, i.e. very near the asymptotic gain that is equal to $10\log(Rd_f) = 10\log 5 = 7$ dB.

With 15 $w(d)$ coefficients, the bound [3.185] practically merges with the bit error rate once the signal to noise ratio exceeds 3.5 dB, that is, for a probability of error of around 10^{-4}.

For low signal to noise ratios, the bound [3.185] no longer makes it possible to reliably evaluate the performance of the code.

3.9. Recursive convolutional codes

Recursive convolutional codes that have been scarcely discussed in books before the invention of turbo codes were considered by coding specialists not to have particular advantages compared to non-systematic (non-recursive) codes.

To carry out concatenated codes it is *a priori* necessary to have elementary codes with good distance properties. This has led the inventors of turbocodes, Berrou *et al.*, to take an interest in systematic codes, for their good behavior with weak signal to noise ratios. However, aware of their short free distance, they sought to define a new family of codes, systematic but with the same free distances as non-systematic codes. They then rediscovered the recursive convolutional codes in their systematic version.

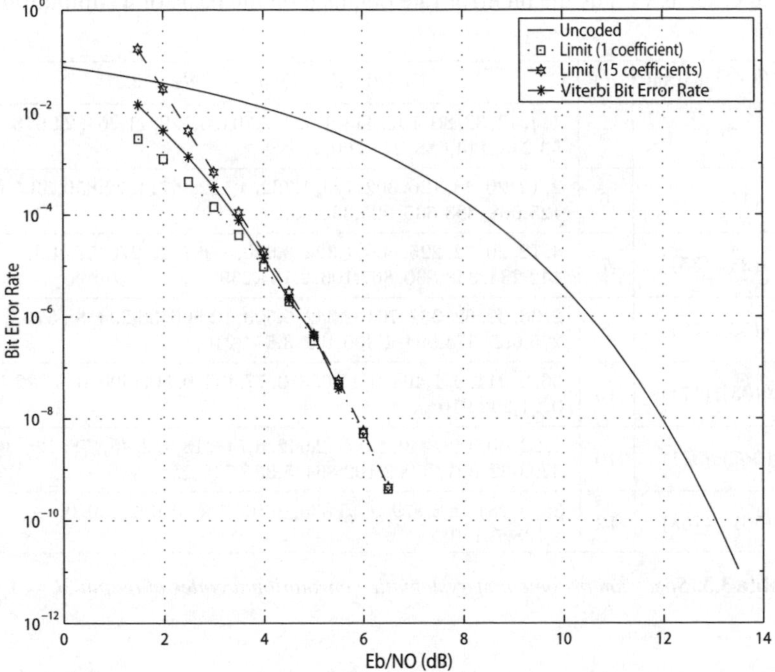

Figure 3.13. *Performance of the convolutional code with parameters 133–171 and output 1/2*

Below we will consider systematic recursive convolutional codes which are used for the construction of the turbocodes.

Let us consider a non-systematic convolutional code with output $R = 1/2$ and memory m (constraint length $\nu = m + 1$).

The encoder delivers two sequences represented by their transforms in $D, C_1(D)$ and $C_2(D)$:

$$C_1(D) = G^1(D)\, d(D)$$
$$C_2(D) = G^2(D)\, d(D) \qquad [3.221]$$

where $d(D)$ is the transform in D associated to the sequence to be coded d_k and where $G^1(D)$ and $G^2(D)$ are the two generator polynomials of the code.

By standardizing relations [3.221] with respect to the polynomial $G^1(D)$ or the polynomial $G^2(D)$, it is possible to define two systematic codes. For example, if we standardize with respect to $G^1(D)$, we obtain:

$$\tilde{C_1}(D) = d(D)$$
$$\tilde{C_2}(D) = (G^2(D)/G^1(D))\, d(D) \qquad [3.222]$$

The systematic nature results from the first equation, while the second equation implies the recursive character of the code specified in this manner. We will reconsider this last point later on.

Let us now introduce the series a_k defined by its transform in D:

$$A(D) = d(D) \,/\, G^1(D) \qquad [3.223]$$

Relations [3.222] then become:

$$\tilde{C_1}(D) = G^1(D)\, A(D)$$
$$\tilde{C_2}(D) = G^2(D)\, A(D) \qquad [3.224]$$

Relations [3.221] and [3.224] are similar; they represent the action of the same code: the non-systematic and non-recursive code with generator polynomials $G^1(D)$ and $G^2(D)$ for two different sequences. In the case of [3.221] the binary registers (memories) receive the information sequence d_k, whereas for [3.224] it is a_k that feeds the shift register of the code.

Thus, two codes with generator polynomials $[G^1(D), G^2(D)]$ (non-systematic non-recursive code) and $[1, G^2(D)/G^1(D)]$ (systematic recursive code) have a lattice diagram with the same structure. The coded sequences delivered by the two coders are identical and, consequently, the two encoders have the same free distance. The same coded sequence delivered by the two encoders corresponds to identical sequences d_k and a_k, but the same coded sequence does not correspond to the same information sequence at the encoder input.

Relation [3.223] is also written:

$$d(D) = G^1(D)A(D) \quad [3.225]$$

which in the temporal field becomes:

$$d_k = \sum_{j=0}^{m} g_j^1 a_{k-j} \quad [3.226]$$

noting g_j^1 the coefficients of the generator polynomial $G^1(D)$. If we take into account the fact that $g_0^1 = 1$, it follows:

$$a_k = d_k + \sum_{j=1}^{m} g_j^1 a_{k-j} \quad [3.227]$$

At the moment $t = k$, a_k is calculated, recursively, according to the information symbol d_k presented at the encoder input and according to the symbols $a_{k-1}, \dots,$ a_{k-m} that are available at this moment in the shift register of the encoder.

To illustrate this, Figure 3.14 represents two systematic recursive encoders constructed on the basis of the convolution encoder from example 3.1. It follows from relation [3.226] that the encoder of a systematic recursive code can be produced using a binary register with register outputs sealed off towards its input. A systematic recursive encoder thus operates similarly to a pseudo-random generator. Let us suppose that the initial state of the encoder is zero and its input is fed with a binary sequence whose first symbol is at 1, the following symbols being all equal to 0. The encoder will pass through a succession of states different from the "all at zero" state.

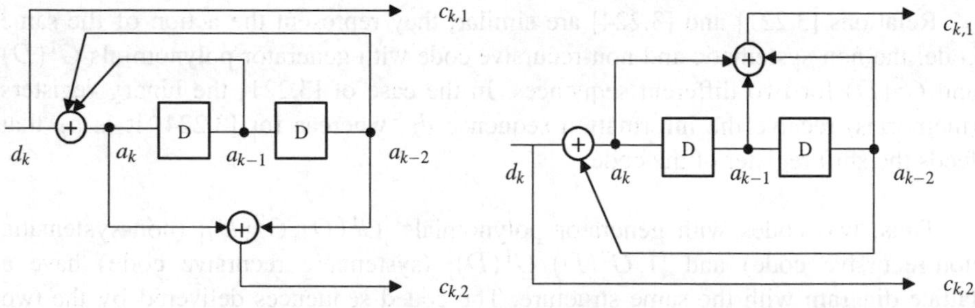

Figure 3.14. *Systematic recursive convolution encoders constructed on the basis of the convolution encoder from example 3.1 ($g^1 = 7_{octal}, g^2 = 5_{octal}$)*

Let us consider a systematic recursive code with memory $m = 4$, for which the generator polynomials are:

$$G^1(D) = 1 + D + D^2 + D^4$$
$$G^2(D) = 1 \quad [3.228]$$

The diagram of the corresponding encoder is represented in Figure 3.15.

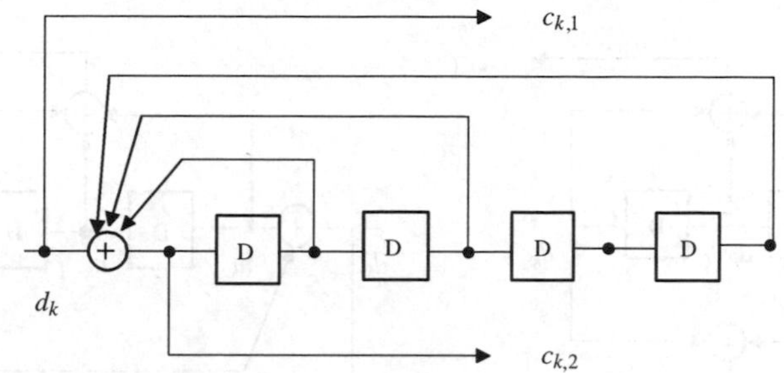

Figure 3.15. *Systematic recursive encoder with polynomials* $G^1(D) = 1 + D + D^2 + D^4$ *and* $G^2(D) = 1$

In Table 3.4 we represented the succession of the encoder states when the sequence of binary information units consists of 1 followed by an infinite number of zeros. The study of table 3.4 reveals a period of $L = 7$ for the encoder.

Input	State of the encoder			
d_k	a_k	a_{k-1}	a_{k-2}	a_{k-3}
1	1	0	0	0
0	1	1	0	0
0	0	1	1	0
0	1	0	1	1
0	0	1	0	1
0	0	0	1	0
0	0	0	1	0
0	0	0	0	1
0	1	0	0	0

Table 3.4. *Sequence of the states entered by the systematic recursive encoder* $G^1(D) = 1 + D + D^2 + D^4$ *and* $G^2(D) = 1$

It is easily understood that the problem of trellis termination (reset to the zero encoder state) is *a priori* more difficult in the case of a recursive code than for a non-recursive code. Indeed, for a recursive code it is not enough to place a sequence of

m symbols at zero at the end of the sequence presented at the input of the encoder to force its state to zero. For a systematic recursive code, the closing lattice symbols will depend on the state of the encoder at the end of the sequence to be coded. In our example, if this state is (1000), we would need to use a sequence of four symbols (1101) to close the lattice, whereas if this state is (0001) it would be enough to use a single symbol equal to 1.

Distance spectrum for systematic recursive convolutional codes

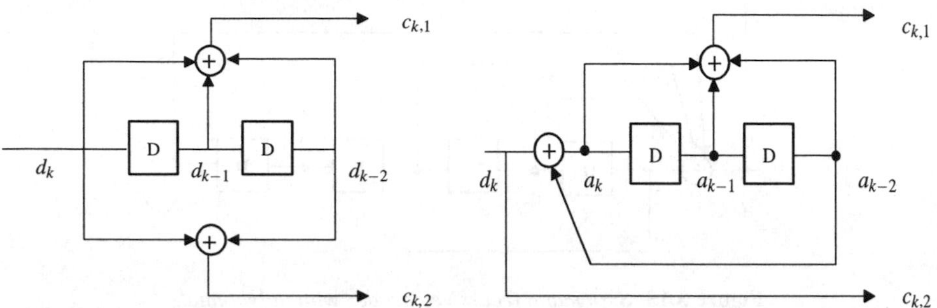

Figure 3.16. *On the left: non-systematic code $G^1(D) = 1 + D + D^2, G^2(D) = 1 + D^2$. On the right: systematic recursive code $1, G^1(D)/G^2(D)$*

To begin, let us start with an example. Let us consider the non-systematic codes and the systematic recursive codes whose encoders are represented in Figure 3.16.

According to relation [3.29], the non-systematic code has the following transfer function:

$$T(D, N) = D^5 N \,/\, (1 - 2DN) \qquad [3.229]$$

The coefficients of serial development of $T(D, N = 1)$ with respect to the variable D are the $n(d)$ coefficients, $n(d)$ representing the number of paths with Hamming weight d:

$$T(D, 1) = D^5 \,/\, (1 - 2D) = \sum_{d=d_f}^{\infty} n(d) D^d \qquad [3.230]$$

Similarly, the coefficients of serial development with respect to D, of the partial derivative with respect to N, of $T(D, N)$ taken at the point $N = 1$ are the $w(d)$ coefficients:

$$\left.\frac{\partial T(D, N)}{\partial N}\right|_{N=1} = D^5 \,/\, (1 - 2D)^2 = \sum_{d=d_f}^{\infty} w(d) D^d \qquad [3.231]$$

d	5	6	7	8	9	10	11	12	13	14	15	16
$n(d)$	1	2	4	8	16	32	64	128	256	512	1,024	2,048
$w(d)$	1	4	12	32	80	192	448	1,024	2,304	5,120	11,264	24,576

Table 3.5. *$n(d)$ and $w(d)$ for the non-recursive non-systematic convolutional code with generator polynomials $G^1(D) = 7_{octal}$ and $G^2(D) = 5_{octal}$*

The quantities $n(d)$ and $w(d)$ for $d \geq d_f (d_f = 5)$ are given in Table 3.5.

The calculation of the transfer function of the systematic recursive code is done in the same manner as for the non-systematic code. In Figure 3.17 we have represented the state transition diagram of the recursive code, the state $a = (00)$ being divided into two states a_e and a_s.

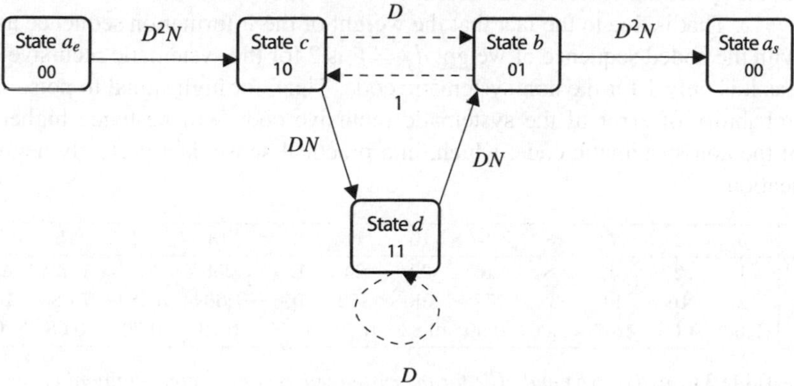

Figure 3.17. *State transition diagram of the systematic recursive convolutional code whose encoder is represented in Figure 3.16*

Compared to the diagram in Figure 3.7 we can notice that the labels of branches are different only with regard to the exhibitor of N. Let us also note that the same sequence of information symbols d_k does not deliver the same coded sequence for the non-systematic encoder and the systematic recursive encoder.

The transfer function of the systematic recursive code is easily obtained from Figure 3.17:

$$T(D, N) = D^5 N^2 (DN^2 - D + 1) \,/\, (1 - D^2 N^2 + D^2 - 2D) \qquad [3.232]$$

We proceed in the same way as in the case of the non-systematic non recursive code to evaluate the quantities $n(d)$ and $w(d)$.

$$T(D,1) = D^5/(1-2D) = \sum_{d=d_f}^{\infty} n(d)D^d \qquad [3.233]$$

$$\left.\frac{\partial T(D,N)}{\partial N}\right|_{N=1} = 2D^5(1-D-D^2)/(1-2D)^2 = \sum_{d=d_f}^{\infty} w(d)D^d \qquad [3.234]$$

Table 3.6 contains the quantities $n(d), w(d)$ and $p(d)$ where $p(d)$ is defined as the relationship between the spectral coefficient of the systematic recursive code (SRC) and the spectral coefficient of the non-systematic code (NSC):

$$\rho(d) \quad = \quad w^{\text{CRS}}(d) \ / \ w^{\text{CNS}}(d) \qquad [3.235]$$

Examining Tables 3.5 and 3.6 we can verify that the first term of the distance spectrum of the systematic recursive code is the double of that of the non-systematic code: $p(d_f) = 2$. That is due to the fact that the weight of the information sequence associated with the coded sequence of weight $d_f = 5$ is 2 for the systematic recursive code, whereas it is only 1 for the non-systematic code. Thus, for high signal to noise ratios, the probability of error of the systematic recursive code will be twice higher than that of the non-systematic code, which, in a practical sense, is a perfectly negligible degradation.

d	5	6	7	8	9	10	11	12	13	14	15	16
$n(d)$	1	2	4	8	16	32	64	128	256	512	1,024	2,048
$w(d)$	2	16	14	32	72	160	352	768	1,664	3,584	7,680	16,384
$\rho(d)$	2.00	4.00	1.67	1.00	0.90	0.83	0.79	0.75	0.72	0.70	0.68	0.67

Table 3.6. *$n(d), w(d)$ and $p(d)$ for the systematic recursive convolutional code whose encoder is represented in Figure 3.16*

Reading these tables we can also note that the coefficients of the distance spectrum of the systematic recursive code are lower than those of the non-systematic code since $d \geq 9$. Thus, for weak signal to noise ratios, the performances of the systematic recursive code will be better than those of the non-systematic code.

The introduction of recursiveness thus made it possible to introduce a new systematic code with performance equivalent to that of the non-systematic code for strong signal to noise ratios, and slightly improved performance for weak signal to noise ratios. This extremely interesting result that we have verified on the example of a particular code, is verified for all systematic recursive codes.

Table 3.7 provides the distance spectra of non-systematic codes and systematic recursive codes with an output of $R = 1/2$ and for lengths of constraint ranging

ν	G^1 G^2 In octal	d_f	$n(d)\ d = d_f, d_f+1, d_f+2, \ldots$ $w_{CRS}(d)\ d = d_f, d_f+1, d_f+2, \ldots$ $w_{CNS}(d)\ d = d_f, d_f+1, d_f+2, \ldots$ $\rho(d)\ d = d_f, d_f+1, d_f+2, \ldots$
3	(5) (7)	5	1, 2, 4, 8, 16, 32, 64, 128, 256, 1,024, 2,048, 4,096, 8,192, 16,384, ... 2, 6, 14, 32, 72, 160, 352, 768, 1,664, 3,584, 7,680, 16,384, 34,816, 73,728, 155,648, ... 1, 4, 12, 32, 80, 192, 448, 1,024, 2,304, 5,120, 11,264, 24,576, 53,248, 114,688, 245,760, ... 2.0, 1.5, 1.17, 1.0, 0.9, 0.83, 0.79, 0.75, 0.72, 0.70, 0.68, 0.67, 0.65, 0.64, 0.63, ...
4	(15) (17)	6	1, 3, 5, 11, 25, 55, 121, 267, 589, 1,299, 2,865, 6,319, 13,937, 30,739, 67,797, ... 4, 9, 20, 51, 124, 303, 728, 1,739, 4,134, 9,771, 22,990, 53,885, 125,858, 293,049, 680,440, ... 2, 7, 18, 49, 130, 333, 836, 2,069, 5,060, 12,255, 29,444, 70,267, 166,726, 393,635, 925,334, ... 2.0, 1.29, 1.11, 1.04, 0.95, 0.91, 0.87, 0.84, 0.82, 0.80, 0.78, 0.77, 0.76, 0.75, 0.74, ...
5	(23) (35)	7	2, 3, 4, 16, 37, 68, 176, 432, 925, 2,156, 5,153, 11,696, 26,868, 62,885, 145,085, ... 8, 12, 16, 84, 213, 406, 1,156, 3,104, 7,021, 17,372, 44,427, 106,518, 257,200, 634,556, 1,537,281, ... 4, 12, 20, 72, 225, 500, 1,324, 3,680, 8,967, 22,270, 57,403, 142,234, 348,830, 867,106, 2,134,239, ... 2.0, 1.0, 0.8, 1.17, 0.95, 0.81, 0.87, 0.84, 0.78, 0.78, 0.77, 0.75, 0.74, 0.73, 0.72, ...
6	(53) (75)	8	1, 8, 7, 12, 48, 95, 281, 605, 1,272, 3,334, 7,615, 18,131, 43,197, 99,210, 237,248, ... 4, 34, 32, 62, 288, 604, 1,884, 4,430, 9,926, 27,794, 67,380, 168,606, 424,768, 1,025,664, 2,570,672, ... 2, 36, 32, 62, 332, 701, 2,342, 5,503, 12,506, 36,234, 88,576, 225,685, 574,994, 1,400,192, 3,554,210, ... 2.0, 0.94, 1.0, 1.0, 0.87, 0.86, 0.80, 0.81, 0.79, 0.77, 0.76, 0.75, 0.74, 0.73, 0.72, ...
7	(133) (20)	10	11, 0, 38, 0, 193, 0, 1,331, 0, 7,275, 0, 40,406, 0, 234,969, 0, 1,337,714, ... 60, 0, 223, 0, 1,368, 0, 10,963, 0, 66,171, 0, 408,918, 0, 2,619,965, 0, 16,222,096, ... 36, 0, 211, 0, 1,404, 0, 11,633, 0, 77,433, 0, 502,690, 0, 3,322,763, 0, 21,292,910, ... 1.67, 1.06, 0.97, 0.94, 0.86, 0.81, 0.79, 0.76, ...
8	(247) (19)	10	1, 6, 12, 26, 52, 132, 317, 730, 1, 823, 4, 446, 10, 739, 25, 358, 60,773, 146,396, 350,399, ... 6, 28, 70, 182, 360, 984, 2,530, 6,156, 16,308, 41,924, 107,014, 265,396, 666,098, 1,677,978, 4,189,876, ... 2, 22, 60, 148, 340, 1,008, 2,642, 6,748, 18,312, 48,478, 126,364, 320,062, 821,350, 2,102,864, 5,335,734, ... 3.0, 1.27, 1.17, 1.23, 1.06, 0.98, 0.96, 0.91, 0.89, 0.87, 0.85, 0.83, 0.81, 0.80, 0.79, ...
9	(561) (753)	12	11, 0, 50, 0, 286, 0, 1,630, 0, 9,639, 0, 55,152, 0, 292,950, 0, ... 67, 0, 349, 0, 2,295, 0, 14,575, 0, 96,680, 0, 606,538, 0, 3,504,145, 0, ... 33, 0, 281, 0, 2,179, 0, 15,035, 0, 105,166, 0, 692,330, 0, 4,138,761, 0, ... 2.03, 1.24, 1.05, 0.96, 0.92, 0.88, 0.85, ...

Table 3.7. *Distance spectrum of non-systematic convolutional codes and of recursive systematic code with output $R = 1/2$ for lengths of constraint ν ranging from 3 to 9*

from 3 to 9. In this table we have also expressed the relationship between the spectral coefficients of systematic recursive codes and the corresponding non-systematic codes. This ratio is initially greater than 1, then it passes below 1, once d is a few units greater than d_f.

To complete the presentation of systematic recursive convolutional codes in Tables 3.8, 3.9, 3.10 and 3.11, we have presented the distance spectrum of some perforated codes for lengths of constraint of 4, 5 and 6 and outputs ranging between 2/3 and 5/6. By way of comparison we have also indicated in these tables the distance spectrum of non-systematic non recursive perforated codes with the same parameters.

Acknowledgements

Tables 3.1, 3.2, 3.6, 3.7, 3.8, 3.9 and 3.10 are borrowed from Punya Thitimajshima, Systematic recursive convolutional codes and their application to parallel concatenation, doctorate thesis in electronics at the Bretagne Occidentale University, December 1993.

Initial convolutional code			**Perforated convolutional code**			
ν	G^1 G^2 In octal	d_f	M	R	d_f	$n(d)\ d = d_f, d_f+1, d_f+2, \ldots$ $w_{CPR}(d)\ d = d_f, d_f+1, d_f+2, \ldots$ $w_{CPNR}(d)\ d = d_f, d_f+1, d_f+2, \ldots$ $\rho(d) = w_{CPR}(d)/w_{CPNR}(d)\ d = d_f, d_f+1, d_f+2, \ldots$
4	(15) (17)	6	1 1 1 0	2/3	4	3, 11, 35, 114, 378, 1,253, 4,147, 13,725, 45,428, 150,362, ... 10, 33, 146, 538, 2,046, 7,595, 27,914, 101,509, 366,222, 1,312,170, ... 10, 43, 200, 826, 3,314, 12,857, 48,834, 182,373, 672,324, 2,452,626, ... 1.0, 0.77, 0.73, 0.65, 0.62, 0.60, 0.57, 0.56, 0.55, 0.54, ...
5	(23) (35)	7	1 1 1 0	2/3	4	1, 0, 27, 0, 345, 0, 4,515, 0, 59,058, 0, ... 3, 0, 106, 0, 1,841, 0, 30,027, 0, 471,718, 0, ... 1, 0, 124, 0, 2,721, 0, 659, 0, 858,436, 0, ... 3.0, 0.86, 0.68, 0.60, 0.55, ...
6	(53) (75)	8	1 0 1 1	2/3	6	19, 0, 220, 0, 3, 089, 0, 42, 725, 0, 586, 592, 0, ... 74, 0, 1,146, 0, 20,251, 0, 337,067, 0, 5,411,831, 0, ... 96, 0, 1,094, 0, 35,936, 0, 637,895, 0, 10,640, 725, 0, ... 0.77, 0.56, 0.53, 0.51, ...

Table 3.8. *Distance spectrum of non-systematic convolutional codes and of recursive systematic code with output $R = 1/2$ for lengths of constraint ν ranging from 3 to 9*

It is interesting to note that there are several systematic recursive perforated codes whose coefficients $w(d)$ are always lower than those of non-systematic non-recursive perforated codes with the same parameters. This comment is particularly true for perforated codes with high output. Thus, it is possible to find perforated recursive systematic codes whose performance, in terms of error probability, will always be better than that of non-systematic non-recursive perforated codes. To illustrate this point we have traced the bit error rate for the non-recursive perforated code with generator polynomials 133–171 (in octal) and of its recursive version, with outputs of 2/3 and 4/5 in Figure 3.18, and with outputs of 3/4 and 5/6 in Figure 3.19.

Initial convolutional code			**Perforated convolutional code**			
ν	G^1 G^2 In octal	d_f	M	R	d_f	$n(d)\ d = d_f, d_f+1, d_f+2, \ldots$ $w_{CPR}(d)\ d = d_f, d_f+1, d_f+2, \ldots$ $w_{CPNR}(d)\ d = d_f, d_f+1, d_f+2, \ldots$ $\rho(d) = w_{CPR}(d)/w_{CPNR}(d)\ d = d_f, d_f+1, d_f+2, \ldots$
4	(15)	6	1 1 0	3/4	4	29, 0, 532, 0, 9,853, 0, 182,372, 0, 3,375,764, 0, ...
	(17)		1 0 1			93, 0, 2,456, 0, 59,503, 0, 1,361,142, 0, 30,003,290, 0, ...
						124, 0, 4,504, 0, 124,337, 0, 3,059,796, 0, 70,674,219, 0, ...
						0.75, 0.55, 0.48, 0.45, 0.43, ...
5	(23)	7	1 0 1	3/4	3	1, 2, 23, 124, 576, 2,847, 14,147, 69,954, 346,050, 1,711,749, ...
	(35)		1 1 0			2, 6, 86, 584, 3,086, 17,278, 96,394, 528,024, 2,865,512, 15,430,036, ...
						1, 7, 125, 936, 5, 915, 36, 580, 216, 612, 1, 246, 685, 7, 035, 254, 39, 092, 197, ...
						2.0, 0.86, 0.69, 0.52, 0.47, 0.45, 0.42, 0.41, 0.40, ...
6	(53)	8	1 0 0	3/4	4	1, 15, 65, 321, 1,661, 8,388, 42,560, 215,586, 1, 091,757, 5,533,847, ...
	(75)		1 1 1			3, 58, 301, 1,734, 10,150, 57,422, 323,730, 1,800,528, 9,926,855, 54,442,646, ...
						3, 85, 490, 3,198, 20,557, 123,312, 724,657, 4,177,616, 23,720,184, 133,193,880, ...
						1.0, 0.68, 0.61, 0.54, 0.49, 0.47, 0.45, 0.43, 0.42, 0.41, ...

Table 3.9. *Distance spectrum for perforated convolution recursive and non-recursive codes with output of 3/4 and lengths of constraint of 4, 5 and 6*

Initial convolutional code				**Perforated convolutional code**		
ν	G^1 G^2 In octal	d_f	M	R	d_f	$n(d)$ $d = d_f, d_f+1, d_f+2, \ldots$ $w_{CPR}(d)$ $d = d_f, d_f+1, d_f+2, \ldots$ $w_{CPNR}(d)$ $d = d_f, d_f+1, d_f+2, \ldots$ $\rho(d) = w_{CPR}(d)/w_{CPNR}(d)$ $d = d_f, d_f+1, d_f+2, \ldots$
4	(15)	6	1 0 1 1	4/5	3	5, 36, 200, 1,060, 5,795, 31,599, 171,969, 936,526, 5,099,930, 27,771,195, ...
	(17)		1 1 0 0			13, 120, 805, 5,125, 32,599, 202,213, 1,234,855, 7,456,754, 44,587,183, 264,479,172, ...
						14, 194, 1, 579, 11,257, 76,930, 502,739, 3,192,644, 19,869,572, 121,718,261, 736,426,298, ...
						0.93, 0.62, 0.51, 0.46, 0.42, 0.40, 0.39, 0.38, 0.37, 0.36, ...
5	(23)	7	1 0 1 0	4/5	3	8, 50, 421, 3,290, 22,488, 155,980, 1,058,726, 7,128,484, 47,398,486, 313,148,273, ...
	(35)		1 1 0 1			11, 78, 753, 6,890, 51,597, 3,849,852, 729,430, 19,106,443, 130,719,110, 884,972,639, ...
						0.73, 0.64, 0.56, 0.48, 0.44, 0.41, 0.39, 0.37, 0.36, 0.35, ...
6	(53)	8	1 0 0 0	4/5	4	7, 54, 307, 2,005, 12,962, 83,111, 532,859, 3,417,085, 21,921,778, 140,627,199, ...
	(75)		1 1 1 1			23, 224, 1,493, 11,367, 83,962, 604,061, 4,297,152, 30,280,003, 211,707,389, 1,470,048,693, ...
						40, 381, 3,251, 27,123, 213,366, 1,619,872, 11,986,282, 87,121,461, 624,743,990, 4,429,930,822, ...
						0.58, 0.59, 0.46, 0.42, 0.39, 0.37, 0.36, 0.35, 0.34, 0.33, ...

Table 3.10. *Distance spectrum for perforated convolution recursive and non-recursive codes with output of 4/5 and lengths of constraint of 4, 5 and 6*

Initial convolutional code			Perforated convolutional code			
ν	G^1 G^2 In octal	d_f	M	R	d_f	$n(d)\ d = d_f, d_f+1, d_f+2, \ldots$ $w_{CPR}(d)\ d = d_f, d_f+1, d_f+2, \ldots$ $w_{CPNR}(d)\ d = d_f, d_f+1, d_f+2, \ldots$ $\rho(d) = w_{CPR}(d)/w_{CPNR}(d)\ d = d_f, d_f+1, d_f+2, \ldots$
4	(15)	6	1 0 1 0 0	5/6	3	15, 96, 601, 3,835, 24,365, 154,829, 984,015, 6,253,538, 39,742, 549,252, 571,973, ...
	(17)		1 1 0 1 1			40, 333, 2,559, 19,373, 142,498, 1,028,859, 7,322,715, 51,517,991, 359,064,087, 2,483, 109,821, ...
						63, 697, 6,367, 52,924, 415,068, 3,139,106, 23,134,480, 167,262,204, 1,191,612,583, 8,390,366,646, ...
						0.64, 0.48, 0.40, 0.37, 0.34, 0.33, 0.32, 0.31, 0.30, 0.30, ...
5	(23)	7	1 0 1 1 1	5/6	3	5, 37, 309, 2,276, 16,553, 121,552, 893,147, 6,560,388, 48,185,069, 353,907,864, ...
	(35)		1 1 0 0 0			13, 128, 1,340, 11,681, 98,362, 822,267, 6,774,771, 55,136,003, 444,440,714, 3,554,300,708, ...
						20, 265, 3,248, 32,299, 297,308, 2,629,391, 22,591,098, 190,034,783, 1,572,790,875, 12,851,680,889, ...
						0.65, 0.48, 0.41, 0.36, 0.33, 0.31, 0.30, 0.29, 0.28, 0.28, ...
6	(53)	8	1 0 0 0 0	5/6	4	19, 171, 1,251, 9,573, 75,097, 84,394, 4,543,202, 35,354,659, 275,053,493, ...
	(75)		1 1 1 1 1			62, 727, 6,354, 56,387, 505,451, 4,420,332, 38,136,726, 305, 026, 118, 2,387, 410,245, ...
						100, 1,592, 17,441, 166,331, 1,591,180, 14,610,169, 130,823,755, 1,152,346,496, 10,010,105,849, ...
						0.62, 0.46, 0.36, 0.34, 0.32, 0.30, 0.29, 0.27, 0.24, ...

Table 3.11. *Distance spectrum for perforated convolution recursive and non-recursive codes with output of 5/6 and lengths of constraint of 4, 5 and 6*

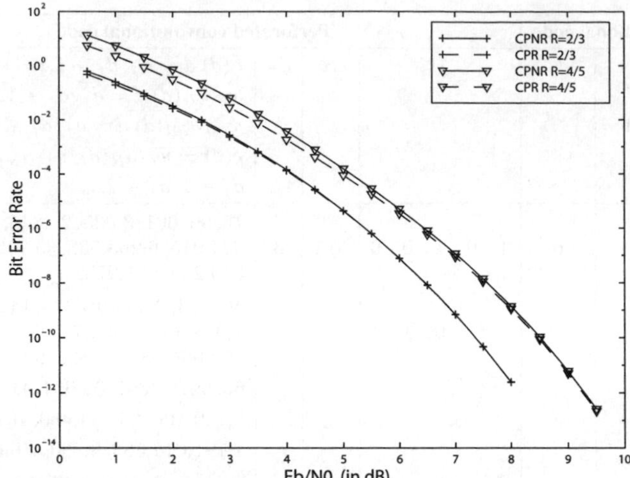

Figure 3.18. *Bit error rates of the non-recursive perforated code with generator polynomials 133–171 (in octal) and of its recursive version for outputs of 2/3 and 4/5*

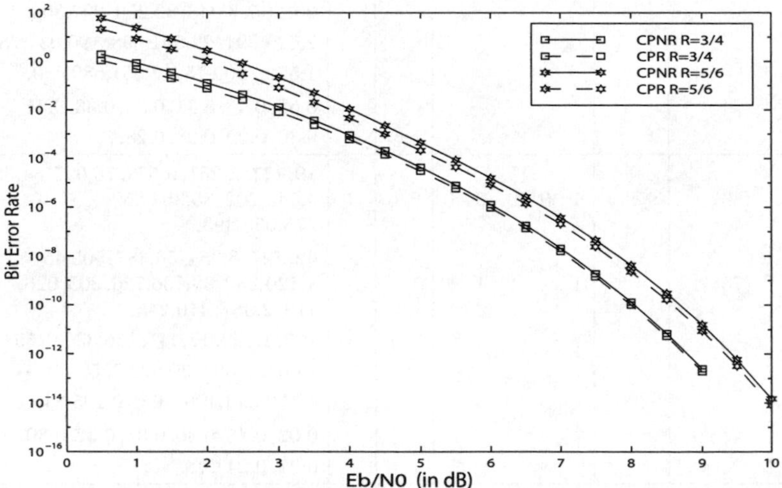

Figure 3.19. *Bit error rates of the non-systematic non-recursive perforated code with generator polynomials 133–171 (in octal) and of its systematic recursive version, for outputs of 3/4 and 5/6*

Chapter 4

Coded Modulations

In traditional transmission systems, information symbols are protected by coding and then a carrier intervenes to modulate. The functions of modulation and coding, and, consequently, of demodulation and decoding, are treated independently. The first examples of codes that combine modulation and coding and demodulation and decoding – and which have led to the concept of *coded modulation* – are the "Trellis Codes" introduced by Gottfried Ungerboeck in 1976. To transmit n bit/symbol with a 2-dimensional modulation, the trellis-coded modulations (TCM) use a constellation with 2^{n+1} points. The redundancy does not cause an expansion of band occupancy, but merely an increase in the size of the constellation. This chapter deals with the most important aspects of coded modulation.

4.1. Hamming distance and Euclidean distance

Binary codes presented in the preceding chapters had been selected for their properties of minimum Hamming distance. However, it is known that for a transmission on a Gaussian channel, the criterion of choosing a modulation scheme is that of Euclidean distance. Indeed, if the signal-to-noise ratio in the channel is sufficiently high, the modulation diagram whose minimum Euclidean distance is the largest will have the weakest probability of error. That leads to the decoding being "with soft decisions", i.e., that the "demodulator" calculates the Euclidean distance (metric) for each decision, while the decoder looks for the codeword with the best metric.

Chapter written by Ezio BIGLIERI.

We can see that if the modulation is binary, the criterion of Hamming distance is equivalent to that of Euclidean distance. Indeed, let us consider an antipodal binary modulation, whose signals $\pm s(t)$ have the same energy $\mathcal{E}$. The square of the Euclidean distance between these two signals is $4\mathcal{E}$. Euclidean distance and Hamming distance are thus proportional. Let us now consider a 4PSK modulation, where the four signals $s_i(t)$, $i = 1, 2, 3, 4$, have the same energy $\mathcal{E}$ and are coded as follows in Table 4.1.

Signal	Phase	Binary characters
$s_1(t)$	0	00
$s_2(t)$	$\pi/2$	01
$s_3(t)$	π	11
$s_4(t)$	$3\pi/2$	10

Table 4.1. *Coding of four signals of a 4PSK modulation*

We can easily see that for this modulation with binary symbols the Euclidean distance also corresponds to the Hamming distance. Indeed, the signals that differ in only one binary character and that thus have a Hamming distance equal to 1 (for example, $s_1(t)$ and $s_2(t)$) have a Euclidean distance of $2\mathcal{E}$, while signals that differ in two binary characters and thus have a Hamming distance equal to 2 (for example, $s_1(t)$ and $s_3(t)$) have a Euclidean distance of $4\mathcal{E}$.

This result cannot be generalized; it is enough to examine the 8PSK modulation to realize that Euclidean distance and Hamming distance are not always dependant. Therefore, the use of a code with a large minimum Hamming distance associated to an 8PSK modulation can lead to a diagram whose minimum Euclidean distance is not the largest. To obtain a system associating coding and powerful modulation for a Gaussian channel it is necessary to code directly in the signals space, i.e. to use a code whose alphabet is formed by signals (or vectors that represents them) and to work directly with the Euclidean distance to optimize the system. It will be said then that we consider the modulator and the encoder jointly: they are joined together in a single block. In the same manner we will not separate the decoder of demodulator functions: we will perform the demodulation and decoding simultaneously.

Introduction to trellis-coded modulation (TCM)

Let us consider a digital transmission via a channel with additive white Gaussian noise. If a signal $\boldsymbol{x}$ is transmitted, the received signal is:

$$\boldsymbol{r} = \boldsymbol{x} + \boldsymbol{n}$$

where $\boldsymbol{n}$ represents a vector whose K components are independent zero-mean random variables with the same variance $N_0/2$. The signal $\boldsymbol{x}$ has K components $s_0, \ldots, s_{K-1}$

whose values belong to a set $\mathcal{S}'$, called *the elementary constellation*, comprising M' signals $\boldsymbol{s}_1, \ldots, \boldsymbol{s}_{M'}$. If the elementary signals have equal probability, the average energy of the transmitted signal will be:

$$\mathcal{E}' = \frac{1}{M'} \sum_{i=1}^{M'} |\boldsymbol{s}_i|^2$$

Now let us consider the transmission of a codeword, comprising the sequence $\boldsymbol{x} = (s_0, \ldots, s_{K-1})$ of K elementary signals. The receiver that minimizes the average probability of error for the sequence at first observes the received sequence $\boldsymbol{r} = (r_0, \ldots, r_{K-1})$, then decides that $\boldsymbol{x} = \hat{\boldsymbol{x}}$ is transmitted if the square of the Euclidean distance:

$$\delta^2 \triangleq \sum_{i=0}^{K-1} |r_i - s_i|^2$$

is minimized taking $s_i = \hat{s}_i, i = 0, \ldots, K-1$. This is equivalent to choosing the codeword $(\hat{s}_0, \ldots, \hat{s}_{K-1})$ closest to the received sequence. As we saw in the previous chapters, the probability of error for the sequence, as well as the probability of error per symbol, is upper bounded for high signal-to-noise ratios by a decreasing function of the ratio $\delta^2_{\text{min}}/N_0$, where δ^2_{min} is the minimum squared Euclidean distance between two sets of signals compatible with the code. Without coding this distance is equal to the minimum distance between the signals of the set $\mathcal{S}'$ (with coding it will be larger).

Coding in signal space consists of choosing the transmitted sequences among the subsets of $\mathcal{S}'^K$. Proceeding in this manner the rate of transmission is reduced, because the number of sequences that we can use is also reduced. To avoid this disadvantage, we can increase the size of $\mathcal{S}'$ by choosing a constellation $\mathcal{S} \supset \mathcal{S}'$ whose size is $M > M'$. We will thus choose M'^K sequences that are subsets of $\mathcal{S}^K$. If this choice is made well, these sequences will have a large minimum distance between them.

We then obtain a minimum distance δ_{free} between two sequences that is larger than the minimum distance δ_{min} between the signals of $\mathcal{S}'$. The reception with maximum likelihood would thus bring a "distance gain": $\delta^2_{\text{free}}/\delta^2_{\text{min}}$.

In order to avoid a reduction of the bit rate, the size of the constellation has been increased by choosing $\mathcal{S}$ instead of $\mathcal{S}'$. This increase can also involve an increase in the energy needed for the transmission ($\mathcal{E}$ instead of $\mathcal{E}'$), and consequently a "loss of energy" $\mathcal{E}/\mathcal{E}'$. We thus define an asymptotic coding gain equal to:

$$\gamma \triangleq \frac{\delta^2_{\text{free}}/\mathcal{E}}{\delta^2_{\text{min}}/\mathcal{E}'} \quad [4.1]$$

where $\mathcal{E}$ and $\mathcal{E}'$ are average energies of the signals transmitted with and without coding.

4.2. Trellis code

The most interesting representation of a code from the point of view of the decoder is the trellis representation. It is a graph that represents the words of code as paths passing through nodes called *encoder states*. The paths consist of a succession of branches where each branch is associated to elementary signal transmitted through the channel. For example, the code with repetition of length 3 will be represented by the trellis in Figure 4.1. The two codewords (s_0, s_0, s_0) and (s_1, s_1, s_1) are represented by two paths of the trellis. The parity code (4,3), whose codewords comprise all the sequences of vectors s_0 and s_1 with an even number of signals s_1, will be represented by the trellis in Figure 4.2. We see here that the encoder has two states, the first corresponding to an even number of s_0 and the second to an odd number. The transmission of a signal s_1 changes the state encoder.

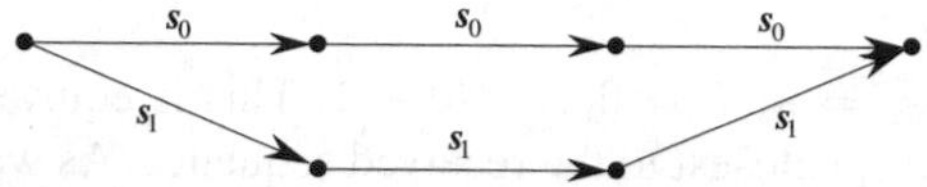

Figure 4.1. *Trellis representation of the code with repetition* $(3, 1)$

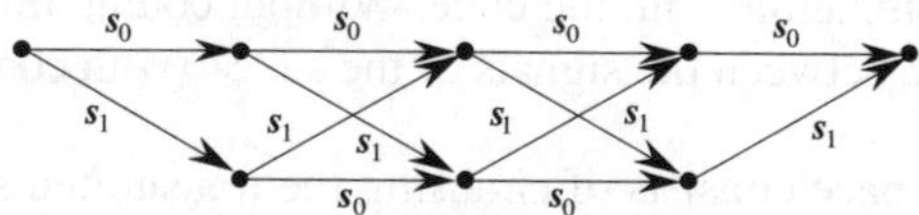

Figure 4.2. *Trellis representation of the parity code* $(4, 3)$

In this context, we may say that a modulation without coding, such as 4PSK modulation, is described by a trellis in a single state. Indeed, the state of the encoder (here of the modulator) does not change from one signal to another. We can also envisage codes whose trellis has "parallel transitions", that is, for which the production of a new symbol does not modify the state of the encoder. An example of a trellis with parallel transitions is represented in Figure 4.3. Here, for each state the encoder can transmit four different signals: s_0, s_1, s_2, s_3.

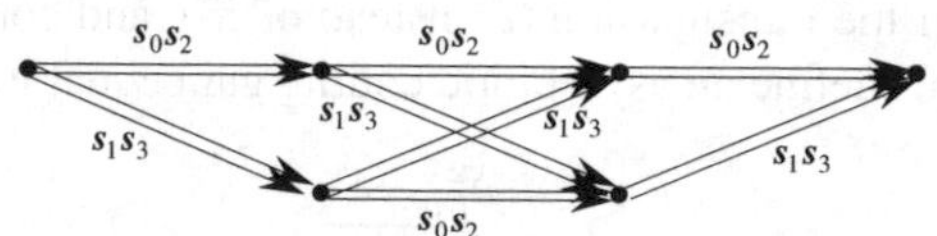

Figure 4.3. *Trellis representation of a code with parallel transitions*

4.3. Decoding

The trellis of a coded modulation, two very simple examples of which we have provided above, describes the correlation between the symbols transmitted through the channel. The process of demodulation/decoding can be understood if we use the same trellis as the one describing coding.

Since there is a bijective correspondence between the sequences of transmitted signals and the paths of the trellis, decoding with maximum likelihood is performed for the channel with additive white Gaussian noise, looking for the path of the trellis that has the shortest Euclidean distance from the received sequence. If we transmit a succession of coded symbols of length K and observe the sequence $r_0, r_1, \ldots, r_{K-1}$ at the output of the channel, the decoder looks for the sequence $s_0, s_1, \ldots, s_{K-1}$ that minimizes:

$$\sum_{i=0}^{K-1} |r_i - s_i|^2$$

That can be done using the Viterbi algorithm.

The branch metrics to be used are obtained as follows. If the branch of the trellis carries the label $\boldsymbol{s}$, then in the discrete time i the metric associated to this branch is $|r_i - \boldsymbol{s}|^2$ if there are no parallel transitions. If two states are connected by parallel transitions and if the branches have the labels, $\boldsymbol{s}', \boldsymbol{s}'', \ldots$ pertaining to the set $\mathcal{S}$, then for the purposes of decoding the trellis will be transformed into a new trellis where the two states are connected by only one branch whose metric would be:

$$\min_{\boldsymbol{s} \in \mathcal{S}} |r_i - \boldsymbol{s}|^2$$

If there are parallel transitions, the decoder first chooses the signal among $\boldsymbol{s}', \boldsymbol{s}'', \ldots,$ which has the smallest distance from r_i (this is a "demodulation"); then it calculates the metric on the basis of the selected signal.

4.4. Some examples of TCM

Let us consider some examples of TCM diagrams and evaluate their coding gain. Let there first be a transmission of 2 bits per symbol. Without coding, a constellation with $M' = 4$ signals would be enough. Let us now examine TCM diagrams with $M = 2M' = 8$ signals, i.e. where coding is obtained thanks to a doubling of the constellation size.

With PSK signals and $M' = 4$ we obtain:

$$\frac{\delta^2_{\min}}{\mathcal{E}'} = 2$$

a value which we will use as a reference to calculate the coding gain of TCM diagrams based on the PSKM. Let us consider TCM diagrams using the 8PSK constellation, whose signals are labeled $\{0, 1, 2, \dots, 7\}$ as indicated in Figure 4.4. We have:

$$\mathcal{E}' = \frac{\delta'^2}{4\sin^2 \pi/8}$$

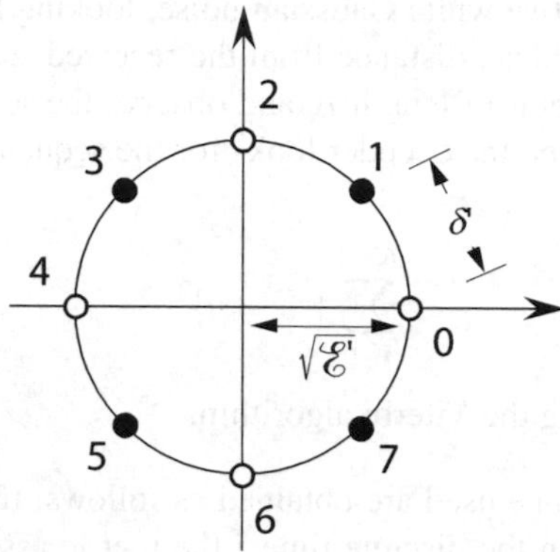

Figure 4.4. *8PSK constellation used in a TCM diagram*

Two states

Let us first consider a diagram with two states (see Figure 4.5). If the encoder is in the state S_1, it uses the sub-constellation $\{0, 2, 4, 6\}$. If it is in the state S_2, it then uses the sub-constellation $\{1, 3, 5, 7\}$. The free distance of this TCM diagram is equal to the shortest distance between the signals associated to parallel transitions (error events of length 1) or between a pair of paths diverging in a node and converging again a few moments after (error events of length 1). The pair of paths that yields the minimum distance called *free distance* is indicated in bold in Figure 4.5. If $\delta(i, j)$ indicates the Euclidean distance between signals i and j, we obtain:

$$\frac{\delta^2_{\text{free}}}{\mathcal{E}} = \frac{1}{\mathcal{E}}[\delta^2(0.2) + \delta^2(0.1)] = 2 + 4\sin^2\frac{\pi}{8} = 2.586$$

There will thus be an asymptotic coding gain with respect to the 4PSK equal to:

$$\gamma = \frac{2.586}{2} = 1.293 \Rightarrow 1.1 \text{ dB}$$

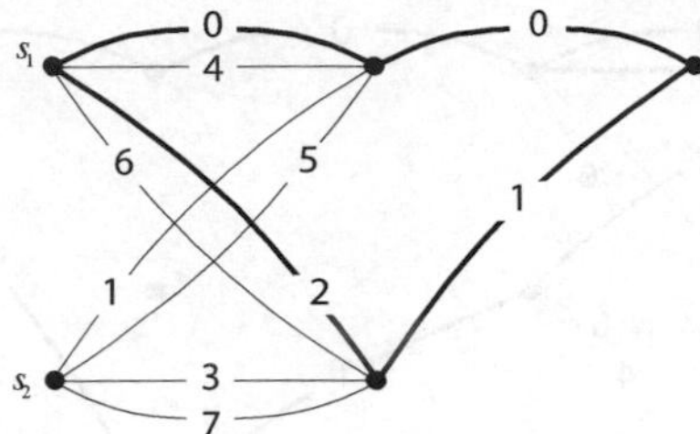

Figure 4.5. *TCM diagram with a trellis with two states, $M' = 4$, and $M = 8$*

Four states

A more complex structure of the TCM trellis would yield a larger coding gain. Still using the constellation in Figure 4.4, let us consider a trellis with 4 states (see Figure 4.6). We can associate the sub-constellation $\{0, 2, 4, 6\}$ to the states S_1 and S_3, and $\{1, 3, 5, 7\}$ to the states S_2 and S_4. In this case the error event that yields the free distance δ_{free} has a length of 1 (parallel transition) and is illustrated in Figure 4.6 in bold. We obtain:

$$\frac{\delta^2_{\text{free}}}{\mathcal{E}} = \delta^2(0.4) = 4$$

It follows that the asymptotic coding gain is:

$$\gamma = \frac{4}{2} = 2 \Rightarrow 3 \text{ dB}$$

Eight states

A way of increasing the coding gain further consists in choosing a trellis with eight states (see Figure 4.7). To simplify Figure 4.7, the four symbols associated to the branches that diverge from a node are indicated at the node level. The first symbol of the four is associated to the transition represented at the top of the figure, the second to the transition immediately below, etc. The error event corresponding to δ_{free} is represented in bold. We obtain:

$$\frac{\delta^2_{\text{free}}}{\mathcal{E}} = \frac{1}{\mathcal{E}}[\delta^2(0.6) + \delta^2(0.7) + \delta^2(0.6)] = 2 + 4\sin^2\frac{\pi}{8} + 2 = 4.586$$

and thus:

$$\gamma = \frac{4.586}{2} = 2.293 \Rightarrow 3.6 \,\text{dB}$$

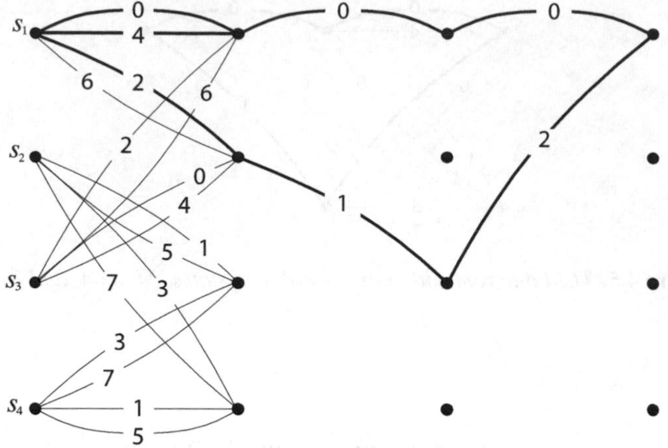

Figure 4.6. *TCM diagram with a trellis with 4 states, $M' = 4$, and $M = 8$*

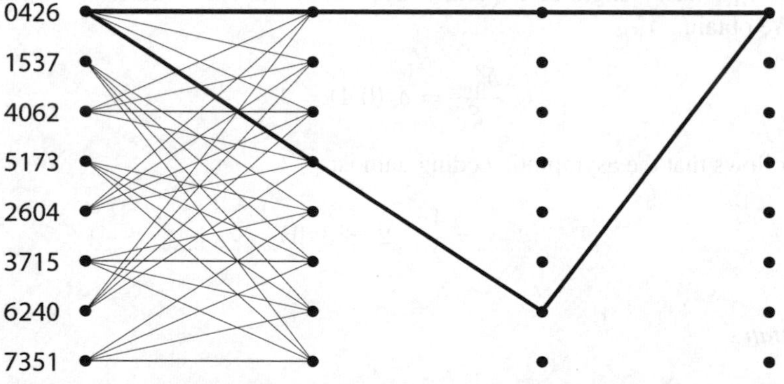

Figure 4.7. *TCM diagram with a trellis with 8 states, $M' = 4$, and $M = 8$*

Coding gains of TCM diagrams

The values of δ_{free} obtained using TCM diagrams of two-dimensional constellations (PSK and QAM) are illustrated in Figure 4.8. The free distances are expressed in dB with respect to the value $\delta^2_{\min} = 2$ correspondent to the non-coded 4PSK. The free distances of several diagrams are given according to R_s/W, spectral output expressed in bits/Hz, for a bandwidth equal to the "Shannon band" $1/T$. We can see that considerable coding gains can be obtained using TCM diagrams with a small number of states: 4, 8 and 16.

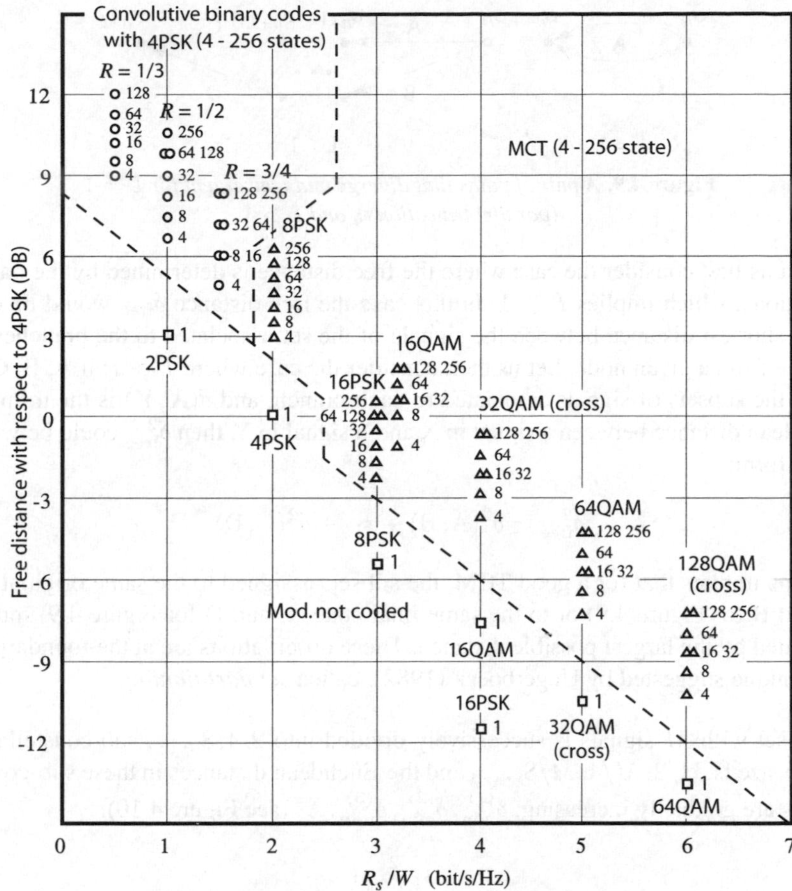

Figure 4.8. *Free distance according to the spectral output for several TCM diagrams based on two-dimensional constellations*

4.5. Choice of a TCM diagram

We can describe a coded modulation diagram on the basis of a trellis by associating each of its branches to an elementary signal. This choice must be made while guaranteeing the largest minimum Euclidean distance.

Partition of a constellation

Let us consider the calculation of the free distance δ_{free}, i.e. the Euclidean distance between a pair of paths that diverge in a node then converge again after L moments (see Figure 4.9).

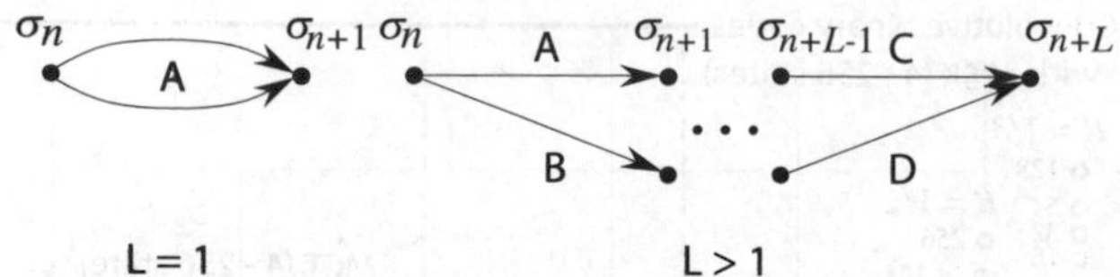

Figure 4.9. *A pair of paths that diverge and convergent for $L = 1$ (parallel transitions), and $L > 1$*

Let us first consider the case where the free distance is determined by the parallel transitions, which implies $L = 1$. In this case the free distance δ_{free} would be equal to the shortest distance between the signals of the set associated to the branches that diverge from a given node. Let us then consider the case where $L > 1$; if A, B, C and D are the subsets of signals associated to each branch, and $\delta(X, Y)$ is the minimum Euclidean distance between a signal in X and a signal in Y, then δ^2_{free} could be written in the form:

$$\delta^2_{\text{free}} = \delta^2(\text{A}, \text{B}) + \cdots + \delta^2(\text{C}, \text{D})$$

This implies that for a good TCM, the subsets assigned to the same original state (A and B for Figure 4.9) or to the same final state (C and D for Figure 4.9) must be separated by the largest possible distance. These observations are at the foundation of a technique suggested by Ungerboeck (1982), called *set partitioning*.

A set with M signals is successively divided into $2, 4, 8, \ldots,$ sub-constellations whose size is $M/2, M/4, M/8, \ldots,$ and the Euclidean distances in these sub-constellations are gradually increasing: $\delta^{(1)}_{\min}, \delta^{(2)}_{\min}, \delta^{(3)}_{\min}, \ldots$ (see Figure 4.10).

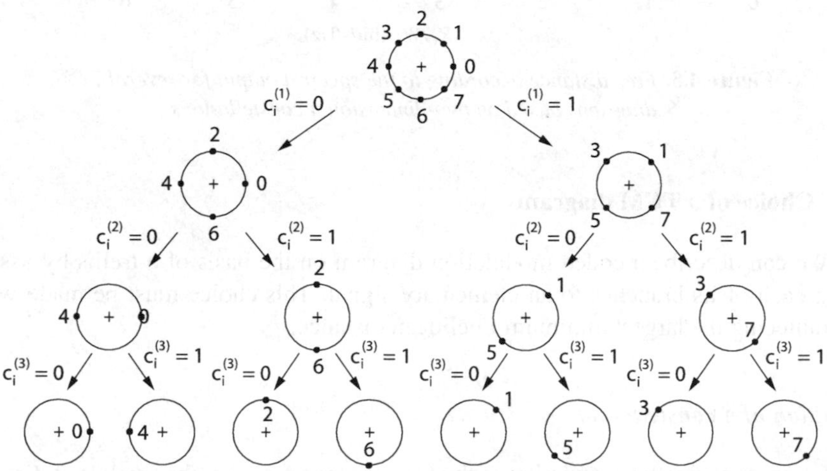

Figure 4.10. *Set partitioning 8PSK*

We will impose certain conditions, called *Ungerboeck conditions* (see conditions 4.1 and 4.2).

CONDITION 4.1 (CONDITION U1). The parallel transitions are associated to signals belonging to the same sub-constellation.

CONDITION 4.2 (CONDITION U2). The branches diverging from the same state or converging towards the same state are associated to the signals belonging to the same sub-constellation of a level superior to that corresponding to condition 4.1.

The two conditions above, plus the condition of symmetry 4.3, lead to the best diagrams of coded modulation.

CONDITION 4.3 (CONDITION U3). All the signals are used with the same frequency.

4.6. TCM representations

We consider here TCM encoders built on the basis of a convolution encoder and a modulation without memory. The source binary symbols are grouped in blocks of m bits, noted $b_i^{(1)}, \ldots, b_i^{(m)}$, and presented at the output of a convolution encoder whose rate of coding is $m/(m+1)$. The latter determines the trellis structure of the TCM diagram (and, in particular, the number of its states). The modulation without memory that follows the encoder generates a bijective correspondence between the binary coded $(m+1)$-tuples and the signals of a constellation with $M = 2^{m+1}$ signals (see Figure 4.11).

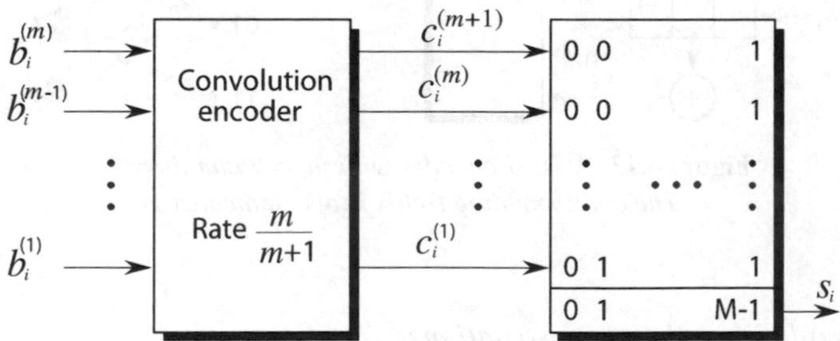

Figure 4.11. *General diagram of a TCM encoder*

It is usually advisable to modify this representation in the manner indicated in Figure 4.12, if certain bits are uncoded. The convolutional code has a rate $\tilde{m}/(\tilde{m}+1)$. The presence of uncoded bits generates parallel transitions; each branch in the trellis of the code is now associated to $2^{m-\tilde{m}}$ signals. The correspondence between the coded bits and the signals of the sub-constellation is indicated in Figure 4.10.

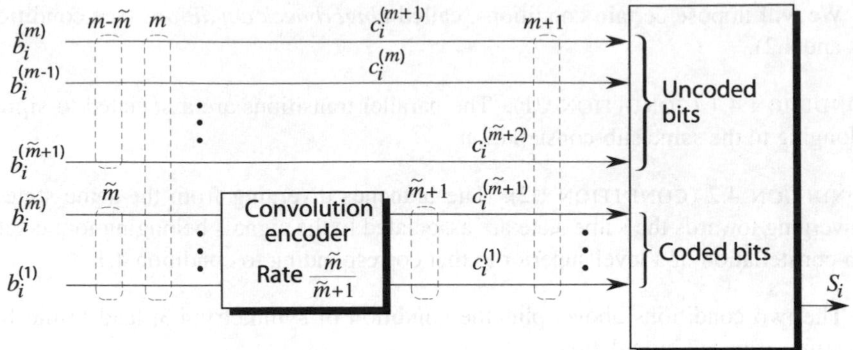

Figure 4.12. *TCM encoder where the uncoded bits are indicated explicitly*

EXAMPLE 4.1. Figure 4.13 shows a TCM encoder and the associated trellis. Here $m = 2$ and $\tilde{m} = 1$, therefore, the nodes of the trellis (corresponding to the encoder states) are connected by parallel transitions, each one associated to two signals. The structure of the trellis is determined by the code.

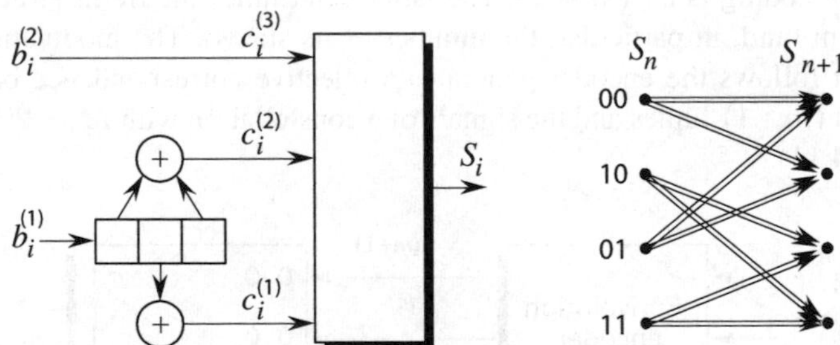

Figure 4.13. *A TCM encoder with $m = 2$ and $\tilde{m} = 1$. The corresponding trellis is also indicated*

TCM with multidimensional constellations

We saw that for a given constellation the performance of a TCM diagram can be improved by increasing the number of states of the trellis. However, when this number exceeds certain values, the coding gain increases by very little. We can then choose to change the elementary constellation. One of the possibilities is to pass from two-dimensional constellations to multidimensional constellations.

Let us consider the constellations generated by concatenating several two-dimensional symbols, such as PSK or QAM symbols. A constellation with $2N$ dimensions can be generated taking N times the Cartesian product of a two-dimensional constellation by itself. If N two-dimensional signals are transmitted in an interval of time of duration T and if each one of them has a duration T/N, we then obtain a constellation with $2N$ dimensions.

EXAMPLE 4.2. A TCM diagram with 4 dimensions can be obtained by concatenating two 4PSK signals. The new constellation will be known as $2 \times$ 4PSK. With the labels indicated in Figure 4.14, the $4^2 = 16$ signals with 4 dimensions are:

$$\{00, 01, 02, 03, 10, 11, 12, 13, 20, 21, 22, 23, 30, 31, 32, 33\}$$

This constellation has the same minimum Euclidean distance as the 4PSK, i.e.:

$$\delta^2_{\min} = \delta^2(00, 01) = \delta^2(0, 1) = 2$$

The following sub-constellation with 8 signals:

$$S = \{00, 02, 11, 13, 20, 22, 31, 33\}$$

has a squared minimum distance equal to 4. If S is divided into four subsets:

$$\{00, 22\} \quad \{20, 02\} \quad \{13, 31\} \quad \{11, 33\}$$

for a trellis with four states, the TCM diagram represented in Figure 4.14 has a squared free distance equal to 8.

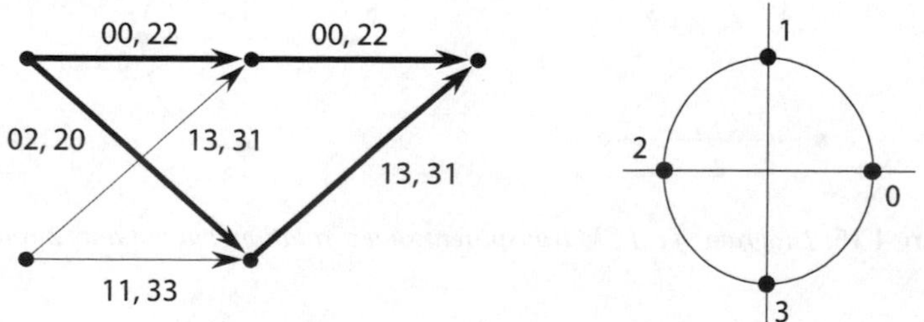

Figure 4.14. *A TCM diagram with two states based on a constellation $2 \times$ 4PSK. The error event that generates the free distance is indicated in bold*

4.7. TCM transparent to rotations

We consider here a channel with *phase offset*, and we are interested in the choice of the TCM for this channel.

Let us consider the example of a PSKM transmission with coherent detection. This detection mode is based on the estimate of the phase of the carrier before demodulation. Several techniques to estimate this phase are based on the elimination of phase jumps ("data noise"). These techniques restore a carrier with a phase ambiguity depending on the constellation used: for example, the phase ambiguity is $\pi/2$ for the QAM modulations and $2\pi/M$ for PSK modulations with M phase states. This ambiguity can be modeled by a random phase jump that takes the values $k2\pi/M, k = 0, 1, \ldots, M - 1$.

To solve this ambiguity, that is, to eliminate this phase displacement, differential coding and decoding are often used. However, when a TCM diagram is used in a transmission system we should ensure that it is transparent with phase rotations by multiples of $2\pi/M$. That means that any coded TCM sequence when affected by a rotation by a multiple of $2\pi/M$, must remain a TCM sequence. Otherwise, a phase rotation can generate a long succession of errors, because, even if there is no noise, the TCM decoder will not recognize the sequence as a TCM sequence.

EXAMPLE 4.3. Let us consider, for example, the trellis in Figure 4.15. It is supposed here that the transmitted sequence is the one, in which all elements are 0. A rotation of π causes the reception of the sequence, in which all elements are 2, which is a TCM sequence. However, a rotation by $\pi/2$ generates the sequence, in which all elements are 1: the latter will not be recognized as a TCM sequence by the Viterbi algorithm (see section 4.2).

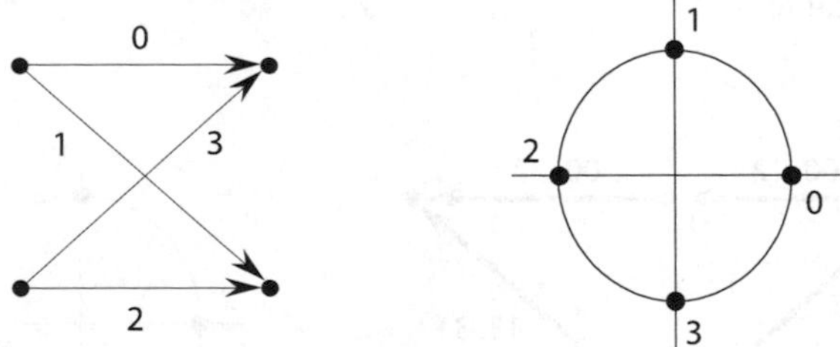

Figure 4.15. *Diagram of a TCM transparent to a π rotation, but not a $\pi/2$ rotation*

The receiver can solve the problem of the ambiguity of phase in several ways. One of them consists of introducing a pilot sequence into the flow of transmitted symbols. A second way consists of using a code whose words are not transparent to rotations. In this case an error of phase could be detected, and thus corrected, by a decoder able to distinguish a TCM sequence, even if it is affected by channel noise, from a sequence invalidated by phase rotation. Finally, a third solution that we will examine consists of using a transparent TCM diagram with rotations. Each sequence having undergone a rotation by a multiple of $2\pi/M$ remains a TCM sequence; the decoder will thus not be affected by this rotation.

If we want a phase offset not to affect a TCM diagram, it is necessary that:

1) the TCM diagram be transparent to phase rotations introduced by the carrier phase estimation circuit;

2) after any rotation, a coded TCM sequence be decoded by the same TCM sequence.

The first of these two properties is geometrical: coded sequences that can be interpreted as a set of points in a Euclidean space with an infinite number of dimensions must be invariant with respect to a certain finite set of rotations. The second property is rather a structural property of the encoder, because it relates to the input-output correspondence determined by the encoder.

4.7.1. *Partitions transparent to rotations*

The first fundamental principle in the construction of a code transparent to rotations is to have a transparent partition.

Let S be a constellation of signals with $2N$ dimensions, $\{Y_1, \ldots, Y_K\}$ be its partition in K subsets, and let us consider the rotations around the point of origin in two-dimensional Euclidean space. Rotations in space with $2N$ dimensions are obtained by considering a separate rotation in each two-dimensional subspace. Let us then consider all of the rotations that leave S unchanged and note this set $R(S)$. If $R(S)$ leaves the partition invariant, i.e. if the effect of each element of $R(S)$ on the partition is simply a permutation of its elements, then the partition is known as *transparent to rotations*.

EXAMPLE 4.4. Let us consider an 8PSK constellation and its partition into 4 signal subsets (see Figure 4.16), and let $R(S)$ be the set of rotations multiples of $\pi/4$, noted $\rho_0, \rho_{\pi/4}, \rho_{\pi/2}$, etc. This partition is transparent to rotations. For example, $\rho_{\pi/4}$ corresponds to the permutation $(Y_1 Y_3 Y_2 Y_4)$, $\rho_{\pi/2}$ to the permutation $(Y_1 Y_2)(Y_3 Y_4)$, ρ_π to the identity permutation, etc.

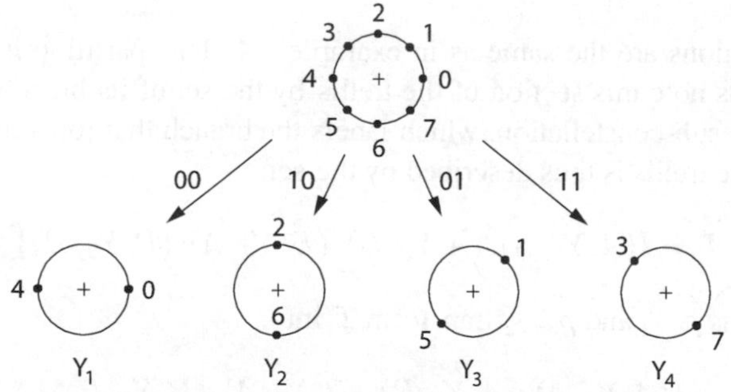

Figure 4.16. *Partition of the 8PSK transparent to rotations*

EXAMPLE 4.5. Let us consider the set of signals of dimension 4 from example 4.2 ($2 \times$ 4PSK), and its partition into four sub-constellations:

$$Y = \{Y_1, Y_2, Y_3, Y_4, Y_5, Y_6, Y_7, Y_8\} \quad [4.2]$$

$$= \{\{00, 22\}, \{11, 33\}, \{02, 20\}, \{13, 31\},$$

$$\{01, 23\}, \{12, 30\}, \{03, 21\}, \{10, 32\}\} \quad [4.3]$$

The elements of $R(\mathcal{S})$ are the rotation pairs, each one a multiple of $\pi/2$, noted $\rho_0, \rho_{\pi/2}, \rho_\pi$ and $\rho_{3\pi/2}$. For example, we can see that the effect of $\rho_{\pi/2}$ on the signal xy, where $x, y = 0, 1, 2, 3$, is to change it into a signal $(x+1)(y+1)$, where the addition is done by *modulo* 4. This partition is transparent to rotations.

4.7.2. *Transparent trellis with rotations*

Now let us consider the effect of a rotation of phase on coded TCM sequences. If the partition Y of $\mathcal{X}$ is transparent to rotations, the TCM becomes transparent to rotations, if for any rotation $\rho \in R(\mathcal{S})$ the sequences of sub-constellations are transformed into sequences of sub-constellations compatible with the code.

Now let us examine a section of the trellis representing the TCM. If all the sub-constellations that label the branches of the trellis are affected by the same rotation ρ, a new section of the trellis is obtained. However, so that the TCM is transparent to rotations, this new section of the trellis must be identical to the initial section (i.e., without rotation) except for a permutation of its states.

EXAMPLE 4.6. Let us consider a section of a trellis in two states (see Figure 4.17) having a base partition:

$$Y = \{Y_1, Y_2, Y_3, Y_4\}$$

where the notations are the same as in example 4.4. This partition is transparent to rotations. Let us note this section of the trellis by the set of its branches (s_i, Y_j, s_k), where Y_j is the sub-constellation, which labels the branch that joins the state s_j with the state s_k. The trellis is thus described by the set:

$$\mathcal{T} = \{(A, Y_1, A), (A, Y_3, B), (B, Y_4, A), (B, Y_2, B)\}$$

The rotations $\rho_{\pi/2}$ and $\rho_{3\pi/2}$ transform $\mathcal{T}$ into:

$$\{(A, Y_2, A), (A, Y_4, B), (B, Y_3, A), (B, Y_1, B)\}$$

which corresponds to the permutation (A, B) of the states of $\mathcal{T}$. Similarly, ρ_0 and ρ_π correspond to the identity permutation.

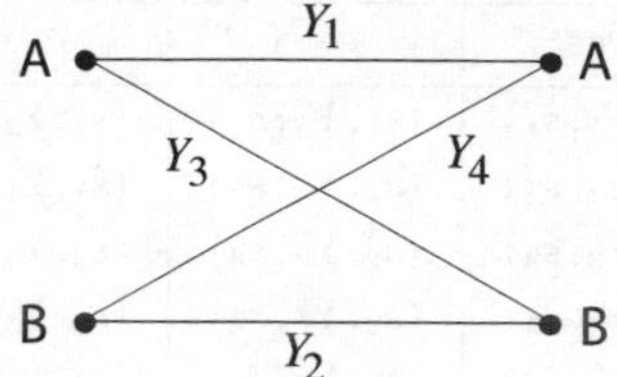

Figure 4.17. *Section of a trellis with two states*

In conclusion, the TCM code is transparent to rotations.

It may be that a TCM satisfies the conditions of rotational transparency only with respect to one subset of $R(\mathcal{S})$, and not the entire $R(\mathcal{S})$. In this case we say that $\mathcal{S}$ is *partially* transparent to rotations.

EXAMPLE 4.7. Let us consider the TCM in Figure 4.18. It has 8 states and its sub-constellations correspond to the partition $Y = \{Y_1, Y_2, Y_3, Y_4\}$ from example 4.6. This partition, as we know, is transparent to rotations. However, the TCM is not transparent. To demonstrate it, let us consider the effect of a rotation by $\pi/4$ (see Table 4.2); we can see that $\rho_{\pi/4}$ generates a single simple permutation of the states of the trellis. Indeed, let us take, for example, the branch (s_1, Y_3, s_1); in the initial trellis there are no states of the type (s_i, Y_3, s_i). This TCM is, however, *partially* invariant: indeed, it is transparent to rotations multiples of $\pi/2$. For example, the effect of $\rho_{\pi/2}$ is described in Table 4.2: it causes the following permutation of its states: $(s_1 s_8)(s_2 s_7)(s_3 s_6)(s_4 s_5)$.

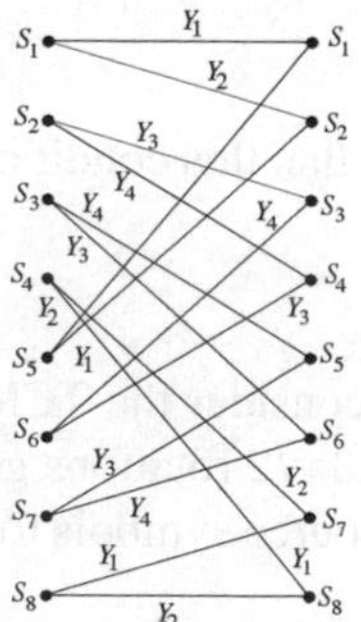

Figure 4.18. *TCM with 8 states non-transparent to rotations*

4.7.3. *Transparent encoder*

Let us now consider the transparency of the encoder, i.e. the property according to which any rotation of a TCM sequence corresponds to the same information

ρ_0	$\rho_{\pi/4}$	$\rho_{\pi/2}$
(s_1, Y_1, s_1)	(s_1, Y_3, s_1)	(s_1, Y_2, s_1)
(s_1, Y_2, s_2)	(s_1, Y_4, s_2)	(s_1, Y_1, s_2)
(s_2, Y_3, s_3)	(s_2, Y_2, s_3)	(s_2, Y_4, s_3)
(s_2, Y_4, s_4)	(s_2, Y_1, s_4)	(s_2, Y_3, s_4)
(s_3, Y_4, s_5)	(s_3, Y_1, s_5)	(s_3, Y_3, s_5)
(s_3, Y_3, s_6)	(s_3, Y_2, s_6)	(s_3, Y_4, s_6)
(s_4, Y_2, s_7)	(s_4, Y_4, s_7)	(s_4, Y_1, s_7)
(s_4, Y_1, s_8)	(s_4, Y_3, s_8)	(s_4, Y_2, s_8)
(s_5, Y_2, s_1)	(s_5, Y_4, s_1)	(s_5, Y_1, s_1)
(s_5, Y_1, s_2)	(s_5, Y_3, s_2)	(s_5, Y_2, s_2)
(s_6, Y_4, s_3)	(s_6, Y_1, s_3)	(s_6, Y_3, s_3)
(s_6, Y_3, s_4)	(s_6, Y_2, s_4)	(s_6, Y_4, s_4)
(s_7, Y_3, s_5)	(s_7, Y_2, s_5)	(s_7, Y_4, s_5)
(s_7, Y_4, s_6)	(s_7, Y_1, s_6)	(s_7, Y_3, s_6)
(s_8, Y_1, s_7)	(s_8, Y_3, s_7)	(s_8, Y_2, s_7)
(s_8, Y_2, s_8)	(s_8, Y_4, s_8)	(s_8, Y_1, s_8)

Table 4.2. *Effect of $\pi/4$ and $\pi/2$ rotations on the TCM of example 4.7*

sequence. If $\boldsymbol{u}$ is a sequence of source symbols and $\boldsymbol{y}$ is a corresponding sequence of sub-constellations, then any rotation $\rho(\boldsymbol{y})$ for which the TCM is transparent, must correspond to the same sequence $\boldsymbol{u}$.

We will be able to observe that this condition is sometimes satisfied by introducing a differential encoder.

EXAMPLE 4.8. Let us again consider the TCM with 8 states and the 8PSK of example 4.7. We saw that $\pi/2$ and $3\pi/2$ rotations generate the permutation $(Y_1 Y_2)(Y_3 Y_4)$. If the encoder associates the source symbols to the sub-constellations according to the rule:

$$00 \Rightarrow Y_1 \qquad 10 \Rightarrow Y_2 \qquad 01 \Rightarrow Y_3 \qquad 11 \Rightarrow Y_4$$

then the only effect of $\rho_{\pi/2}$ and $\rho_{3\pi/2}$ is to change the first bit of the pair of binary source symbols, while ρ_0 and ρ_π change nothing. Therefore, if the first bit is subjected to differential coding, the TCM will be transparent to rotations multiple of $\pi/2$.

4.7.4. *General considerations*

We can say in general that rotational transparency constraints may lead to a reduction of the coding gain of the two-dimensional TCM. The use of a decoder transparent to rotations may require the use of a non-linear convolutional code. The loss of performance caused by transparency constraints is generally weaker for multidimensional constellations.

4.8. TCM error probability

This section is dedicated to the calculation of the probability of error of a TCM. The assumption is that the transmission is carried out via a channel with additive white Gaussian noise, and that the reception is at maximum probability. The reader will not be surprised, if, asymptotically for a high signal-to-noise ratio, the upper and lower bounds of the probability of error decrease as δ_{free} increases. This proves that the Euclidean free distance is the most relevant parameter for a comparison of TCM in the Gaussian channel with a high signal-to-noise ratio. This also explains why the parameter γ, the ratio between the free distance between the coded constellation and the minimum distance from the uncoded constellation, is called *asymptotic coding gain.*

Since there are no techniques that optimally choose a TCM, this choice will be based on a study of a class of codes. It is thus very important to have a fast and effective algorithm to calculate the free distance and the probability of error.

4.8.1. *Upper bound of the probability of an error event*

A convolutional code whose rate is $m/(m+1)$ simultaneously receives m binary symbols from source $\boldsymbol{b}_i$, and transforms them into $\boldsymbol{c}_i$ blocks with $m+1$ binary symbols, each constituting the entry to a non-linear system without memory (see Figure 4.11). This system generates at its output the elementary signals s_i. As of now, the $(m+1)$-tuple binary $\boldsymbol{c}_i$ will be called *label* of the signal s_i.

There is a bijective correspondence between s_i and its label c_i. Thus, two sequences of L signals can be described equivalently by the two sequences of their labels, i.e.:

$$\boldsymbol{c}_k, \boldsymbol{c}_{k+1}, \ldots, \boldsymbol{c}_{k+L-1}$$

and:

$$\boldsymbol{c}_k' = \boldsymbol{c}_k \oplus \boldsymbol{e}_k, \quad \boldsymbol{c}_{k+1}' = \boldsymbol{c}_{k+1} \oplus \boldsymbol{e}_{k+1}, \ldots, \boldsymbol{c}_{k+L-1}' = \boldsymbol{c}_{k+L-1} \oplus \boldsymbol{e}_{k+L-1}$$

where $e_i, i = k, \ldots, k+L-1$, form a series of binary vectors, that we will call *error vectors*, and $\oplus$ indicates *modulo*-2 addition.

Let $\boldsymbol{X}_L$ and $\hat{\boldsymbol{X}}_L$ be two sequences of L elementary signals. An *error event* of length L occurs when the demodulator instead of the transmitted sequence $\boldsymbol{X}_L$ chooses a different sequence $\hat{\boldsymbol{X}}_L$ corresponding to a path in the trellis that diverges from the correct path at a certain moment, and converges with it exactly L moments later. The probability of error will thus be obtained by adding in $L, L = 1, 2, \ldots,$ the probabilities of the error events of length L, i.e. the joint probabilities that $\boldsymbol{X}_L$ is transmitted and that $\hat{\boldsymbol{X}}_L$ is decided.

The union bound gives the following inequality for the probability of an error event:

$$P(e) \leq \frac{1}{N_\sigma} \sum_{L=1}^{\infty} \sum_{\boldsymbol{X}_L} \sum_{\hat{\boldsymbol{X}}_L \neq \boldsymbol{X}_L} P\{\boldsymbol{X}_L\} P\{\boldsymbol{X}_L \to \hat{\boldsymbol{X}}_L\} \qquad [4.4]$$

Division by N_σ, the number of states of the code trellis, makes it possible to obtain the probability of error *per state*.

Once again exploiting the bijective correspondence between the output symbols and the labels, if $\boldsymbol{C}_L$ indicates a series of labels $\boldsymbol{c}_i$ of length L, and $\boldsymbol{E}_L$ designates a series (still of length L) of error vector $\boldsymbol{e}_i$, we can write [4.4]:

$$\begin{aligned} P(e) &\leq \frac{1}{N_\sigma} \sum_{L=1}^{\infty} \sum_{\boldsymbol{C}_L} P\{\boldsymbol{C}_L\} \sum_{\hat{\boldsymbol{C}}_L \neq \boldsymbol{C}_L} P\{\boldsymbol{C}_L \to \hat{\boldsymbol{C}}_L\} \\ &= \frac{1}{N_\sigma} \sum_{L=1}^{\infty} \sum_{\boldsymbol{C}_L} P\{\boldsymbol{C}_L\} \sum_{\boldsymbol{E}_L \neq \boldsymbol{0}} P\{\boldsymbol{C}_L \to \boldsymbol{C}_L \oplus \boldsymbol{E}_L\} \\ &= \frac{1}{N_\sigma} \sum_{L=1}^{\infty} \sum_{\boldsymbol{E}_L \neq \boldsymbol{0}} P\{\boldsymbol{E}_L\} \end{aligned} \qquad [4.5]$$

where:

$$P\{\boldsymbol{E}_L\} \triangleq \sum_{\boldsymbol{C}_L} P\{\boldsymbol{C}_L\} P\{\boldsymbol{C}_L \to \boldsymbol{C}_L \oplus \boldsymbol{E}_L\} \qquad [4.6]$$

expresses the pair-wise error probability of particular error events of length L generated by the sequence of errors $\boldsymbol{E}_L$. The pair-wise error probability that appears in the last equation can be calculated exactly. However, we will not do it and will rather use a bound that leads to the Bhattacharyya bound.

Let $f(\boldsymbol{c})$ be the signal whose label is $\boldsymbol{c}$, and $f(\boldsymbol{C}_L)$ be the sequence of signals whose label is $\boldsymbol{C}_L$. We then have:

$$\begin{aligned} P\{\boldsymbol{C}_L \to \hat{\boldsymbol{C}}_L\} &= \frac{1}{2}\operatorname{erfc}\left(\frac{|f(\boldsymbol{C}_L) - f(\hat{\boldsymbol{C}}_L)|}{2\sqrt{N_0}}\right) \\ &\leq \frac{1}{2}\exp\left\{-\frac{1}{4N_0}|f(\boldsymbol{C}_L) - f(\hat{\boldsymbol{C}}_L)|^2\right\} \\ &= \frac{1}{2}\exp\left\{-\frac{1}{4N_0}\sum_{n=1}^{L}|f(\boldsymbol{c}_n) - f(\hat{\boldsymbol{c}}_n)|^2\right\} \end{aligned} \quad [4.7]$$

Let us now define the function:

$$W(\boldsymbol{E}_L) \triangleq \sum_{\boldsymbol{C}_L} P\{\boldsymbol{C}_L\}\mathrm{e}^{-|f(\boldsymbol{C}_L) - f(\boldsymbol{C}_L \oplus \boldsymbol{E}_L)|^2/4N_0} \quad [4.8]$$

If we observe that $P\{\boldsymbol{C}_L\} = P\{\boldsymbol{X}_L\}$, equation [4.4] can be written in the following form:

$$P(e) \leq \frac{1}{2N_\sigma}\sum_{L=1}^{\infty}\sum_{\boldsymbol{E}_L \neq \boldsymbol{0}} W(\boldsymbol{E}_L) \quad [4.9]$$

The expression [4.9] shows that $P(e)$ is upper bounded by a sum of the functions of $\boldsymbol{E}_L$ vectors that generate error events, extended to all their possible lengths. We will thus have to enumerate these vectors. Before continuing, let us note that a technique often used for the calculation of the probability of error (in particular, for TCM codes with a great number of states, or for the transmission in channels that are not Gaussian) is to include in [4.9] a finite number of terms chosen among those with a small value of L. Since we expect that these error events generate minimum distances, they will contribute the most to the probability of error events. This technique will have to be used with caution, because the truncation of the series associated to the upper bound does not necessarily yield an upper bound.

4.8.1.1. *Enumeration of error events*

We can enumerate all the vectors of error by using the transfer function of a diagram of the error states, i.e. a graph whose branch labels are matrices $N_\sigma \times N_\sigma$, where N_σ is the number of states of the trellis. In particular, let us recall that under our assumptions all the source symbols have the same probability 2^{-m}, and let us define the matrix $N_\sigma \times N_\sigma$ "of the error weights" $\boldsymbol{G}(\boldsymbol{e}_n)$ in the following manner. The component i, j of $\boldsymbol{G}(\boldsymbol{e}_n)$ is equal to zero if there is no transition between states i and j of the trellis; otherwise it has the expression:

$$[\boldsymbol{G}(\boldsymbol{e}_n)]_{i,j} = 2^{-m}\sum_{\boldsymbol{c}_{i\to j}} Z^{|f(\boldsymbol{c}_{i\to j}) - f(\boldsymbol{c}_{i\to j}\oplus\boldsymbol{e}_n)|^2} \quad [4.10]$$

where $\boldsymbol{c}_{i \to j}$ are the vectors of the labels generated by the transition from state i to the state j (the sum takes into account the parallel transitions that can exist between these two states).

To each sequence $\boldsymbol{E}_L = \boldsymbol{e}_1, \ldots, \boldsymbol{e}_L$ of labels in the diagram of error state there corresponds a sequence of L matrices of error weights $\boldsymbol{G}(\boldsymbol{e}_1), \ldots, \boldsymbol{G}(\boldsymbol{e}_L)$, with:

$$W(\boldsymbol{E}_L) = \mathbf{1}' \left[\prod_{n=1}^{L} \boldsymbol{G}(\boldsymbol{e}_n) \right] \mathbf{1} \Bigg|_{Z = \mathrm{e}^{-1/4N_0}} \qquad [4.11]$$

where $\mathbf{1}$ is the column vector whose N_σ elements take the value 1, and $\mathbf{1}'$ is its transpose. Consequently, if we call $\boldsymbol{A}$ a matrix $N_\sigma \times N_\sigma$, then $\mathbf{1}'\boldsymbol{A}1$ will be the sum of all the elements of $\boldsymbol{A}$. The element i, j of the matrix $\prod_{n=1}^{L} \boldsymbol{G}(\boldsymbol{e}_n)$ enumerates the Euclidean distances generated by the transitions from state i towards the state j in exactly L stages.

Now, to calculate $P(e)$ it is necessary to sum $W(\boldsymbol{E}_L)$ for all the possible error sequences $\boldsymbol{E}_L$ using [4.9].

4.8.1.1.1. The error state diagram

Since the convolutional code that generates the TCM is linear, the set of the possible sequences $\boldsymbol{e}_1, \ldots, \boldsymbol{e}_L$ is identical to the set of coded sequences. Therefore, the error sequences can be described using the trellis associated to the encoder and can be enumerated using a diagram of state that is a copy of the one describing the code. We call this diagram the *error state diagram*. Its structure is only determined by the convolutional code, and differs from the code state diagram only in so far as the branch labels, which are now the $\boldsymbol{G}(\boldsymbol{e}_i)$ matrices.

4.8.1.1.2. The bound resulting from the transfer function

Using [4.11] and [4.9] we can write:

$$P(e) \leq \frac{1}{2N_\sigma} T(Z) \Bigg|_{Z = \exp(-1/4N_0)} \qquad [4.12]$$

where:

$$T(Z) = \mathbf{1}'\boldsymbol{G}\mathbf{1} \qquad [4.13]$$

and the matrix:

$$\boldsymbol{G} \triangleq \sum_{L=1}^{\infty} \sum_{\boldsymbol{E}_L \neq \mathbf{0}} \prod_{n=1}^{L} \boldsymbol{G}(\boldsymbol{e}_n) \qquad [4.14]$$

is the transfer function of the error state diagram. We will call $T(Z)$ the scalar transfer function of the error state diagram.

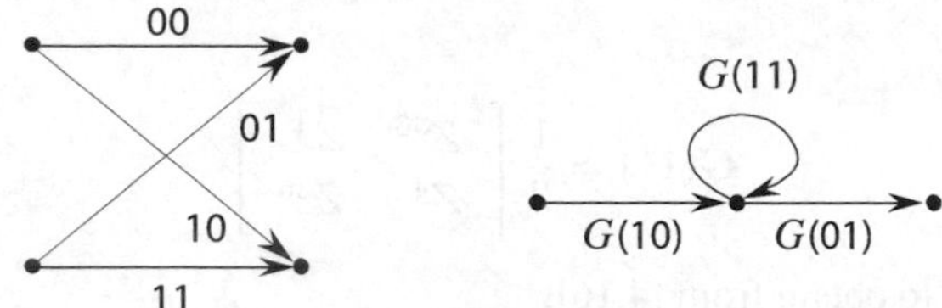

Figure 4.19. *Trellis diagram of a TCM with two states and $m = 1$ and the error state diagram*

EXAMPLE 4.9. Let us consider the TCM whose section of the trellis diagram is represented in Figure 4.19 where $m = 1$ and $M = 4$ (binary source, signals at 4 levels). The error state diagram is also represented in this figure. If we note by $\boldsymbol{e} = (e_2 e_1)$ the error vector and if $\bar{e} = 1 \oplus e$ ($\bar{e}$ is the complement of e), we can write the general shape of the matrix $\boldsymbol{G}(\boldsymbol{e})$ in the following fashion:

$$\boldsymbol{G}(e_2 e_1) = \frac{1}{2} \begin{bmatrix} Z^{\|f(00)-f(e_2 e_1)\|^2} & Z^{\|f(10)-f(\bar{e}_2 e_1)\|^2} \\ Z^{\|f(01)-f(e_2 \bar{e}_1)\|^2} & Z^{\|f(11)-f(\bar{e}_2 \bar{e}_1)\|^2} \end{bmatrix} \qquad [4.15]$$

The transfer function of the error state diagram is:

$$\boldsymbol{G} = \boldsymbol{G}(10)\, [\boldsymbol{I} - \boldsymbol{G}(11)]^{-1}\, \boldsymbol{G}(01) \qquad [4.16]$$

where $\boldsymbol{I}$ symbolizes the identity matrix 2×2.

We can observe that [4.15] and [4.16] can be written without knowing the signals used by the TCM. Indeed, providing the constellation of the signals amounts to providing the four values of the function $f(\cdot)$. In turn, these values yield those of the elements of $\boldsymbol{G}(e_2 e_1)$, for which the transfer function $T(Z)$ is calculated.

First of all let us consider a pulse amplitude modulation (IAM) with four states, with the following correspondence:

$$f(00) = +3 \quad f(01) = 1 \quad f(10) = -1 \quad f(11) = -3$$

In this case we have:

$$\boldsymbol{G}(00) = \frac{1}{2} \begin{bmatrix} 1 & 1 \\ 1 & 1 \end{bmatrix} \qquad [4.17]$$

$$\boldsymbol{G}(01) = \frac{1}{2} \begin{bmatrix} Z^4 & Z^4 \\ Z^4 & Z^4 \end{bmatrix} \qquad [4.18]$$

$$\boldsymbol{G}(10) = \frac{1}{2} \begin{bmatrix} Z^{16} & Z^{16} \\ Z^{16} & Z^{16} \end{bmatrix} \qquad [4.19]$$

and:

$$G(11) = \frac{1}{2}\begin{bmatrix} Z^{36} & Z^4 \\ Z^4 & Z^{36} \end{bmatrix} \tag{4.20}$$

which enables us to obtain from [4.16]:

$$G = \frac{1}{2}\frac{Z^{20}}{1-\frac{1}{2}(Z^4+Z^{36})}\begin{bmatrix} 1 & 1 \\ 1 & 1 \end{bmatrix} \tag{4.21}$$

and finally the transfer function has the expression:

$$T(Z) = \frac{1}{2}\mathbf{1}'\boldsymbol{G}\mathbf{1} = \frac{Z^{20}}{1-\frac{1}{2}(Z^4+Z^{36})} \tag{4.22}$$

If we consider a 4PSK constellation with unitary energy as represented in Figure 4.20, we will have:

$$f(00)=1 \quad f(01)=j \quad f(10)=-1 \quad f(11)=-j$$

and thus:

$$G(00) = \frac{1}{2}\begin{bmatrix} 1 & 1 \\ 1 & 1 \end{bmatrix} \tag{4.23}$$

$$G(01) = \frac{1}{2}\begin{bmatrix} Z^2 & Z^2 \\ Z^2 & Z^2 \end{bmatrix} \tag{4.24}$$

$$G(10) = \frac{1}{2}\begin{bmatrix} Z^4 & Z^4 \\ Z^4 & Z^4 \end{bmatrix} \tag{4.25}$$

and:

$$G(11) = \frac{1}{2}\begin{bmatrix} Z^2 & Z^2 \\ Z^2 & Z^2 \end{bmatrix} \tag{4.26}$$

Finally:

$$G = \frac{1}{2}\frac{Z^6}{1-Z^2}\begin{bmatrix} 1 & 1 \\ 1 & 1 \end{bmatrix} \tag{4.27}$$

and, thus, the transfer function is equal to:

$$T(Z) = \frac{1}{2}\mathbf{1}'\boldsymbol{G}\mathbf{1} = \frac{Z^6}{1-Z^2} \tag{4.28}$$

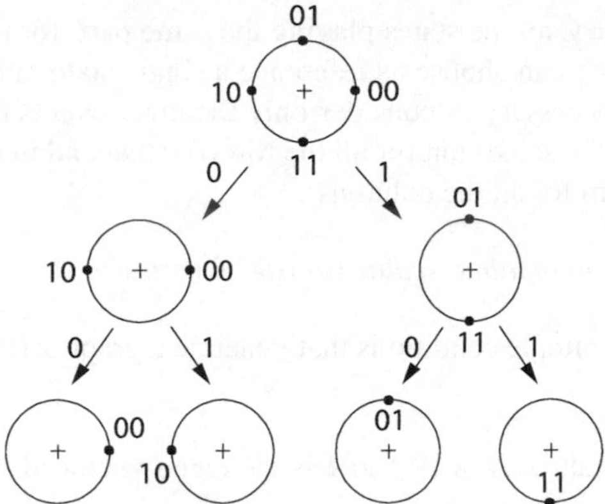

Figure 4.20. *Constellation of signals (4PSK) and its partition*

4.8.1.2. *Interpretation and symmetry*

Examining the matrix $\boldsymbol{G}$ defined in [4.14] we may observe that its element i, j provides us with an upper bound of the probability than an error event starts at node i and finishes at node j. Similarly, $\boldsymbol{G1}$ is a vector whose element i is a bound of the probability of an error event that starts at node i, and $\boldsymbol{1}'\boldsymbol{G}$ is a vector whose element j is a bound of the probability of all the error events that end at node j.

In matrix $\boldsymbol{G}$ we can observe various degrees of symmetry implicated in a TCM. It may so occur that all the elements of the matrix $\boldsymbol{G}$ are equal: such is the case of the example of 4PSK above. This fact can be interpreted by saying that all the paths in the trellis contribute equally to the probability of error (more precisely, we should say that they contribute *equally to the upper bound* of the probability of error). In the context of the analysis of a TCM we will be able to take a single path as reference and calculate the probability of error knowing that the transmitted signal is the one corresponding to this path. A sufficient condition for this to occur is that all the matrices $\boldsymbol{G}(e)$ have equal components. However, this condition is not necessary, as we may by considering the example of IAM with four states: the matrix $\boldsymbol{G}$ has equal elements although the components of $\boldsymbol{G}(11)$ are not identical.

If all the matrices $\boldsymbol{G}(e)$ have equal components, the branches of the error state diagram in the calculation of the bound based on the transfer function can be simply labeled by the common component of these matrices, leading to a *scalar* transfer function. However, to obtain this result it is not necessary that the degree of symmetry of the code be high. All that is needed is to have a weaker symmetry: the sum of all the elements of a row (or column) of $\boldsymbol{G}$ is identical for all the rows (or columns).

With this symmetry, all the states playing the same part, for the calculation of the probability of error we can choose as reference a single state rather than all the pairs of states. It will be necessary to consider only the error events that start in a certain state (when we have the same sum for all the rows) or that end in the same state (when we have the same sum for all the columns).

Algebraic conditions to obtain a scalar transfer function

We will establish simple conditions that generate a graph whose labels are *scalars* and not matrices.

If $\boldsymbol{A}$ is a square matrix $N \times N$ and $\mathbf{1}$ is the eigen-vector of its transpose $\boldsymbol{A}'$, that is:

$$\mathbf{1}'\boldsymbol{A} = \alpha\mathbf{1}'$$

where α is a constant, the sum of the components of any column of $\boldsymbol{A}$ does not depend on the column order. It will be said that $\boldsymbol{A}$ is *uniform by columns*. Similarly, if $\mathbf{1}$ is an eigen-vector of the square matrix $\boldsymbol{B}$, i.e. if:

$$\boldsymbol{B}\mathbf{1} = \beta\mathbf{1}$$

where β is a constant, then the sum of the components of any row does not depend on the row. In this case it is said that $\boldsymbol{B}$ is *uniform by rows*.

However, the product and the sum of two uniform matrices (by lines or by columns) are uniform matrices. For example, if $\boldsymbol{B}_1$ and $\boldsymbol{B}_2$ are uniform by rows with respective eigen-values β_1 and β_2, then $\boldsymbol{B}_3 \triangleq \boldsymbol{B}_1 + \boldsymbol{B}_2$ and $\boldsymbol{B}_4 \triangleq \boldsymbol{B}_1\boldsymbol{B}_2$ verifies the following relations:

$$\boldsymbol{B}_3\mathbf{1} = (\beta_1 + \beta_2)\mathbf{1}$$

and:

$$\boldsymbol{B}_4\mathbf{1} = \beta_1\beta_2\mathbf{1}$$

which shows that $\boldsymbol{B}_3$ and $\boldsymbol{B}_4$ are also uniform by rows, with respective eigen-values $\beta_1 + \beta_2$ and $\beta_1\beta_2$. Moreover, for a matrix $\boldsymbol{A}$ of a uniform order N (by rows or by columns) we have:

$$\mathbf{1}'\boldsymbol{A}\mathbf{1} = N\alpha$$

It follows from the above that if all the matrices $\boldsymbol{G}(e)$ are uniform by rows, or uniform by columns, the transfer function (which is a sum of products of error matrices, as we can see from [4.14]) can be calculated using only scalar labels in the error state diagram. These labels are the sums of the components of a row (or a column), and we say that the TCM is *uniform*. According to the definition of matrices $\boldsymbol{G}(e)$, we can observe that $\boldsymbol{G}(e)$ is uniform by rows if the transitions that diverge from any trellis node carry the same set of labels (the order of transitions is not have important). $\boldsymbol{G}(e)$ is uniform by columns if the transitions leading to any node of the trellis carry the same set of labels.

4.8.1.3. *Asymptotic considerations*

The i, j element of the matrix $\boldsymbol{G}$ is equal to a series of powers of Z. Let $\nu_{ij}(\delta_\ell)Z^{\delta_\ell^2}$ be the general term of the series, where:

$$\nu_{ij}(\delta_\ell) = \frac{1}{M^{L_1}}n_1 + \frac{1}{M^{L_2}}n_2 + \cdots$$

and $n_h, h = 1, 2, \ldots$, is the number of erroneous paths that start at node i at the moment 0 (for example) and after L_h moments finish at the node j, the distance associated with which is δ_ℓ. Since $1/M^{L_h}$ is the probability of a series of symbols of length $L_h, \nu_{ij}(\delta_\ell)$ can be interpreted as the average number of paths competing with the reference path diverging at the node i, and converging at the node j, and being at a distance of δ_ℓ from it. Consequently, the quantity:

$$\nu_i(\delta_\ell) \triangleq \sum_j \nu_{ij}(\delta_\ell)$$

can be interpreted as the average number of paths competing with the reference path, diverging at node i, converging in any node, and being at the distance of δ_ℓ from it. Similarly:

$$N(\delta_\ell) \triangleq \frac{1}{N_\sigma^2}\sum_{i,j} \nu_{ij}(\delta_\ell)$$

is the average number of competition paths at a distance δ_ℓ.

For high signal-to-noise ratios, i.e., when $N_0 \to 0$, the only terms of the matrix $\boldsymbol{G}$ that contribute to the probability of error, while differing from zero to a significant degree, will be of the type $\nu_{ij}(\delta_{\text{free}})Z^{\delta_{\text{free}}^2}$. Thus, asymptotically, we will have:

$$P(e) \sim \frac{1}{2}N(\delta_{\text{free}})\mathrm{e}^{-\delta_{\text{free}}^2/4N_0}$$

4.8.1.4. *A tighter upper bound*

An upper bound tighter than [4.12] can be obtained using a more precise expression that the Bhattacharyya bound in [4.7]. Let us recall that we have exactly:

$$P\{\boldsymbol{C}_L \to \boldsymbol{C}'_L\} = \frac{1}{2}\text{erfc}\left(\frac{|f(\boldsymbol{C}_L) - f(\boldsymbol{C}'_L)|}{2\sqrt{N_0}}\right) \qquad [4.29]$$

Since the minimum value taken by $|f(\boldsymbol{C_L}) - f(\boldsymbol{C'_L})|$ is equal to δ_{free}, using the inequality:

$$\text{erfc}\left(\sqrt{x+y}\right) \leq \text{erfc}\left(\sqrt{x}\right)\text{e}^{y}, \quad x \geq 0,\ y \geq 0 \qquad [4.30]$$

we obtain the bound:

$$P\{\boldsymbol{C_L} \to \boldsymbol{C'_L}\}$$
$$\leq \frac{1}{2}\text{erfc}\left(\frac{\delta_{\text{free}}}{2\sqrt{N_0}}\right)\text{e}^{\delta_{\text{free}}/4N_0} \cdot \exp\left\{-\frac{1}{4N_0}|f(\boldsymbol{C_L}) - f(\boldsymbol{C'_L})|^2\right\} \qquad [4.31]$$

In conclusion, we obtain the following bound for the probability of error:

$$P(e) \leq \frac{1}{2}\text{erfc}\left(\frac{\delta_{\text{free}}}{2\sqrt{N_0}}\right)\text{e}^{\delta^2_{\text{free}}/4N_0}T(Z)\Big|_{Z=\text{e}^{-1/4N_0}} \qquad [4.32]$$

We also have, approximately, for strong signals to noise ratios:

$$P(e) \cong N(\delta_{\text{free}})\frac{1}{2}\text{erfc}\left(\frac{\delta_{\text{free}}}{2\sqrt{N_0}}\right) \qquad [4.33]$$

4.8.1.5. *Bit error probability*

A bound of the probability of error per bit can be obtained by modifying the error matrices. The components of the matrix $\boldsymbol{G}(\boldsymbol{e})$ associated to the transitions from state i to the state j of the error state diagram must be multiplied by the W^{ϵ} factor, where ϵ is the Hamming weight (i.e., the number of "1") of the vector $\boldsymbol{b}$ associated to the transition $i \to j$.

With this new definition of the error matrices, the component i, j of the matrix $\boldsymbol{G}$ can now be expressed as a series of powers of unspecified Z and W. The general term of the series will be $\mu_{pq}(\delta_\ell, \epsilon_h)Z^{\delta^2_\ell}W^{\epsilon_h}$, where $\mu_{pq}(\delta_\ell, \epsilon_h)$ can be interpreted as the average number of paths with ϵ_h errors in bits and at a distance δ_ℓ from any path in the trellis diverging from node i and converging at node j. If we calculate the derivative of these terms with respect to W and if we pose $W = 1$, each one of them would yield the average number of errors in bits by branch generated by the incorrect paths from i to j. If we divide these quantities by m, the number of source bits by transition in the trellis, and if the series is summed, we obtain a bound for the bit error probability in the form:

$$P_b(e) \leq \frac{1}{2m}\frac{\partial}{\partial W}T_2(Z, W)\Big|_{W=1,\ Z=\text{e}^{-1/4N_0}} \qquad [4.34]$$

Another upper bound can be determined using the finer inequality obtained above in [4.7]. We then have:

$$P_b(e) \leq \frac{1}{2m}\text{erfc}\left(\frac{\delta_{\text{free}}}{2\sqrt{N_0}}\right)\exp(\delta^2_{\text{free}}/4N_0)\frac{\partial}{\partial W}T_2(Z, W)\Big|_{W=1,\ Z=\text{e}^{-1/4N_0}} \qquad [4.35]$$

4.8.1.6. *Lower bound of the probability of error*

We can also calculate a lower bound of the probability of an error event. Our calculations are based on the fact that the probability of error for a real decoder is larger than that obtained for an ideal decoder using collateral information brought by a "benevolent genius". The decoder helped by this genius functions as follows. The genius observes a long series of transmitted symbols, or, which is equivalent, the sequence:

$$\boldsymbol{C} = (\boldsymbol{c}_i)_{i=0}^{K-1}$$

of labels, and informs the decoder that the transmitted sequence is $\boldsymbol{C}$ *or* the sequence:

$$\boldsymbol{C}' = (\boldsymbol{c}_i')_{i=0}^{K-1}$$

where $\boldsymbol{C}'$ is randomly selected among the possible transmitted sequences that are at the smallest Euclidean distance from $\boldsymbol{C}$ (not necessarily δ_{free}, because $\boldsymbol{C}$ may not have a sequence $\boldsymbol{C}'$ at the free distance).

The probability of error of this decoder helped by the "genius" is the one obtained on the basis of a binary transmission diagram, in which the only sequences that can be transmitted are $\boldsymbol{C}$ and $\boldsymbol{C}'$:

$$P_G(e \mid \boldsymbol{C}) = \frac{1}{2}\text{erfc}\left(\frac{|f(\boldsymbol{C}) - f(\boldsymbol{C}')|}{2\sqrt{N_0}}\right) \tag{4.36}$$

Let us now consider the probabilities $P_G(e)$. We have:

$$\begin{aligned} P_G(e) &= \sum_{\boldsymbol{C}} P(\boldsymbol{C})\frac{1}{2}\text{erfc}\left(\frac{|f(\boldsymbol{C}) - f(\boldsymbol{C}')|}{2\sqrt{N_0}}\right) \\ &\geq \sum_{\boldsymbol{C}} I(\boldsymbol{C})P(\boldsymbol{C})\frac{1}{2}\text{erfc}\left(\frac{\delta_{\text{free}}}{2\sqrt{N_0}}\right) \end{aligned} \tag{4.37}$$

where $I(\boldsymbol{C}) = 1$ if $\boldsymbol{C}$ admits a sequence a the δ_{free} distance:

$$\min_{\boldsymbol{C}'} d[f(\boldsymbol{C}), f(\boldsymbol{C}')] = \delta_{\text{free}}$$

and $I(\boldsymbol{C}) = 0$ otherwise. In conclusion:

$$P(e) \geq \psi\frac{1}{2}\text{erfc}\left(\frac{\delta_{\text{free}}}{2\sqrt{N_0}}\right)$$

where:

$$\psi = \sum_{\boldsymbol{C}} P(\boldsymbol{C})I(\boldsymbol{C}) \tag{4.38}$$

represent the probability that at every moment a path in the trellis of the code, selected randomly, has another path that diverges from that of this moment, and converges with it later, and such that the Euclidean distance between them be equal δ_{free}. If all the sequences have this property, the following lower bound is obtained:

$$P(e) \geq \frac{1}{2}\text{erfc}\left(\frac{\delta_{\text{free}}}{2\sqrt{N_0}}\right) \qquad [4.39]$$

but that, in general, is not always true. For [4.39] to be valid, it is necessary that all the paths in the trellis be equivalent, and that each of them has a path at the δ_{free} distance. This result is obtained if the elements of each error matrix are equal.

We can finally obtain a lower bound of the probability of error per bit noting that the average fraction of erroneous information bits in the first branch of an error event cannot be lower than $1/m$. Thus:

$$P_b \geq \frac{\psi}{m}\frac{1}{2}\text{erfc}\left(\frac{\delta_{\text{free}}}{2\sqrt{N_0}}\right)$$

4.8.2. *Examples*

Let us now consider some examples of calculations of error probabilities for the TCM. According to the theory developed above, this calculation comprises two stages. The first is the evaluation of the transfer function of the error state diagram with formal labels (here we would have to remember that matrix operations are not commutative if the TCM is not uniform). In the second we replace formal labels with real labels (matrices or scalars) and calculate the matrix $\boldsymbol{G}$.

Code with four states

A TCM with four states is represented in Figure 4.21 with the corresponding diagram of error state. T_α, T_β and T_γ represent respectively the transfer function of the error state diagram from the starting node towards the nodes α, β and γ. We can write:

$$\begin{aligned}
T_\alpha &= \boldsymbol{G}(10) + T_\gamma \boldsymbol{G}(00) \\
T_\beta &= T_\alpha \boldsymbol{G}(11) + T_\beta \boldsymbol{G}(01) \\
T_\gamma &= T_\alpha \boldsymbol{G}(01) + T_\beta \boldsymbol{G}(11) \\
T(Z) &= T_\gamma \boldsymbol{G}(10)
\end{aligned}$$

To simplify, we only examine here the case where the labels are scalar and thus the property of commutation is verified. Defining $g_0 \triangleq \boldsymbol{G}(00)$, $g_1 \triangleq \boldsymbol{G}(01)$, $g_2 \triangleq \boldsymbol{G}(10)$ and $g_3 \triangleq \boldsymbol{G}(11)$, we obtain the following result:

$$T(Z) = \frac{g_2^2(g_1 - g_1^2 + g_3^2)}{(1 - g_0 g_1)(1 - g_1) - g_0 g_3^2} \qquad [4.40]$$

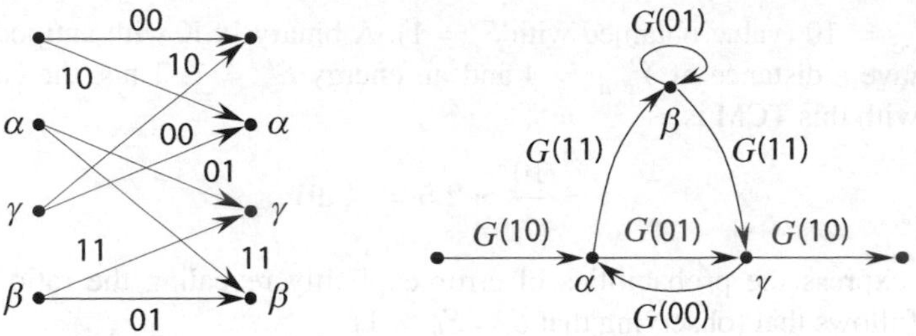

Figure 4.21. *Trellis diagram for a TCM with four states with $m = 1$, and the corresponding error state diagram*

IAM with 4 states

Using equation [4.40] we can obtain an upper bound of the probability of an error event by replacing g_i with the values obtained on the basis of the calculation of the error matrices $\boldsymbol{G}(\cdot)$. We will perform this operation for a constellation IAM4 with:

$$f(00) = +3, \qquad f(01) = +1, \qquad f(10) = -1, \qquad f(11) = -3$$

The matrices $\boldsymbol{G}$ for this constellation have been calculated in [4.17]–[4.20]. Since these matrices are uniform by lines, the transfer function can be obtained on the basis of equation [4.40]:

$$\begin{aligned} T(Z) &= Z^{36} \frac{4 - 3Z^4 + 2Z^{36} + Z^{68}}{4 - 8Z^8 + 3Z^8 - 2Z^{40} - Z^{72}} \\ &= Z^{36} + \frac{5}{4}Z^{40} + \frac{7}{4}Z^{44} + \cdots \end{aligned} \qquad [4.41]$$

We see that we have $\delta^2_{\text{free}} = 36$, value obtained for an average energy $\mathcal{E} = (9+1+1+9)/4 = 5$. A binary IAM ($\pm 1$) without coding would have a minimum distance $\delta^2_{\min} = 4$ and an energy $\mathcal{E}' = 1$, and thus the coding gain of this TCM is:

$$\gamma = \frac{36}{20} = 1.8 \Rightarrow 2.55 \text{ dB}$$

4PSK

With a 4PSK with unitary energy and for:

$$f(00) = 1, \quad f(01) = j, \quad f(10) = -1, \quad f(11) = -j$$

we obtain the $\boldsymbol{G}$ matrices on the basis of the equations [4.23]–[4.26]. Here we have a uniformity, and, thus, the transfer function obtained from [4.40] becomes:

$$T(Z) = \frac{Z^{10}}{1 - 2Z^2} = Z^{10} + 2Z^{12} + 4Z^{14} + \cdots \qquad [4.42]$$

where $\delta^2_{\text{free}} = 10$ (value obtained with $\mathcal{E} = 1$). A binary PSK with antipodal signals ± 1 will have a distance of $\delta^2_{\min} = 4$ and an energy $\mathcal{E}' = 1$. Thus, the coding gain obtained with this TCM is:

$$\gamma = \frac{10}{4} = 2.5 \Rightarrow 4 \text{ dB}$$

If we express the probabilities of error explicitly revealing the ratio $\mathcal{E}_b/N_0$ of [4.42], it follows that (observing that $\mathcal{E} = \mathcal{E}_b = 1$):

$$P(e) \leq \frac{1}{2} \frac{\text{e}^{-5\mathcal{E}b/2N_0}}{1 - 2\text{e}^{-\mathcal{E}_b/2N_0}}$$

The improved upper bound [4.32] yields:

$$P(e) \leq \frac{1}{2} \operatorname{erfc}\left(\sqrt{\frac{5}{2}\frac{\mathcal{E}_b}{N_0}}\right) \cdot \frac{1}{1 - 2\text{e}^{-\mathcal{E}_b/2N_0}}$$

The lower bound [4.39] yields:

$$P(e) \geq \frac{1}{2} \operatorname{erfc}\left(\sqrt{\frac{5}{2}\frac{\mathcal{E}_b}{N_0}}\right)$$

These probabilities of error should be compared with those of 2PSK modulation without coding equal to:

$$P(e) = \frac{1}{2} \operatorname{erfc}\left(\sqrt{\frac{\mathcal{E}_b}{N_0}}\right)$$

These four probabilities of error are represented in Figure 4.22. On the basis of Figure 4.22 we can observe that the lower bound and the improved upper bound are very close to each other and very close to the exact value of the probability of error. Unfortunately that happens only for TCM built on the basis of constellations with a small number of points and for trellis with a low number of states. Moreover, if we compare the probability $P(e)$ for a 2PSK modulation without coding with the two TCM bounds, we explicitly see that the coding gain is very close to 5/2, as has been mentioned previously.

4.8.3. *Calculation of* δ_{free}

The results obtained for the upper and lower bounds of probability of error of a TCM show than δ_{free} plays a central part in determining its performances. Consequently, if we need to retain only one parameter to evaluate the performance of a TCM, it is its free distance δ_{free}. It is thus normal to look for an algorithm to determine this distance.

Use of the error state diagram

The first technique that we will describe for the calculation of δ_{free} is based on the error state diagram. We have already observed that the transfer function $T(Z)$ contains information on the δ_{free} distance. In the preceding examples we saw that the value of δ^2_{free} can be obtained on the basis of the serial development of this function: the smallest exhibitor of Z in this series is δ^2_{free}. However, it should be known that an exact expression of $T(Z)$ cannot be determined.

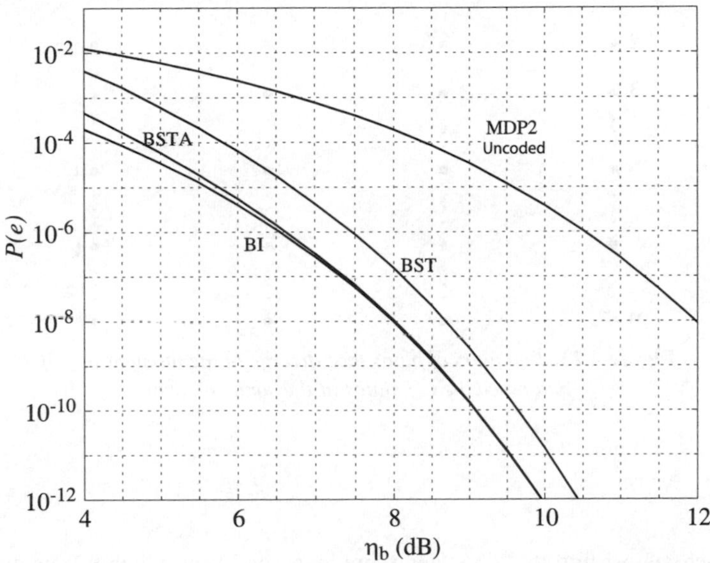

Figure 4.22. *Probabilities of error of a TCM with 4 states with 4PSK modulation. BST: upper bound based on the transfer function. BSTA: improved upper bound based on the transfer function. BI: lower bound. The probability of error of the 2PSK without coding is also represented. Here* $\eta_b = \mathcal{E}_b/N_0$

For this reason we will describe an algorithm for the numerical calculation of δ_{free}. Let us consider the trellis associated to the TCM; each pair of branches in a section of the trellis corresponds to a distance between the signals that label the branches. If there are parallel transitions, each branch will be associated to a sub-constellation. In the latter case, we will use only the minimum distance between two signals: one belonging to the first sub-constellation, the other belonging to the second. The square of the distance between the sequences of signals associated to the two paths in the trellis is obtained by summing up the squares of the distances individually.

The algorithm is based on the update of the components of a $\boldsymbol{D}^{(n)} = (\delta_{ij}^{(n)})$ matrix, which is equal to the squares of the minimum distances between all the pairs of paths that diverge from an initial state and reach the i and j states at a discrete moment n. Two pairs of this type of path are represented in Figure 4.23. We can observe there that the matrix $\boldsymbol{D}^{(n)}$ is symmetric, and that its components on the principal diagonal are the distances between the paths that converge in a single state ("error events").

The algorithm is described below.

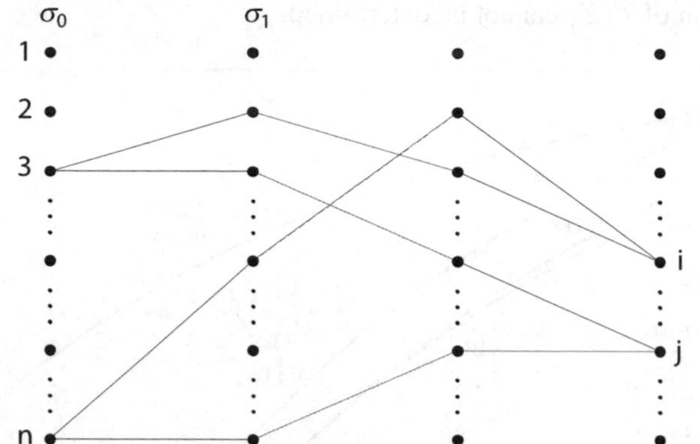

Figure 4.23. *Two pairs of paths that diverge at the moment $n = 0$ and reach the i, j states at the same moment*

Stage 1

For each state i find the 2^m states ("predecessors") by which a transition to i is possible and record them in a table. Let $\delta_{ij} = -1$ for any i and $j \geq i$. If there are parallel transitions, for each i let δ_{ii} be equal to the smallest Euclidean distance between the signals associated with the parallel transitions that lead to the state i.

Stage 2

For each pair of states $(i, j), j \geq i$ find the minimum Euclidean distance between the pairs of paths that diverge from the same (any) starting states and join the same pair of states i, j at the same moment. Two pairs of this kind are shown in Figure 4.24. This distance is $\delta_{ij}^{(1)}$.

Stage 3

For the two states of the pair $(i, j), j > i$ find in the table defined at Stage 1 the 2^m predecessors $i_1, \ldots, i_{2^m}$ and $j_1, \ldots, j_{2^m}$ (see Figure 4.25). In general, there will be 2^{2m} possible paths at the moment $n - 1$ that pass by i and j at the moment n. They pass by the pairs:

$$(i_1, j_1), (i_1, j_2), \ldots, (i_1, j_{2^m})$$

$$\ldots$$

$$(i_{2^m}, j_1), (i_{2^m}, j_2), \ldots, (i_{2^m}, j_{2^m})$$

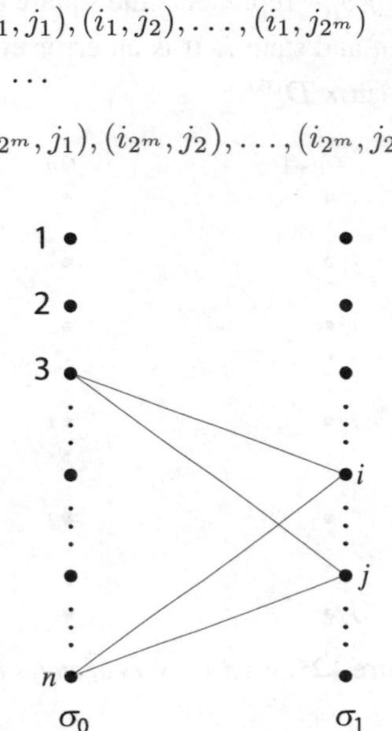

Figure 4.24. *Two pairs of paths that start from two different states and rejoin the same pair of states at the same moment*

The minimum distance between all the paths that pass by (i, j) at the moment n is:

$$\begin{aligned} \delta_{ij}^{(n)} = \min\Big\{ & \delta_{i_1 j_1}^{(n-1)} + \delta^2(i_1 \to i, j_1 \to j), \\ & \delta_{i_1 j_2}^{(n-1)} + \delta^2(i_1 \to i, j_2 \to j), \\ & \ldots \\ & \delta_{i_1 j_{2^m}}^{(n-1)} + \delta^2(i_1 \to i, j_M \to j), \\ & \ldots \\ & \delta_{i_{2^m} j_{2^m}}^{(n-1)} + \delta^2(i_{2^m} \to i, j_{2^m} \to j)\Big\} \end{aligned} \qquad [4.43]$$

In [4.43], the distances $\delta^{(n-1)}$ stem from the calculations at Stage 2, where, for example, $\delta(i_1 \to i, j_1 \to j)$ is the Euclidean distance between the two signals associated to the transitions $i_1 \to i$ and $j_1 \to j$. These can be calculated only once, at the beginning. When one of the already calculated distances $\delta_{\ell m}^{(n-1)}$ is equal to -1, the corresponding term in the second member of [4.43] disappears. Indeed, the value

$\delta_{\ell m}^{(n-1)} = -1$ tells us that there are no pairs of paths that pass by the states ℓ and m at moment $n-1$. When $i = j$, $\delta_{ii}^{(n)}$ represents the square of the distance between two paths that meet at the stage n and state i. It is an error event. If $\delta_{ii}^{(n)} < \delta_{ii}^{(n-1)}$, then $\delta_{ii}^{(n)}$ replaces $\delta_{ii}^{(n-1)}$ in the matrix $\boldsymbol{D}^{(n)}$.

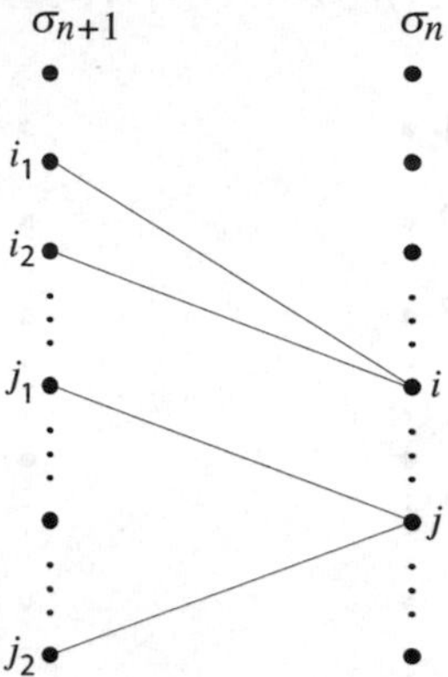

Figure 4.25. *Predecessors of states* i, j

Stage 4

If:

$$\delta_{ij}^{(n)} < \min_i \delta_{ii}^{(n)} \qquad [4.44]$$

for at least one pair (i, j), then we change n to $n+1$ and return to Stage 3. In other words, it is necessary to stop the iterations and to define:

$$\delta_{\text{free}}^2 = \min_i \delta_{ii}^{(n)}$$

4.9. Power spectral density

Let us consider here the calculation of the power spectral density of a TCM. We will determine the sufficient conditions for the spectrum of this TCM signal to be equal to that of an uncoded signal. To simplify, we will consider two-dimensional linear modulations: thus the signal transmitted on the channel has the following form:

$$y(t) = \sum_{n=-\infty}^{\infty} a_n s(t - nT) \qquad [4.45]$$

where $s(t)$ is a signal defined in the interval $(0, T)$ of the Fourier transform $S(f)$, and (a_n) is a series of complex coded symbols produced by the TCM. If the source symbols are independent and have equal probability, it stems from the regular and invariant in time structure of the TCM trellis that the sequence (a_n) is stationary. Under these conditions we obtain the following result. If:

$$\mathbb{E}[a_n] = \mu_a$$

and:

$$\mathbb{E}[a_\ell a_m^*] = \sigma_a^2 \rho_{\ell-m} + |\mu_a|^2$$

it results from this that $\rho_0 = 1$ and $\rho_\infty = 0$, and consequently the power spectral density of $y(t)$ is equal to:

$$\mathcal{G}(f) = \mathcal{G}^{(c)}(f) + \mathcal{G}^{(d)}(f) \qquad [4.46]$$

where $\mathcal{G}^{(c)}(f)$, the continuous part of the spectrum, has an expression:

$$\mathcal{G}^{(c)}(f) = \frac{\sigma_a^2}{T}|S(f)|^2 \sum_{\ell=-\infty}^{\infty} \rho_k \mathrm{e}^{-j2\pi f \ell T} \qquad [4.47]$$

and $\mathcal{G}^{(d)}(f)$, the discrete part of the spectrum (the "striped spectrum") is equal:

$$\mathcal{G}^{(d)}(f) = \frac{|\mu_a|^2}{T^2}|S(f)|^2 \sum_{\ell=-\infty}^{\infty} \delta\left(f - \frac{\ell}{T}\right) \qquad [4.48]$$

If the symbols a_n are not correlated (i.e. $\rho_\ell = \delta_{0,l}$ with δ being the Kronecker symbol) we obtain the following particular result:

$$\mathcal{G}^{(c)}(f) = \frac{\sigma_a^2}{T}|S(f)|^2 \qquad [4.49]$$

We have obtained the power spectral density without coding at the output of a modulator that uses the same signal $s(t)$ as the TCM. We will now examine the conditions for a TCM to have the same power spectral density as an uncoded signal (in particular, it does not lead to an expansion of the bandwidth).

Let us admit that $\mu_a = 0$, which implies that there are no spectral stripes. We are not restrictive, if we make the assumption that $\sigma_a^2 = 1$. Let σ_n be the state of the encoder when the symbol x_n is transmitted, and σ_{n+1} be the following state. The correlation coefficients ρ_ℓ can be expressed in the form:

$$\rho_k = \sum_{a_k} \sum_{\sigma_k} \sum_{\sigma_1} \sum_{a_0} a_k a_0^* P[a_k, \sigma_k, \sigma_1, a_0]$$

with:

$$P[a_k, \sigma_k, \sigma_1, a_0] = P[a_k \mid \sigma_k] \cdot P[\sigma_k, \sigma_1] \cdot P[a_0 \mid \sigma_1]$$

which becomes:

$$\rho_k = \sum_{\sigma_k} \sum_{\sigma_1} \mathbb{E}[a_k \mid \sigma_k] \cdot P[\sigma_k, \sigma_1] \cdot \mathbb{E}[a_0^* \mid \sigma_1]$$

Therefore, for $\rho_\ell = 0, \ell \neq 0$ it is enough that $\mathbb{E}[a_n|\sigma_n] = 0$ or $\mathbb{E}[a_{n-1}|\sigma_n] = 0$, for each σ_n. The first condition is equivalent to saying that for all the encoder states the transmitted symbols have a zero mean. The second condition is equivalent to affirming that for each encoder state the mean of the symbols that make it possible to the encoder to reach this state is zero. This condition is verified for many good TCM.

Let us, finally, consider the spectral stripes. A sufficient condition so that there are none is that $\mu_a = 0$, i.e. the mean of the symbols at the output of the encoder is equal to zero.

4.10. Multi-level coding

Let us consider, for example, the 8PSK signals. The partition of this constellation is represented in Figure 4.10, where δ_0, δ_1, and δ_2 are the Euclidean distances between the signals of the sub-constellations corresponding to the various partition levels. We suppose signals to have a unitary energy, and thus:

$$\delta_0^2 = 4\sin^2 \pi/8 \approx 0.5858 \qquad \delta_1^2 = 2 \qquad \delta_2^2 = 4$$

We observe that the three bits are protected unequally. That is due to the fact that bit 3 is protected from the noise by the Euclidean distance δ_0, bit 2 by the Euclidean distance δ_1, and bit 1 by the Euclidean distance δ_2. The unequal protection of the three bits follows from $\delta_0 < \delta_1 < \delta_2$.

If we want to increase the reliability of the transmission, the use of the same code to protect the three bits is thus not an effective solution. A better solution consists of using three different codes:

– the most powerful code, $\mathcal{C}_0$, to protect bit 3;
– the $\mathcal{C}_1$ code, less powerful than $\mathcal{C}_0$, to protect bit 2;
– the $\mathcal{C}_2$ code, less powerful than $\mathcal{C}_1$, to protect bit 1.

Proceeding thus we generate a coded modulation: indeed, we can see the coded modulation as a technique that combines the Euclidean distances (generated by the constellation of signals) with the Hamming distances (generated by the codes). Therefore, if we note d_0, d_1, and d_2 the minimum Hamming distances of the codes $\mathcal{C}_0, \mathcal{C}_1$, and $\mathcal{C}_2$, we obtain a good coding efficiency choosing:

$$d_0 > d_1 > d_2$$

Figure 4.26 shows the diagram of multi-level coded modulation obtained following this principle. Coded binary symbols are used, by triplet, to choose a signal of the "elementary" 8PSK constellation.

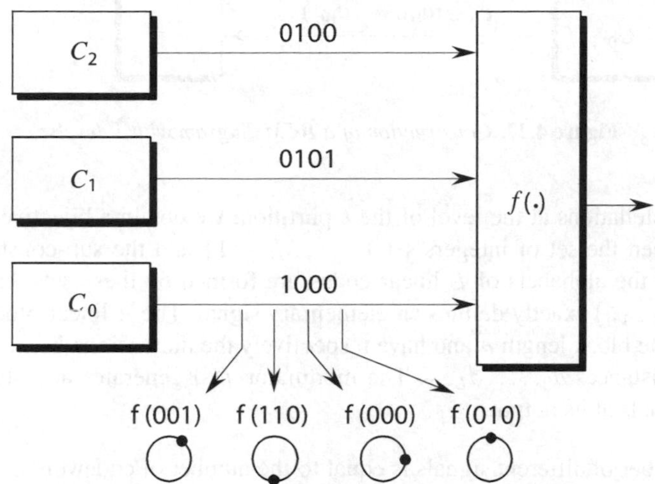

Figure 4.26. *Coded modulation system with three levels*

We observe here that trellis-coded modulation can be seen as a particular case of the multi-level modulation, for which certain bits are coded by a convolutional code, while other bits are uncoded. If we use block codes, the construction that follows is called *block coded modulation* (BCM).

4.10.1. *Block coded modulation*

We will describe the principal properties of BCM. Figure 4.27 shows a general outline with L levels.

The starting point is an "elementary" constellation (8PSK of the example above) with M signals, and a partition with L levels. At the 0 level the elementary constellation is split into M_0 sub-constellations, each with M/M_0 signals. At level 1 each sub-constellation is split into M_1 sub-constellations, each with $M/(M_0M_1)$ signals, etc, until the $L-1$ level with M sub-constellations, each with one signal. If we number

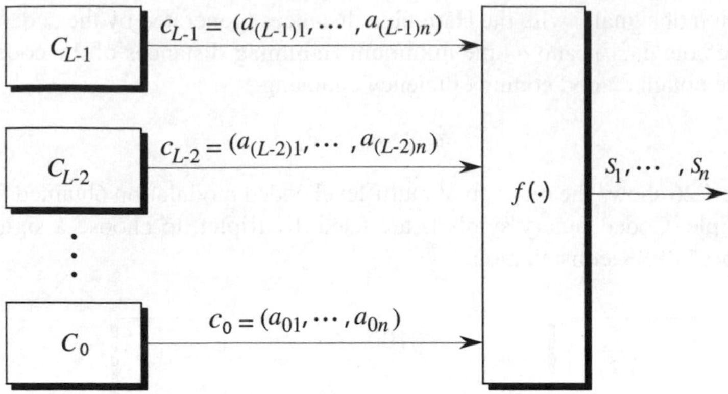

Figure 4.27. *Construction of a BCM diagram with L levels*

the sub-constellations at the level of the ℓ partition, we obtain a bijective correspondence between the set of integers $\{0, 1, \ldots, M_\ell - 1\}$ and the sub-constellations of this level. If the alphabets of L linear codes are formed by these sets, each L-tuple $(a_{0i}, \ldots, a_{(L-1)i})$ exactly defines an elementary signal. The L linear block encoders have the same block length n and have respectively the dimensions $k_0, \ldots, k_{L-1}$ and Hamming distances $d_0, \ldots, d_{L-1}$. The modulator $f(\cdot)$ generates an n-tuple of elementary signals at its output.

The number of different signals is equal to the number of codewords:

$$\mathcal{M} = \prod_{\ell=0}^{L-1} q_i^{k_i}$$

where q_i is the number of symbols in the alphabet of code C_i. The minimum Euclidean distance between the n-tuples of elementary signals is bounded less well by:

$$\delta_{\min}^2 \geq \min\left(\delta_0^2\, d_0, \delta_1^2\, d_1, \ldots, \delta_L^2\, d_L\right), \qquad [4.50]$$

where δ_ℓ is the minimum Euclidean distance between the signals of the sub-constellations of the level ℓ partition. To demonstrate [4.50] it is sufficient to note that at the ℓ level where the minimum Euclidean distance is d_ℓ two words of the C_ℓ code differ in at least d_ℓ positions; consequently, both n-tuples corresponding to the elementary signals will be at least at an Euclidean distance whose square is equal to $\delta_\ell^2 d_\ell$.

OBSERVATION. If we impose the value of the square of the minimum Euclidean distance, we will be able to choose the minimum Hamming distances in the following manner:

$$d_\ell = \lceil \delta_{\min}^2 / \delta_\ell^2 \rceil$$

where $\lceil x \rceil$ is the smallest integer that is not inferior to x.

EXAMPLE 4.10. Let us consider $L = 3, n = 7$ and the 8PSK as an elementary constellation with a correspondence between the binary symbols and the signals of the constellation illustrated in Figure 4.28. Let $\mathcal{C}_2$ be the non-redundant binary code (7,7,1), $\mathcal{C}_1$ be the binary parity code (7,6,2), and $\mathcal{C}_0$ the binary code with repetition (7,1,7). The resulting BCM diagram will have $2^{1+6+7} = 2^{14}$ signals, 14 dimensions, and thus 1 bit per dimension (like the 4PSK). Its minimum Euclidean distance, standardized with respect to average energy $\mathcal{E}$ of the signals, is upper bounded by [4.50]:

$$\frac{d_{\min}^2}{\mathcal{E}} \geq \min\{4 \times 1,\ 2 \times 2,\ 0.586 \times 7\} = 4$$

In this case we can demonstrate that it is exactly equal to 4 (or an asymptotic coding gain of 3 dB with respect to 4PSK).

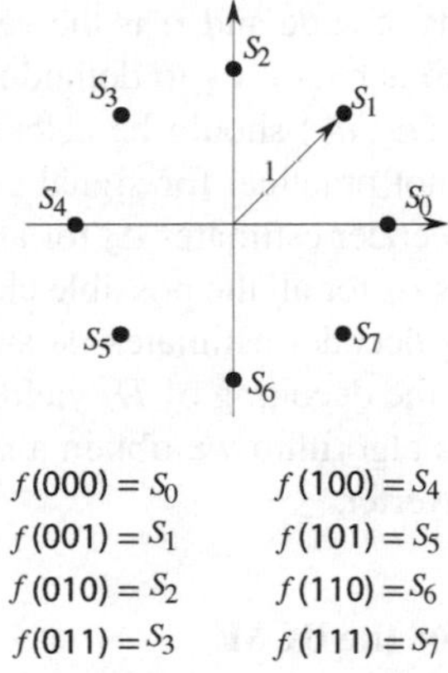

Figure 4.28. *Elementary constellation: 8PSK*

4.10.2. *Decoding of multilevel codes by stages*

We will describe the principle of decoding by stages of a multilevel code. In certain cases (for example, when $L = 2$ and $\mathcal{C}_1$ is a non-redundant code) decoding by stages is optimal. However, in the general case, it is a sub-optimal algorithm, albeit of a much lower complexity than the optimal algorithm.

The basic idea of the stages algorithm is as follows: the codes $\mathcal{C}_0, \ldots, \mathcal{C}_{L-1}$ are decoded the ones after the other. First of all, we decode $\mathcal{C}_0$, the most powerful code; then we decode $\mathcal{C}_1$ supposing that the $\mathcal{C}_0$ code was decoded correctly; then we decode $\mathcal{C}_2$ supposing that the two preceding codes were decoded correctly, etc.

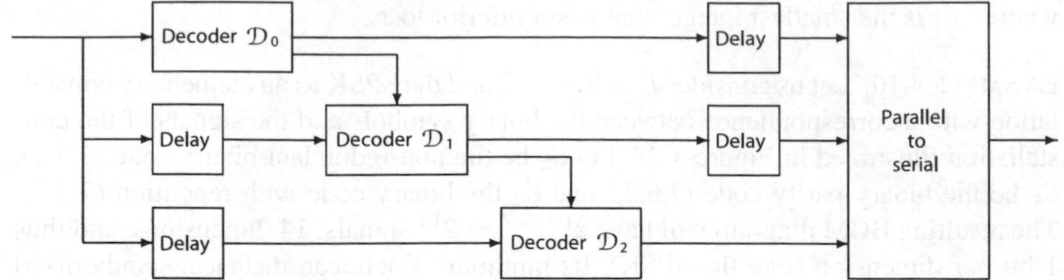

Figure 4.29. *Decoder by stages for a block coded modulation with three levels*

The block diagram of a multilevel decoder is illustrated in Figure 4.29 for $L = 3$. If the received signal is:

$$\boldsymbol{r} = F(\boldsymbol{c}_0, \boldsymbol{c}_1, \boldsymbol{c}_2) + \boldsymbol{n}$$

where $\boldsymbol{c}_0, \boldsymbol{c}_1, \boldsymbol{c}_2$ are the words of code and $\boldsymbol{n}$ is the vector of noise, the receiver must estimate the three words of code $\boldsymbol{c}_0, \boldsymbol{c}_1, \boldsymbol{c}_2$ to demodulate $\boldsymbol{r}$. In theory, the metrics of all the possible vectors $F(\boldsymbol{c}_0, \boldsymbol{c}_1, \boldsymbol{c}_2)$ should be calculated before finding their minimum value. This process is not practical for signal constellations of a large size. In decoding by stages, the $\mathcal{D}_0$ decoder estimates $\boldsymbol{c}_0$ for all the possible choices of $\boldsymbol{c}_1, \boldsymbol{c}_2$; then the $\mathcal{D}_1$ decoder estimates $\boldsymbol{c}_1$ for all the possible choices of $\boldsymbol{c}_2$, supposing a correct choice of $\boldsymbol{c}_0$. Finally, the $\mathcal{D}_2$ decoder estimates $\boldsymbol{c}_2$ supposing a correct choice of $\boldsymbol{c}_0$ and $\boldsymbol{c}_1$. We observe here that the decoding of $\mathcal{D}_\ell$ yields an estimate of k_ℓ source symbols, and at each level of this algorithm we obtain a block of source symbols, which is sent to a parallel/serial converter.

4.11. Probability of error for the BCM

We will now describe a procedure of calculation of the BCM performances. In particular, we will calculate the probabilities of error per symbol and bit. Our description will be based on trellis, where a path corresponds to a series of source symbols, and, also, to a series of elementary signals. This trellis can be used for optimal decoding, i.e. with maximum likelihood (using the Viterbi algorithm), just as for the calculation of the probability of error. The trellis will have n levels where n is the common length of the elementary words of codes, and can be generated by taking the direct product of the L trellis of the codes $\mathcal{C}_0, \ldots, \mathcal{C}_{L-1}$.

Let $\boldsymbol{x} = (s_1, \ldots, s_n)$ be a n-tuple of block coded signals. If we use the union bound with the assumption that the coded symbols have equal probability, we obtain:

$$P(e) = \frac{1}{\mathcal{M}} \sum_{i=1}^{\mathcal{M}} P(e|\boldsymbol{x}_i) \leq \frac{1}{\mathcal{M}} \sum_{i=1}^{\mathcal{M}} \sum_{j \neq i} P[\boldsymbol{x}_i \to \boldsymbol{x}_j] \qquad [4.51]$$

where M is the total number of coded signals, noted $\boldsymbol{x}_1, \ldots, \boldsymbol{x}_{\mathcal{M}}$, and $P[\boldsymbol{x}_i \to \boldsymbol{x}_j]$ is the pair-wise error probability. The exact calculation of this expression requires enumerating all the pairs $\boldsymbol{x}_i \neq \boldsymbol{x}_j$, or, which is equivalent, all the pairs of words of the code $\mathcal{C} = \mathcal{C}_0 \times \ldots \times \mathcal{C}_{L-1}$.

If $\mathcal{T}_0, \mathcal{T}_1, \ldots, \mathcal{T}_{L-1}$ are the trellis of $\mathcal{C}_0, \mathcal{C}_1, \ldots, \mathcal{C}_{L-1}$, the trellis representing the code will be the trellis produced:

$$\mathcal{T} = \mathcal{T}_0 \otimes \mathcal{T}_1 \otimes \ldots \otimes \mathcal{T}_{L-1}$$

Each state of $\mathcal{T}$ is a L-tuple of code states, and each of its branches is labeled by the corresponding value of the vector $\mathbf{c}$. Thus, each word $\boldsymbol{x}$ is represented by a path in $\mathcal{T}$.

If we do not make an assumption on the symmetry of the BCM diagram, we need to consider the product $\mathcal{T}^2$ of the two trellis, the first representing the sequence of symbols transmitted via the channel and the second the sequence of symbols received.

Probability of error per bit

To obtain the probability of error per bit we multiply each branch label of the $\mathcal{T}^2$ trellis by I^α, where I is unspecified and α is the number of bits where the two signals associated with the branch differ. Let $T(Z, I)$ be the new transfer function of the $\mathcal{T}^2$ trellis; in a Gaussian channel the probability of error per symbol will be upper bounded by:

$$P_b(e) \leq \frac{1}{\mathcal{M}} \frac{1}{m} \frac{\partial}{\partial I} T(Z, I) \Big|_{I=1, Z=\exp(-1/4N_0)}$$

where m is the number of information bits associated to each coded symbol.

4.11.1. *Additive Gaussian channel*

If the transmission takes place through channel with additive Gaussian noise of power spectral density $N_0/2$, the Bhattacharyya bound yields:

$$P[\boldsymbol{x} \to \hat{\boldsymbol{x}}] \leq \prod_{\ell=1}^{n} Z^{\delta^2(x_\ell, \hat{x}_\ell)} \qquad [4.52]$$

where $Z = \exp(-1/4N_0)$, and δ represents the Euclidean distance.

If the quantity $Z^{\delta^2(x_\ell, \hat{x}_\ell)}$ labels the branch of the trellis product $\mathcal{T}^2$ associated with the pair $(x_\ell, \hat{x}_\ell)$, the transfer function $T(Z)$ of $\mathcal{T}^2$ enumerates all the possible $P[\boldsymbol{x} \to \hat{\boldsymbol{x}}]$. Among these quantities we also obtain those where $\boldsymbol{x} = \hat{\boldsymbol{x}}$, which number $\mathcal{M}$; we will thus write on the basis of [4.51]:

$$P(e) \leq \frac{1}{\mathcal{M}} [T(Z) - \mathcal{M}] \qquad [4.53]$$

4.11.2. *Calculation of the transfer function*

The calculation of the transfer function of $\mathcal{T}^2$ can be performed by representing each section of the trellis by a matrix. Let us consider the case of additive Gaussian noise, and suppose that the ℓth section of the trellis has ν_1 input nodes and ν_2 output nodes. We will note by s one of its input nodes and by t one of its output nodes, so that the pair (s,t) corresponds to a transition between these two nodes. This section is described by a $\boldsymbol{T}_\ell$ matrix of size $\nu_1 \times \nu_2$ whose element (s,t) is zero if there are no branches connecting s to t, and equal to $Z^{d^2(\boldsymbol{x}_\ell, \hat{\boldsymbol{x}}_\ell)}$ if the transition (s,t) corresponds to the ℓth branch of the trellis associated with the pair $(\boldsymbol{x}_\ell, \hat{\boldsymbol{x}}_\ell)$ of coded signals. By calculating the product $\prod_{\ell=1}^{n} \boldsymbol{T}_\ell$ we obtain the required transfer function.

EXAMPLE 4.11. Let us again consider the BCM diagram described in example 4.10. The trellis $\mathcal{T}_1, \mathcal{T}_2, \mathcal{T}_3$ and $\mathcal{T}$ are represented in Figure 4.30. For the additive Gaussian channel we obtain the following bound for the symbol error probability:

$$\begin{aligned} P(e) \leq\; & 91Z^{8} + 64Z^{8,2} + 448Z^{13,86} + 1001Z^{16} \\ & + 1344Z^{19,52} + 3003Z^{24} + 2240Z^{25,18} \\ & + 2240Z^{30,82} + 3003Z^{32} + 1344Z^{36,48} \\ & + 1001Z^{40} + 448Z^{42,14} + 64Z^{47,8} \\ & + 91Z^{48} + Z^{56} \Big|_{Z=\exp(-\mathcal{E}_b/4N_0)} \end{aligned}$$

whereas the bit error probability is equal to:

$$\begin{aligned} P_b(e) \leq\; & \frac{235}{14}Z^{8} + \frac{184}{7}Z^{8,2} + 200Z^{13,86} \\ & + \frac{2321}{7}Z^{16} + 648Z^{19,52} + \frac{18513}{14}Z^{24} \\ & + 1160Z^{25,18} + 1240Z^{30,82} + \frac{10758}{7}Z^{32} \\ & + 792Z^{36,48} + \frac{7645}{14}Z^{40} + 280Z^{42,14} \\ & + \frac{296}{7}Z^{47,8} + \frac{345}{7}Z^{48} + \frac{1}{2}Z^{56} \Big|_{Z=\exp(-E_b/4N_0)} \end{aligned}$$

These error probabilities are illustrated in Figures 4.31 and 4.32 These figures also show a bound that improves [4.53] using the inequality [4.30].

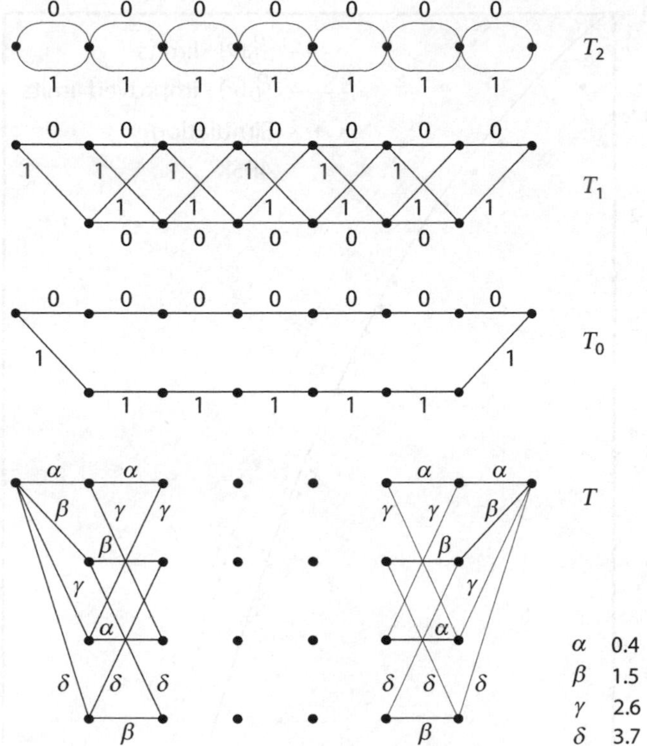

Figure 4.30. *Trellis produced for the BCM diagram of example 4.10*

4.12. Coded modulations for channels with fading

For the Gaussian channel considered in this chapter until now the received signal is affected only by constant attenuation and delay.

We will now consider a channel affected by fading, i.e. whose attenuation is variable in time. This situation corresponds to channels with multiple paths with movements of the transmitter with respect to the receiver. That causes a variation in the time of the amplitude and the phase of the received signal, effects that can seriously deteriorate the performance of a communication system and justify the use of channel coding.

4.12.1. *Modeling of channels with fading*

The propagation by multiple paths takes place when the electromagnetic energy that carries the information signals propagates between the transmitter and the receiver

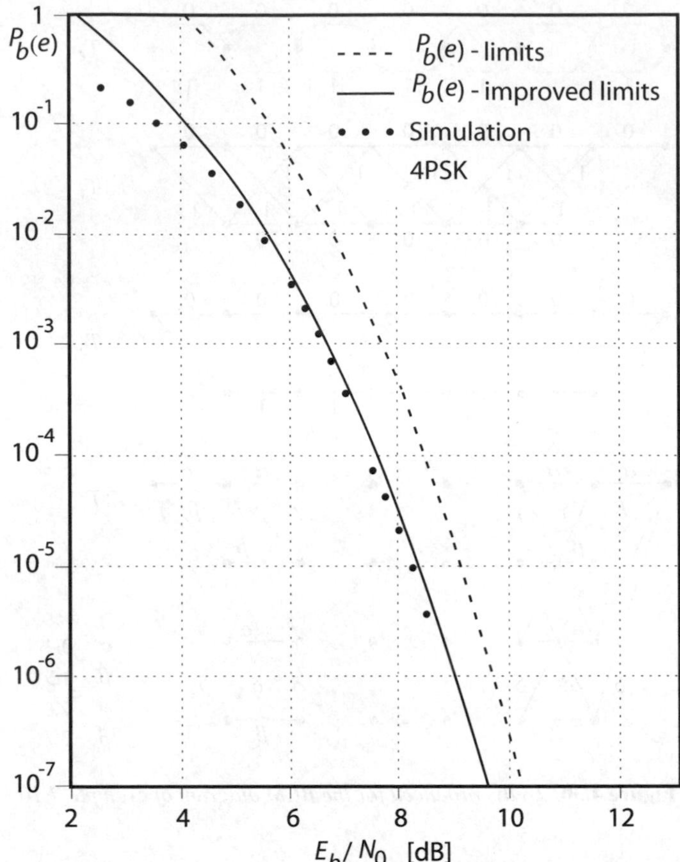

Figure 4.31. *Probability of error per symbol of a BCM diagram in an additive Gaussian channel*

following a multiple-paths model. This situation occurs, for example, in the case of transmissions inside a building, or communications terrestrial radio-mobile. The waves are reflected from fixed or mobile obstacles (buildings, hills, cars, etc.) and lead to multiple paths.

4.12.1.1. *Delay spread*

The components of the signal that passes along several paths, direct or indirect, have different delays. They combine to generate a deformed version of the transmitted signal. If an ideal impulse had been transmitted and if the bandwidth of the channel is rather broad, the received signal will consist of several impulses whose delays and phases are different. These impulses each correspond to a propagation path. We call *delay spread* the difference in time between the first impulse received and the last. If

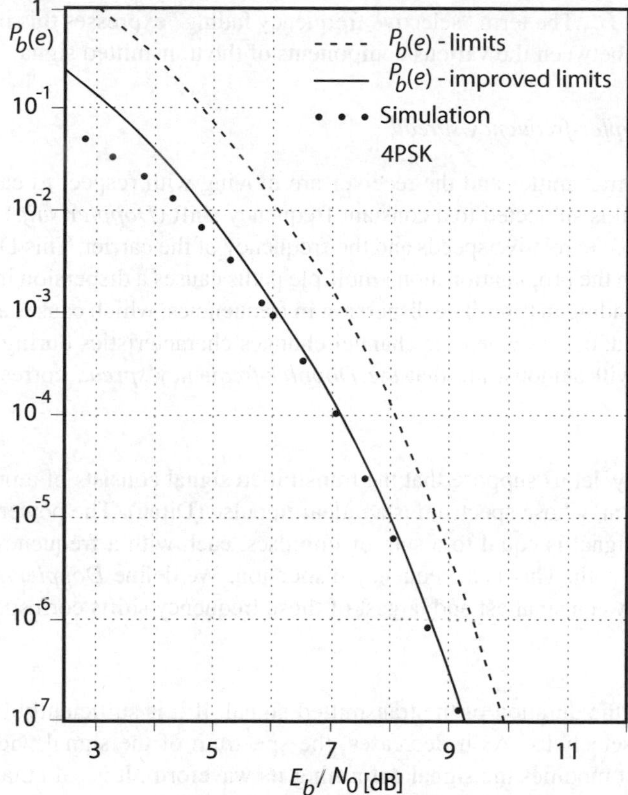

Figure 4.32. *Probability of error per bit of a BCM diagram in an additive Gaussian channel*

the band of the channel is not sufficiently broad, the impulses are widened and superimposed. In the context of this chapter we will say that the *delay spread* introduces a temporal dispersion and selective frequency fading.

If B_x is the bandwidth of the transmitted signal and if it is small compared to the channel bandwidth, the signal is not deformed, and there is no selectivity in frequencies. As B_x increases, the shape of the signal will deteriorate. A measurement of the band of the signal beyond which this deformation becomes considerable is generally given according to the channel coherence band, noted B_c and defined as the inverse of the *delay spread*. The coherence band is equal to the frequency separation of two components of the signal that undergo two independent attenuations. A signal with $B_x \gg B_c$ is prone to selective frequency fading; more precisely, the envelope and the phase of two non-modulated carriers transmitted through a channel with fading will be very different if the instantaneous frequency deviation between these carriers is greater than B_c. The term "selective frequency fading" expresses this independence of correlation between the various components of the transmitted signal.

4.12.1.2. *Doppler-frequency spread*

When the transmitter and the receiver are moving with respect to each other, the received signal is subjected to a constant frequency shift (*Doppler shift*) proportional to the difference in relative speeds and the frequency of the carrier. This Doppler effect combined with the propagation along multiple paths causes a dispersion in frequencies and selective fading in time. The dispersion in frequencies, which causes a widening of the signal band, occurs when the channel changes characteristics during signal propagation. We will demonstrate that the *Doppler-frequency spread* corresponds to the *delay spread.*

To simplify, let us suppose that the transmitted signal consists of a pure carrier of infinite duration, whose spectrum is an ideal impulse (Dirac). The power spectrum of the received signal is equal to a sum of impulses, each with a frequency shift corresponding to a path. This is a frequency dispersion. We define *Doppler spread* as the difference between smallest and largest of these frequency shifts corresponding to the various paths.

Let T_x be the duration of the transmitted signal: if it is sufficiently large, there is no temporal selectivity. As it decreases, the spectrum of the signal widens, and the Doppler effect modifies the signal deforming its waveform. It is said that the channel is selective in time.

The duration of the signal, above which this deformation becomes considerable, is called the *coherence time* of the channel. We note it by T_c, and define it as the inverse of the *Doppler spread.* If T_x is the duration of an impulse, and if it is short to the point that the channel does not change to a significant degree during the transmission, the signal will be received without deformation. Its deformation, on the contrary, becomes perceptible, if T_x is much larger than T_c, the coherence time of the channel.

4.12.1.3. *Classification of channels with fading*

The description above has demonstrated that the two quantities B_c and T_c describe how the channel behaves with respect to the transmitted signal. In particular:

– if $B_x \ll B_c$, there is no selective fading in frequencies, and thus no temporal dispersion. The transfer function of the channel is constant, and the channel is known as *flat*, or *non-selective*, in frequency;

– if $T_x \ll T_c$, there is no selective fading in time, and the channel is called *flat*, or *non-selective*, in time.

4.12.1.4. *Examples of radio channels with fading*

4.12.1.4.1. Propagation along to two paths: effect of movement

Let us consider the situation represented in Figure 4.33. The vehicle moves at a constant speed v; the transmitted signal is non-modulated carrier with the frequency f_0 propagated along two paths, which, for simplicity's sake, we suppose to have the same delay and the same attenuation. The angles of reception of the two paths are noted 0 and γ. The Doppler effect generates the following received signal:

$$y(t) = A\exp\left[j2\pi f_0\left(1-\frac{v}{c}\right)t\right] + A\exp\left[j2\pi f_0\left(1-\frac{v}{c}\cos\gamma\right)t\right] \qquad [4.54]$$

We observe according to [4.54] that the received signal includes a pair of complex sinusoids: this effect can be interpreted as a dilation of the signal spectrum, and thus a frequency dispersion caused by the channel and due to the combined effects of the Doppler shift and the propagation along multiple paths.

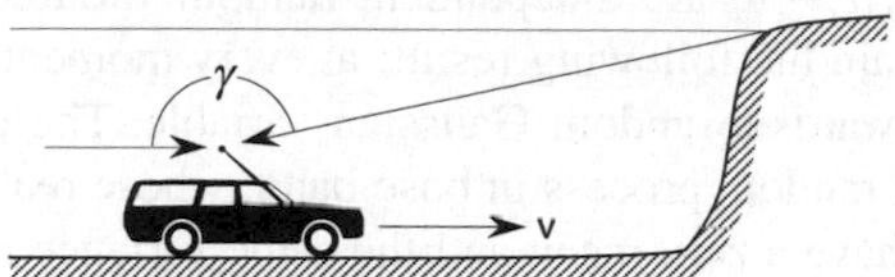

Figure 4.33. *Propagation along two paths: effect of movement*

The equation [4.54] can be written in the form:

$$y(t) = A\left[\exp\left(-j2\pi f_0\frac{v}{c}t\right) + \exp\left(-j2\pi f_0\frac{v}{c}\cos\gamma t\right)\right]e^{j2\pi f_0 t} \qquad [4.55]$$

The absolute value of the term inside the square brackets is the instantaneous envelope of the received signal:

$$R(t) = 2A\left|\cos\left[2\pi\frac{v}{c}f_0\frac{1-\cos\gamma}{2}t\right]\right|$$

The last equation shows an important effect: the envelope of the received signal varies with time as a sinusoid. Its frequency is:

$$\frac{v}{c}f_0\frac{1-\cos\gamma}{2}$$

We have at the same time a selective fading and a frequency dispersion.

4.12.1.4.2. Propagation along multiple paths: effect of movement

Let us suppose now that the transmitted signal is received along N paths, as depicted in Figure 4.34. The receiver is moving with a speed v and A_i, θ_i and γ_i are respectively the amplitude, the phase and the angle of incidence of the vector of the ith path. The received signal contains contributions affected by various Doppler shifts, i.e.:

$$f_i \triangleq f_0 \frac{v}{c} \cos \gamma_i, \qquad i = 1, 2, \ldots, N$$

The received analytical signal can be written in the form:

$$y(t) = \sum_{i=1}^{N} A_i \exp j[2\pi(f_0 - f_i)t + \theta_i] \qquad [4.56]$$

The complex envelope of the received signal is:

$$R(t)e^{j\Theta(t)} = \sum_{i=1}^{N} A_i e^{-j(2\pi f_i t - \theta_i)}$$

If the number N of paths is sufficiently large, we can suppose that the attenuations A_i and the phases $2\pi f_i t - \theta_i$ are independent random variables. Using the central bound theorem we obtain the following result: at every moment, since $N \to \infty$, the resulting sum tends towards a random Gaussian variable. The complex envelope of the received signal is a random process in base band, whose real and imaginary parts are independent; they have a zero mean and the same variance σ^2. In this case, $R(t)$ and $\Theta(t)$ are two independent random processes, where $\Theta(t)$ has a uniform density in the $(0, 2\pi)$ interval, and $R(t)$ has a Rayleigh density, i.e.:

$$f_R(r) = \begin{cases} \frac{r}{\sigma^2} e^{-r^2/2\sigma^2}, & 0 \leq r < \infty \\ 0, & r < 0 \end{cases} \qquad [4.57]$$

We can also modify this channel model supposing that, as it often happens in practice, there exists a path, whose power is much greater than that of other paths. We can thus express the complex envelope of the received signal in the form:

$$R(t)e^{j\Theta(t)} = u(t)e^{j\alpha(t)} + v(t)e^{j\beta(t)}$$

where $u(t)$ has a Rayleigh probability density function, $\alpha(t)$ is uniform in $(0, 2\pi)$, and $v(t)$ and $\beta(t)$ are non-random signals. In this model, $R(t)$ has a Rice probability density function:

$$f_R(r) = \frac{r}{\sigma^2} \exp\left\{ -\frac{r^2 + v^2}{2\sigma^2} \right\} I_0\left(\frac{rv}{\sigma^2}\right) \qquad [4.58]$$

where $r \geq 0$, and $I_0(\cdot)$ is related to the modified Bessel of the order zero and first type. Here $R(t)$ and $\Theta(t)$ are not independent.

In [4.58] v is the envelope of the fixed path component, while $2\sigma^2$ is the power of the Rayleigh component. The "Rice factor":

$$K = \frac{v^2}{2\sigma^2}$$

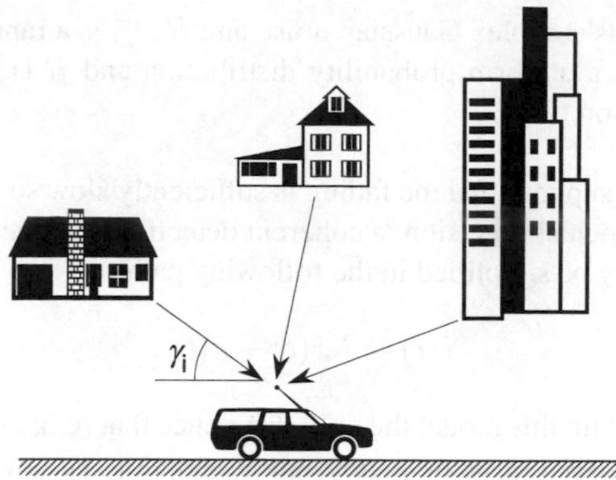

Figure 4.34. *Propagation along multiple paths: effect of movement*

is equal to the relationship between the power of the fixed component and the power of the Rayleigh component. When $K \to 0$, i.e., the power of the fixed path tends towards zero, since $I_0(0) = 1$ the density of Rice probability becomes a Rayleigh density. In addition, if $K \to \infty$, i.e., if the power of the fixed path is dominating with respect to the power of other random paths, the Rice density becomes equal to a Gaussian density.

4.12.2. *Rayleigh fading channel: Euclidean distance and Hamming distance*

The models of channel that we have examined are based on a narrow band transmission, which is equivalent to supposing that the duration of a symbol is much larger than the *delay spread* caused by the propagation by multiple paths. If such is the case, all the frequency components of the transmitted signal undergo the same attenuation and the same phase offset; the channel is therefore flat in frequency. If, moreover, the channel changes very slowly with respect to the duration of a symbol (very slow movement of the transmitter and the receiver) the fading $R(t)\exp[j\Theta(t)]$ remains almost constant during the transmission of a symbol.

This flat in frequency and slow in time model of fading will be studied more thoroughly later on. If the fading is non-selective, we can model it as a multiplicative process, whereas if it is slow it can be modeled by a random variable with a constant value for the duration of the symbol. If we also admit the presence of Gaussian noise, and if $\tilde{x}(t)$ is the complex envelope of the signal transmitted during the time interval $(0, T)$, then the complex envelope of the signal received at the output of a channel affected by a fading that is slow and flat in frequency can be written in the form:

$$\tilde{r}(t) = R\mathrm{e}^{j\Theta}\tilde{x}(t) + \tilde{n}(t) \quad [4.59]$$

where $\tilde{n}(t)$ represents complex Gaussian noise and $R\,e^{j\Theta}$ is a random Gaussian variable, with Θ having a uniform probability distribution and R is a Rice or Rayleigh probability distribution.

If, moreover, we suppose that the fading is sufficiently slow so that phase Θ can be estimated with a sufficient precision, a coherent detection becomes possible, and thus the model [4.59] may be simplified in the following path:

$$\tilde{r}(t) = R\tilde{x}(t) + \tilde{n}(t) \quad [4.60]$$

We can see that with this model the only difference that remains with the Gaussian channel is that R, instead of being a constant attenuation, is now a random variable, whose value balances the amplitude and, consequently, the power of the received signal. Let us suppose then that the value of R (which we will call here the *state of the channel*) is known exactly by the receiver. Optimal detection would consist in this case of minimizing the Euclidean distance:

$$\int_0^T [r(t) - Rx(t)]^2\,dt \quad \text{or} \quad |\boldsymbol{r} - R\boldsymbol{x}|^2 \quad [4.61]$$

with respect to the transmitted signal $x(t)$ (or to the transmitted vector $\boldsymbol{x}$).

Let us consider the transmission of a coded sequence $\boldsymbol{x} = (s_1, s_2, \ldots, s_n)$, whose components are elementary signals belonging to a constellation $\mathcal{S}$. Here we do not distinguish between block codes and convolutional codes (with *soft* decoding) or coded modulation. Moreover, we suppose that due to perfect interleaving (i.e., of infinite depth), the random variables representing the fading that affects the s_k signals are independent. We are able to write for the components of the received sequence $(r_1, r_2, \ldots, r_n)$:

$$r_k = R_k s_k + n_k \quad [4.62]$$

where R_k are independent and, with the assumption of white noise, the noise components n_k are also independent.

Coherent detection of the coded sequence is based on the search for the sequence $\boldsymbol{x} = (s_1, \ldots, s_n)$ that minimizes the distance:

$$\sum_{k=1}^{n} |r_k - R_k s_k|^2 \quad [4.63]$$

The pair-wise error probability can be written in the form:

$$P\{\boldsymbol{x} \to \hat{\boldsymbol{x}}\} = P(X < 0) \quad [4.64]$$

where:

$$X \triangleq \sum_{k=1}^{n} \left[|r_k - R_k \hat{s}_k|^2 - |r_k - R_k s_k|^2 \right]$$
$$= \sum_{k=1}^{n} \left[|R_k(s_k - \hat{s}_k) + n_k|^2 - |n_k|^2 \right]$$
$$= \sum_{k=1}^{n} \left[R_k^2 |s_k - \hat{s}_k|^2 + 2R_k(n_k,\, s_k - \hat{s}_k) \right] \qquad [4.65]$$

Using the Chernoff bound, valid for any random continuous variable X:

$$P(X < 0) \leq \min_{z>0} \Phi_X(z) \qquad [4.66]$$

where $\Phi(z)$ is the bilateral Laplace transform of the probability density of X, i.e.:

$$\Phi_X(z) \triangleq \mathbb{E}\left[\mathrm{e}^{-zX} \right] \qquad [4.67]$$

With the preceding assumptions, all the terms in the sum (4.65) are independent, and noting that R_k observations are independent and distributed identically we obtain:

$$\Phi_X(z) = \prod_{k=1}^{n} \mathrm{E}_{R_k} \left[\exp z(N_0 z - 1) R_k^2 |s_k - \hat{s}_k|^2 \right] \qquad [4.68]$$
$$= \prod_{k \in \mathcal{K}} \mathrm{E}_{R_k} \left[\exp z(N_0 z - 1) R_k^2 |s_k - \hat{s}_k|^2 \right] \qquad [4.69]$$

where the last equality differs from the one above, since the set where k takes its values is reduced from $\{1, \ldots, n\}$ to $\mathcal{K}$, the set of k, such that $s_k \neq \hat{s}_k$. That can be done, because for values of k that yield $s_k = \hat{s}_k$, the exponentials in [4.69] take the value of 1 and thus do not contribute to $\Phi_X(z)$ at all.

The number of elements in the set $\mathcal{K}$ is equal to the Hamming distance between $\boldsymbol{x}$ and $\hat{\boldsymbol{x}}$, i.e. the number of components, for which $\boldsymbol{x}$ and $\hat{\boldsymbol{x}}$ differ. We note this Hamming distance by $d_H(\boldsymbol{x}, \hat{\boldsymbol{x}})$.

The $\Phi_X(z)$ function then takes the value:

$$\Phi_X(z) = \prod_{k \in \mathcal{K}} \frac{1}{1 - z(N_0 z - 1)|s_k - \hat{s}_k|^2} \qquad [4.70]$$

Since the choice $z = 1/2N_0$ minimizes each term of the product, and thus $\Phi_X(z)$ for real z, using [4.66] we obtain the upper bound:

$$P\{\boldsymbol{x} \to \hat{\boldsymbol{x}}\} \leq \prod_{k \in \hat{\mathcal{K}}} \frac{1}{1 + |s_k - \hat{s}_k|^2 / 4N_0} \qquad [4.71]$$

EXAMPLE 4.12. Let us calculate the upper Chernoff bound for the probability of error considering a block code with a rate of R_c. Let us suppose that we use an antipodal binary constellation with waveforms whose energy is $\mathcal{E}$, and that the demodulation is coherent with a perfect knowledge of the state of the channel. Here we use [4.71], noting that for $\hat{s}_k \neq s_k$ we have:

$$|s_k - \hat{s}_k|^2 = 4\bar{\mathcal{E}} = 4R_c\bar{\mathcal{E}}_b$$

where $\bar{\mathcal{E}}_b$ is the average energy per bit. For two words of code $\boldsymbol{x}, \hat{\boldsymbol{x}}$ of Hamming distance $d_H(\boldsymbol{x}, \hat{\boldsymbol{x}})$ we obtain:

$$P\{\boldsymbol{x} \to \hat{\boldsymbol{x}}\} \leq \left(\frac{1}{1 + R_c\bar{\mathcal{E}}_b/N_0}\right)^{d_H(\boldsymbol{x}, \hat{\boldsymbol{x}})}$$

and thus, for a linear code:

$$P(e) = P(e \mid \boldsymbol{x}) \leq \sum_d A_d \left(\frac{1}{1 + R_c\bar{\mathcal{E}}_b/N_0}\right)^d$$

where the exponent of the sum takes its values in the set of non-zero Hamming weights of the code, and A_d is the number of words of code whose Hamming weight is d. We can see that for sufficiently high signal-to-noise ratios the term dominating in the expression of $P(e)$ is the one whose exhibitor is $d_{\min}$, the minimum Hamming distance of the code. The fact that the probability of error is inversely proportional to the signal-to-noise ratio to the power of $d_{\min}$ can be seen as a *diversity of code* of the order $d_{\min}$. In this context, the various diagrams of diversity can be interpreted as the manifestations of a code with repetition, whose Hamming distance is equal to the diversity.

How should a code for the Rayleigh channel be chosen?

We can take a bound of [4.71] writing:

$$P\{\boldsymbol{x} \to \hat{\mathbf{x}}\} \leq \prod_{k \in \mathcal{K}} \frac{1}{|s_k - \hat{s}_k|^2/4N_0}$$

$$= \frac{1}{[\delta^2(\boldsymbol{x}, \hat{\boldsymbol{x}})/4N_0]^{d_H(\boldsymbol{x}, \hat{\boldsymbol{x}})}} \qquad [4.72]$$

(which is close to the Chernoff bound for sufficiently small values of N_0). The quantity:

$$\delta^2(\boldsymbol{x}, \hat{\boldsymbol{x}}) \triangleq \left[\prod_{k \in \mathcal{K}} |s_k - \hat{s}_k|^2\right]^{1/d_H(\boldsymbol{x}, \hat{\boldsymbol{x}})}$$

is the geometric mean of the squares of Euclidean distances between the components of $\boldsymbol{x}$ and $\hat{\boldsymbol{x}}$. The latter result shows that the probability of error is (approximately) inversely proportional to the product of the squared Euclidean distances between the components of $\boldsymbol{x}$, $\hat{\boldsymbol{x}}$ that are different, and to a power of the signal-to-noise ratio whose exhibitor is the Hamming distance between $\boldsymbol{x}$ and $\hat{\boldsymbol{x}}$.

We know that the bound by union of the probability of error of a system with coding can be obtained by summing the probabilities of error per pair associated with all the various "error events". For small values of power spectral density of the N_0 noise, i.e., for strong signal-to-noise ratios, a small number of terms contribute a dominant share to the bound. Within the framework of this discussion this corresponds to error events whose Hamming distance $d_H(\boldsymbol{x}, \hat{\boldsymbol{x}})$ is minimum. We note this quantity as L_c to underline the fact that it corresponds to the diversity brought by the code. We have:

$$P\{\boldsymbol{x} \rightarrow \hat{\boldsymbol{x}}\} \overset{\sim}{\leqslant} \frac{\nu}{[\delta^2(\boldsymbol{x},\, \hat{\boldsymbol{x}})/4N_0]^{L_c}} \qquad [4.73]$$

where ν is the number of error events that dominate the bound. For error events having the same Hamming distance, the values taken by $\delta^2(\boldsymbol{x}, \hat{\boldsymbol{x}})$ and by ν are also considerable. This observation can be used to choose the codes for the Rayleigh channel with a high signal-to-noise ratio. The Euclidean distance, which plays a central part in the choice of a code for the Gaussian channel, plays a secondary part here, and we can verify that, in general, codes optimized for the Gaussian channel will not be optimal for the Rayleigh channel.

We will also be able to note that for "conventional" systems that separate binary modulation from binary coding, the Hamming distance is proportional to the Euclidean distance and, thus, a system optimized for the Gaussian channel will also be optimal for the Rayleigh channel. This solution offers the advantage of being robust, i.e. to be powerful for the Rayleigh channel as well as for the Gaussian channel.

4.13. Bit interleaved coded modulation (BICM)

We observed at the end of the previous section that a code powerful for the Gaussian channel and the Rayleigh channel must lead to a large Euclidean distance and have a large Hamming distance. We can obtain this result using bit interleaved coded modulations. Such a system can be obtained by carrying out a code diversity equal to the number of bits, rather than the number of signals in an error event, as is the case for trellis-coded modulations. It is initially necessary to interleave the bits at the output of the binary encoder, and to use a suitable metric in the *soft* decoder.

The result is that for certain channels we obtain no advantage by combining coding and modulation.

This solution is robust, because changes of behavior of the physical channel do not affect the performances of the coded system.

The performance of BICM depend separately on the Euclidean distance between the signals of the constellation used and the Hamming distance of the selected code. The metric to be used here differs from the usual metric: with TCM the metric associated to the transmitted signal s is $p(r|s)$, while for BICM the metric is:

$$\sum_{s \in \mathcal{S}_i(b)} p(r \mid s) \qquad [4.74]$$

where $\mathcal{S}_i(b)$ is the subset of signals in the constellation $\mathcal{S}$ whose binary label is $b(b \in \{0, 1\})$ in the ith position. We see then that the performance of BICM will depend on the labeling used: in particular, Gray coding is more powerful than coding coming from the Ungerboeck partition.

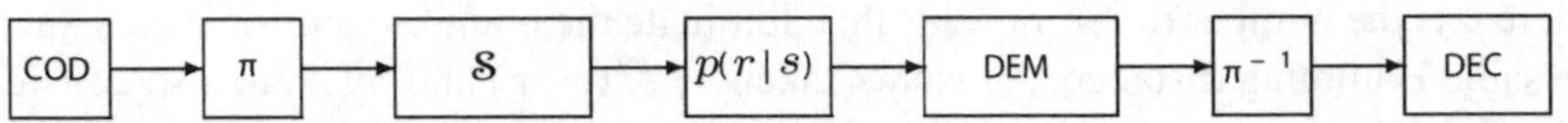

Figure 4.35. *Block diagram of a transmission system with traditional coded modulation and BICM. For the traditional coded modulation π represents the interleaver at the signal level, while in the case of BICM π is the interlacer at the bit level*

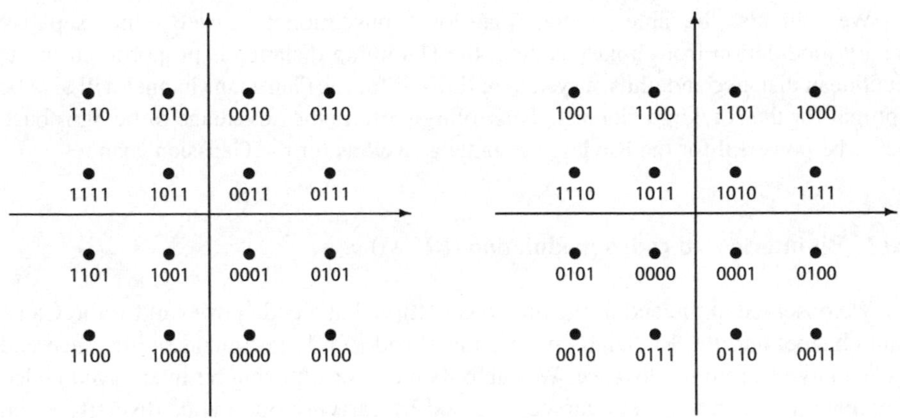

Figure 4.36. *16QAM signals with UNGERBOECK and GRAY coding*

BICM increases the Hamming distance to the detriment of the Euclidean distance: see Table 4.3.

Encoder Memory	BICM		TCM	
	δ^2_{free}	d_H	δ^2_{free}	d_H
2	1.2	3	2.0	1
3	1.6	4	2.4	2
4	1.6	4	2.8	2
5	2.4	6	3.2	2
6	2.4	6	3.6	3
7	3.2	8	3.6	3
8	3.2	8	4.0	3

Table 4.3. *Euclidean and Hamming distances of certain BICM and TCM for a 16QAM constellation and a rate of transmission of 3 bits per pair of dimensions (average energy being standardized to 1)*

4.14. Bibliography

[BEN 99] BENEDETTO S., BIGLIERI E., *Digital Transmission Principles with Wireless Applications*. Plenum, New York, 1999.

[BIG 84] BIGLIERI E., "High level modulation and coding for nonlinear satellite channels", *IEEE Trans. Commun.*, Vol. COM-32, No. 5, May 1984.

[BIG 91] BIGLIERI E., DIVSALAR D., MCLANE P.J., SIMON M.K, *Introduction to Trellis-Coded Modulation with Applications*. Macmillan, New York, 1991.

[BIG 98] BIGLIERI E., PROAKIS J., SHAMAI S. (SHITZ), "Fading channels: information-theoretic and communications aspects", *IEEE Trans. Inform. Theory*, Vol. 44, No. 6, pp. 2619–2693, Oct. 1998.

[CAI 98] CAIRE G., TARICCO G., BIGLIERI E., "Bit-interleaved coded modulation", *IEEE Trans. Inform. Theory*, Vol. 44, No. 3, pp. 927–946, Mar. 1998.

[FOR 98] FORNEY, JR., G.D. UNGERBOECK G., "Modulation and coding for linear Gaussian channels" *IEEE Trans. Inform. Theory*, Vol. 44, No. 6, pp. 2384–2415, Oct. 1998.

[UNG 82] UNGERBOECK G., "Channel coding with multilevel/phase signals", *IEEE Trans. Inform. Theory*, Vol. IT-28, pp. 55–67, Jan. 1982.

[WEI 82] WEI L.-F., "Rotationally invariant convolutional channel encoding with expanded signalspace, Part II: nonlinear codes", *IEEE J. Select. Areas Commun.*, Vol. 2, pp. 672–686, Sept. 1984.

[WEI 84] WEI L.-F., "Trellis-coded modulation using multidimensional constellations" *IEEE Trans. Inform. Theory*, Vol. IT-33, pp. 483–501, July 1982.

[WEI 89] WEI L.-F., "Rotationally invariant trellis-coded modulations with multidimensional M-PSK" *IEEEJ. Select. Areas Commun.*, Vol. 7, pp. 1285–1295, Dec. 1989.

Chapter 5

Turbocodes

5.1. History of turbocodes

The invention of turbocodes does not derive from a linear and limpid theory, much less a beautiful mathematical development. It is the product of a long search, whose origin is to be found in the intuitions and work of some European researchers: Gerard Battail, Joachim Hagenauer and Peter Hoeher who, at the end of the 1980s [BAT 87, BAT 89, HAG 89a, HAG 89b], announced the promise of probabilistic treatment in communication systems. Others before, in particular in the United States, such as Michael Tanner [TAN 81] and Robert Gallager [GAL 62], had earlier come up with coding and decoding processes that were the precursors of turbocodes.

In the laboratories of the Ecole Nationale Supérieure de Telecommunications de Bretagne (Brittany National Telecommunications Graduate School) some sought the simplest way possible to translate the Viterbi algorithm with soft output (SOVA: *Soft-output Viterbi Algorithm*) proposed in [BAT 87], in MOS transistors. A suitable solution [BER 93a] was found after two years, which made it possible for the researchers to form an opinion on probabilistic decoding. Thus, they observed following Battail and Hagenauer that a decoder with soft input and output could be regarded as an amplifier of signal-to-noise ratio, which encouraged them to implement concepts commonly used in amplifiers, in particular, negative feedback. We must, however, note that this parallel with amplifiers only makes sense if the values considered at the input and the output of the decoder provide information on

---

Chapter written by Claude BERROU, Catherine DOUILLARD, Michel JÉZÉQUEL and Annie PICART.

the same data, i.e. in practice, if the code is systematic, which was not the case for convolutional codes used hitherto.

The development of turbocodes passed through many very pragmatic stages, just as the introduction of neologisms, such as "parallel concatenation" or "extrinsic information", now integrated in the jargon of the information theory. Here in a few words are the reflections that marked out this work.

5.1.1. *Concatenation*

With the simplified version of the SOVA it became possible to cascade the "signal-to-noise ratio amplifiers" and to carry out the experiments reported in [HAG 89b], namely to decode a classical (i.e. serial) concatenation of two normal (i.e. non-systematic, non-recursive) convolutional codes, or even more than two. Concatenation is a simple means of obtaining high distances and thus large asymptotic gains [FOR 66], but performance with a low signal-to-noise ratio is degraded by the obligation to distribute redundant energy between the various constituent codes. This apparent antagonism between increased distance and good behavior with strong noise seemed to be impossible to circumvent in the search for good corrector codes.

Figure 5.1 represents the first diagram of concatenated coding and corresponding decoding, developed to highlight the contribution (approximately 1.5 dB) of pondering the internal code at the decoder output.

5.1.2. *Negative feedback in the decoder*

The use of information in the receiver in Figure 5.1 is far from optimal. Indeed, the first elementary decoder benefits only from Y_1 redundancy symbols produced by the internal encoder. The second decoder in turn benefits from Y_2 redundancy symbols and the work of the decoder that precedes it. This dissymmetry in the use of received information suggests re-injecting the result of the operation of the outer decoder into the inner decoder in a form to be defined. This re-injection of the output into the input is similar to the principle of the turbo engine, which gave its prefix to the turbocode[1], although would have been more rigorous to speak only of turbodecoding, since no negative feedback intervenes in concatenated coding.

1. This can also be written as turbo-code as well as turbo code, according to the local custom.

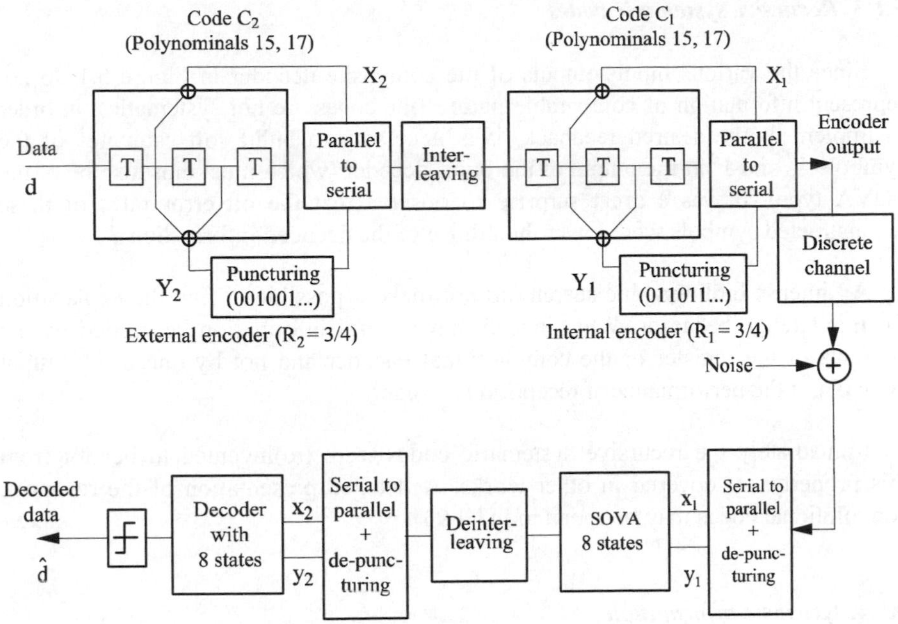

Figure 5.1. *Serial concatenation of two convolutional codes with 3/4 (external code) and 2/3 (internal code) outputs, leading to a total output of 1/2. Decoding of the internal code using a Viterbi algorithm with soft output (SOVA). It is on the basis of this diagram suggested by G. Battail [BAT 89, BAT 87], J. Hagenauer and P. Hoeher [HAG 89a, HAG 89b] that the turbocodes have been developed through successive improvements*

Digital information processing has at least one great disadvantage: it does not easily deal with the technique of negative feedback, which is simple to implement in analog circuits. Due to delays (trellis, interleaving, etc.), an iterative procedure must be used[2]. This procedure increases the latency of the decoder, but the constant progress of micro-electronics makes it possible today to do that which would not have been reasonable a little while ago. Two material solutions are possible depending on the flow of processed information. If this flow is low, a single decoding processor functioning at high clock frequency can carry out all the iterations necessary with a tolerable added delay. If the flow is high, a cascade of decoding modules can be implemented as a monolith to allow *pipe line* processing at a high speed (in this case, which is typically that of the diffusion, the problems of latency are generally less crucial).

2. Perhaps one day analog electronics will remove this handicap [LOE 98].

5.1.3. *Recursive systematic codes*

Since the various inputs/outputs of the composite decoder in Figure 5.1 do not represent information of comparable nature (the codes are not systematic), in order to implement the desired feedback, it is necessary to build soft estimates of the symbols X^2 and Y^2 at the output of the outer decoder (which must thus also be of the SOVA type). It was a great surprise to observe that the bit error rates of these reconstructed symbols were lower than those of the decoded information $\hat{d}$.

An intense bibliographic search did not make it possible to find the explanation for this strange behavior. Why, indeed, have useful information be carried by the contents of the register of the convolutional encoder and not by one of its output symbols, if the performance at reception is worse?

Immediately the recursive systematic codes were (re)invented to benefit from this property not covered in other works. A detailed presentation of the recursive convolutional codes may be found in [THI 93].

5.1.4. *Extrinsic information*

The SOVA decoder of a systematic code provides a good estimate of the Logarithm of Likelihood Ratio (LLR) with respect to d_i unit coded to the moment i, that is naturally seen as the sum of two contributions. The first, intrinsic information resulting from the transmission channel directly linked to d_i, is already available before any decoding; the second, extrinsic information, is brought by the decoding of the link (convolutional, parity, etc.) that exists between the data d_i and other symbols of the codeword.

In a turbo decoder it is the extrinsic information, in the form of a probability or a LLR, that must be exchanged between the various (typically two) processors seeking to converge towards the same decision on the transmitted codeword.

Indeed, intrinsic information, which is already exploited by each of these processors, should not be used a second time as new information, for fear of missing the errors (an instability, as would be said in electronics).

This principle, in a certain fashion, has already been posed by Gallager in [GAL 62]. It has also been used by Lodge *et al.* [LOD 93].

5.1.5. *Parallel concatenation*

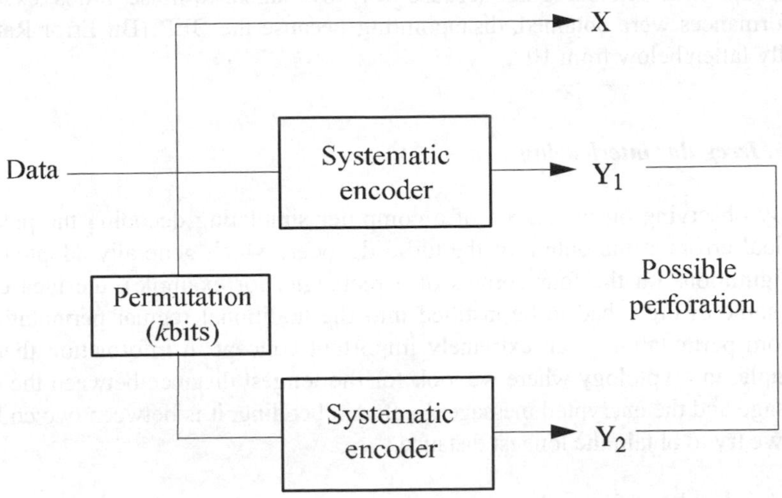

Figure 5.2. *Parallel concatenation: a symmetric structure of coding (and decoding) that makes it possible to obtain, with a given coding output, more redundancy symbols than serial concatenation, and thus a better diversity*

It is not by analogy with the code-products that the idea of concatenation known as parallel was born. More trivially, it was necessary to simplify the problems of clock distribution in the broadband integrated circuit, which was the aim of the study, and traditional (serial) concatenation obliges to consider different clocks for internal and outer decoders. Parallel concatenation (Figure 5.2) simplifies the architecture of the system because the two encoders and the two decoders associated with it function with the same clock, which is the data clock. However, what is the general principle of correct error coding? Distribute the energy available for transmission between various redundant symbols in such a way that the decoder can best benefit from a diversity effect. Moreover, the lower the output of coding is, the more the effect of diversity is important. With the same total coding output, the external encoder of a parallel concatenation functions with a lower output than that of serial concatenation. For example, to obtain a total output of 1/2, serial concatenation can associate two elementary codes with outputs 3/4 and 2/3, as in Figure 5.1, whereas parallel concatenation associates two codes with the same output 2/3. The decoder with a low signal-to-noise ratio is then favored by this addition of diversity. That explains why the threshold of convergence (i.e. the signal-to-noise ratio, at which the corresponding decoder starts to correct the majority of errors) is more favorable when concatenation is parallel.

The results of the first simulations were very encouraging and disappointing at the same time: encouraging because for low signal-to-noise ratios exceptional performances were obtained, disappointing because the BER (Bit Error Rates) had hardly fallen below from 10^{-4}.

5.1.6. *Irregular interleaving*

By observing on the screen of a computer simulating decoding the patterns of residual errors at the output of the turbo decoder, which generally adopted regular configurations (at the four corners of a rectangle, for example), the idea emerged that some disorder had to be instilled into the traditional regular permutation. The random permutation is an extremely important concept in information theory, for example, in cryptology where we look for the longest distance between the original message and the encrypted message. In channel coding, it is between two codewords that we try to obtain the longest distance.

This problem of irregular permutation is conceptually captivating because it uses algebra, geometry and coding. With a serial concatenated encoder, regular interleaving can be enough to obtain a large minimum distance, but – as has already been highlighted – the threshold of convergence of the algorithm of "turbo" decoding is further removed than that of parallel concatenation. Perhaps, the composite code, which will join together the advantageous properties of each of the two concatenation paradigms will be found quickly.

5.2. A simple and convincing illustration of the turbo effect

It is only interesting to apply retroactive processing to a reception chain if the data to be estimated there is correlated and/or redundant. Let us consider the simple case of a phase modulation at 2 items[3] (PSK2), preceded by a convolutional coding with output 1/2 (Figure 5.3). This slightly magical example is proposed by Narayanan and Stüber [NAR 99].

The receiver performs a demodulation operation followed by corrector decoding. The demodulator provides estimates x_i and y_i of the transmitted symbols, treated as independent, although they come from the same encoder, which the demodulator does not take into account. That is an important loss of information.

3. We prefer speaking of constellations with M points than of constellations with M states. The two names coincide only if the point is addressed exactly by the contents of a coding register.

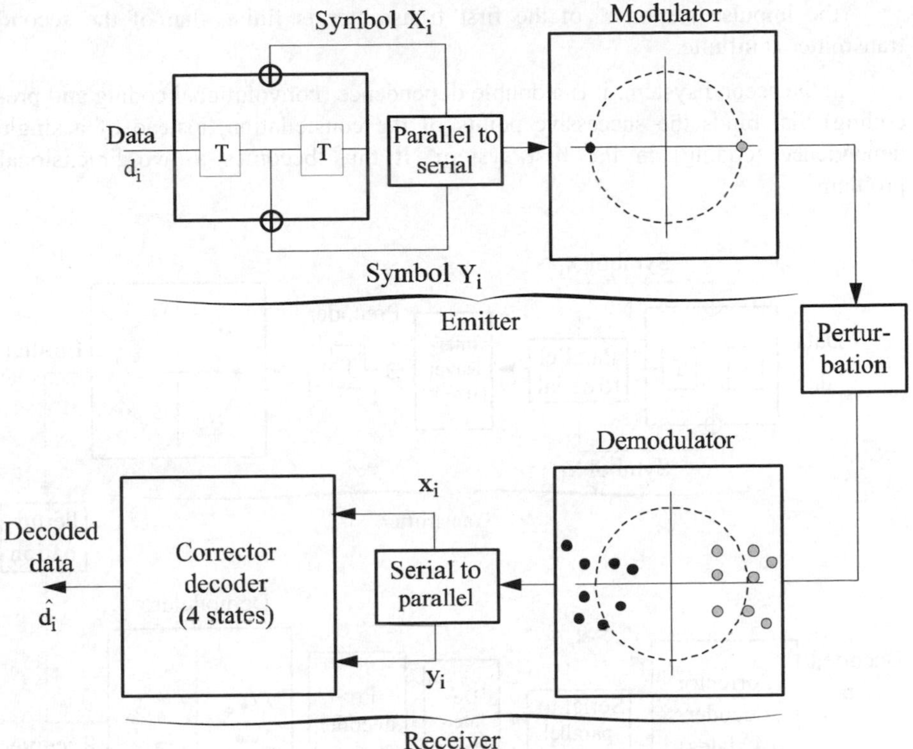

Figure 5.3. *A simple diagram of convolutional coding and modulation at 2 points*

Let us modify the transmission chain intercalating between encoder and modulator an interleaver of size *k* and a recursive pre-coder (Figure 5.4).

In the receiver, a trellis pre-decoder with 2 states and a de-interleaver are also inserted between the demodulator and the corrector decoder.

We have thus quite simply replaced PSK2 modulation by a differential modulation at 2 points and introduced a temporal permutation by means of the interleaver.

The modifications made to the transmission diagram are rather simple. However, they confer remarkable properties on it:

– the number of states for the transmitter of the first system is 4; for the second, it is 2^{k+3};

– the impulse response[4] of the first transmitter is finite; that of the second transmitter is infinite;

– in the second system, it is a double dependence (convolutional coding and pre-coding) that binds the successive points of the constellation, instead of a single dependence (coding) in the first system. It thus becomes a two-dimensional problem.

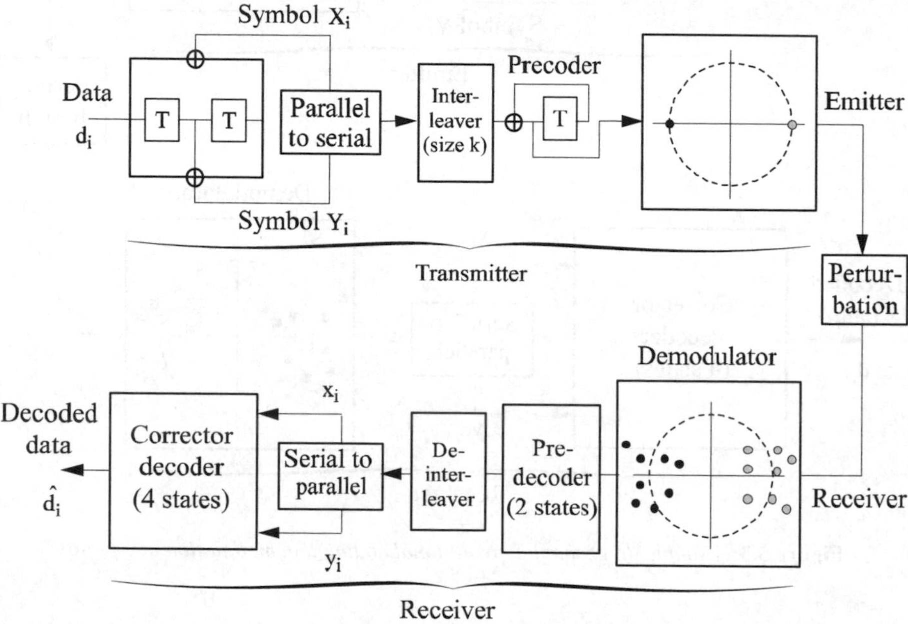

Figure 5.4. *Differential modulation with traditional convolutional coding and an interleaver*

These three properties constitute the three keys of "turbo" processing. Let us cover them one by one.

The number of states of a system is characteristic of its complexity, not necessarily of its performance. However, by judiciously choosing a permutation of interleaving, the trellis representative of the transmitter, which can be gigantic (for example a trellis with 2^{103} states for k = 100), can be good, i.e. the minimum Hamming distance between the series associated to two competing paths of the trellis can be large.

4. The impulse response of a linear binary system is the signal delivered at its output when its input is fed by the "all-zero" sequence except for one position.

The impulse response of the modified system is infinite because of the recursive character of the pre-coder. In what is that a remarkable property, for this transmitter in particular and for any coding system in general? Let us suppose that the binary sequence at the transmitter input has infinite length. If the receiver makes a mistake in its decision, it will be mistaken at least by two binary values (only one difference between its decisions and the good sequence would correspond to a transmitted signal indefinitely different from the good signal). These two false values will at least be delivered to a downstream processor. It is clear that if this processor is authorized to act retroactively on the decisions of the receiver, the information feedback will be more negative (antagonist) that in the case of a single error. The downstream processor using its own information will thus say "no" twice rather than once, questioning the decisions of the receiver more strongly.

The argument is also valid if instead of considering the information feedback of a downstream processor towards the receiver, we implement retroactive processing inside the receiver itself. In the case of Figure 5.4, for example, it is the corrector decoder, which in the event of an error opposes its "dissent" to the demodulator at least twice.

Lastly, and the latter point is not unrelated to the previous one, the successive points of the constellation are determined by a register with feedback. There is a recursive relation between the current data (U_i) and the contents (A_i) of the pre-coder's register: $A_i = A_{i-1} + U_i \pmod 2$ (Figure 5.5). Of course, without information on the current data U_i the pre-decoder can only deliver estimates a_i independent from each other for the transmitted points. But if the pre-decoder receives the feedback u_i on the U_i data from the corrector decoder, it itself becomes a corrector. Indeed, at every moment i it has an information pair (a_i, u_i) similar to that, which a systematic encoder with output 1/2 would provide, and that without having to add redundancy to the transmission! The double dependence of the symbols transmitted by the modulator is used profitably in the receiver by the feedback of the decoder towards the pre-decoder. This is the "turbo" effect.

Figure 5.6 provides the performance (BER according to the signal-to-noise ratio E_b/N_0) of the system with additive white Gaussian noise (AWGN). We notice the extreme simplicity of the receiver: a demodulator, a pre-decoder with 2 states, a de-interleaver and a decoder with 4 states. Even if, strictly speaking, it is necessary to multiply this complexity by 7 (the number of iterations carried out), that remains very reasonable compared to the traditional concatenated system (convolutional code with 64 states and Reed-Solomon code correcting $t = 8$ symbols), whose correction capacity appears lower than the BER considered.

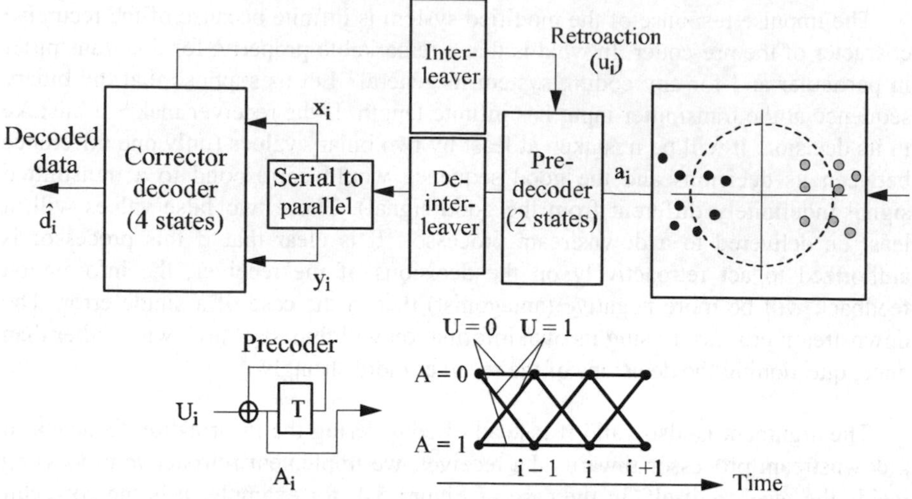

Figure 5.5. *The pre-decoder becomes corrector through feedback*

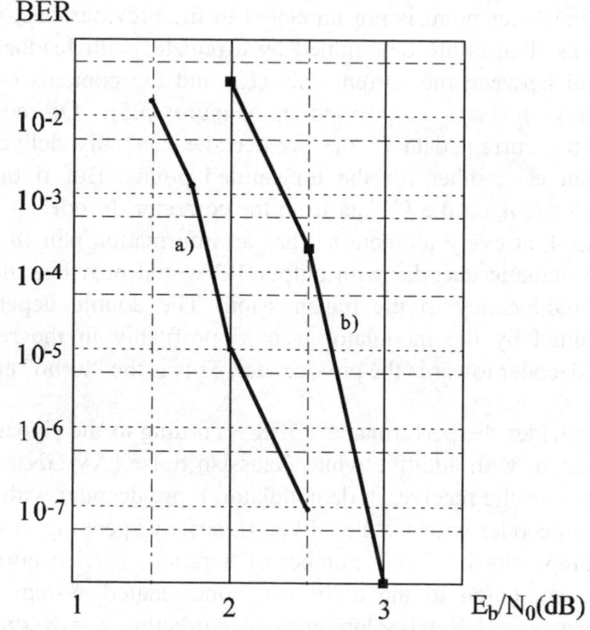

Figure 5.6. *(a) BER obtained with the system in Figure 5.4 and turbo processing (MAP algorithm, 7 iterations, k = 3392). (b) BER obtained with a convolutional encoder with 64 states concatenated with a Reed-Solomon code (204, 188, t = 8) and a PSK2 modulation. Simulations carried out with AWGN, according to [NAR 99]*

The principle of this modulator-encoder couple and the associated receiver can easily be extended to constellations with a larger number of points to so increase spectral efficiency, and by adopting codes with 8 or 16 states, its performance can be rendered close to the theoretical limit.

The example that we chose to illustrate the "turbo" effect is significant in more than one way. First of all, it shows that excellent performance can be obtained with a simple system. That goes against the pessimism exhibited until the recent years regarding the possibility of reconciling theory and practice of the Shannon paradigm. Then, it obviously attests the need for considering any reception problem "in two senses". If we removed feedback by u_i from Figure 5.5, we would lose approximately 4 dB in the resulting link for a BER of 10^{-5}! Finally, it highlights all the importance of the interleaving function, at the same time powerful, because it increases the minimum distance from the transmitter[5], and simple, because its implementation (writing and reading from memory) is rudimentary.

The downstream information feedback in a reception chain, which is the guiding principle of turbo decoding, has been generalized to various types of processing, such as detection and equalization [DOU 95, GLA 97], multi-user detection [ALE 99], reception with multiple antennae [LEK 00], etc., and proves to be essential each time we wish to exploit all the information available in a sequence of probabilistic treatments.

5.3. Turbocodes

5.3.1. *Coding*

Random codes have always, since the precursory work of Shannon [SHA 48], constituted a reference for correct error coding. Systematic random coding of a block of k bits of information, leading to a codeword of length n, at the first stage and once and for all, consists of randomly drawing and memorizing k binary "markers" of $n - k$ bits, whose memorizing address is noted i ($1 \leq i \leq k$). The redundancy associated to any information block is then formed by the modulo 2 summing up of all the markers whose address i is such that the ith bit of information equals "1". The codeword, finally, consists of the concatenation of k information bits and $n - k$ redundancy bits. The output R of the code is k/n.

5. The minimal distance of the system in Figure 5.4 without interlacing is 10 (5, for the code with generators 5,7 in octal, times 2 for the pre-coder). The asymptotic gain would thus only be 7 dB, whereas Figure 5.6 makes it possible to observe a gain already higher than 8 dB for a BER of 10^{-6}.

This very simple construction of the codeword is based on the property of linearity of addition and leads to increased minimum distances for sufficiently large values of $n - k$. Since two codewords differ in at least one bit of information and the redundancy bits are drawn randomly, the average minimum distance is $1+\frac{n-k}{2}$. However, the minimum distance of this code being a random variable, its various manifestations may be lower than this average size. A realistic approximation of the effective minimum distance is $\frac{n-k}{4}$ (an approximation deduced from the Gilbert-Varshamov limit). A way of building an almost random encoder is represented in Figure 5.7. It is a multiple parallel concatenation of Recursive Systematic Circular Convolutional (RSCC) codes [BER 99a]. The sequence of k binary data is coded N times by N RSCC encoders, each time in a different order. The Π_j permutations are drawn randomly (except for the first one, which can be the identity permutation). Each elementary encoder produces $\frac{k}{N}$ redundancy symbols (N being a divider of k), the total output of the concatenation being 1/2.

The proportion of input sequences of a recursive encoder constructed on the basis of a pseudo-random generator with memory ν, initially positioned in the state 0, that replace the register in the same state at the end of coding, is

$$p_1 = 2^{-\nu} \qquad [5.1]$$

because there are 2^{ν} possible return states. These sequences, called RTZ (Return to Zero) [POD 95], are linear combinations of the minimum RTZ sequence, given by the recursiveness polynomial of the generator ($1 + D + D^3$ in the case of Figure 5.7).

The proportion of *RTZ* sequences for the multidimensional encoder is lowered to:

$$p_N = 2^{-N\nu} \qquad [5.2]$$

because it is necessary that the sequence remains RTZ for the N encoders after each randomly drawn permutation.

Other sequences, of $1 - p_N$ proportion, produce codewords which have a minimum distance at least equal to

$$d_{\min} = \frac{k}{2N} \qquad [5.3]$$

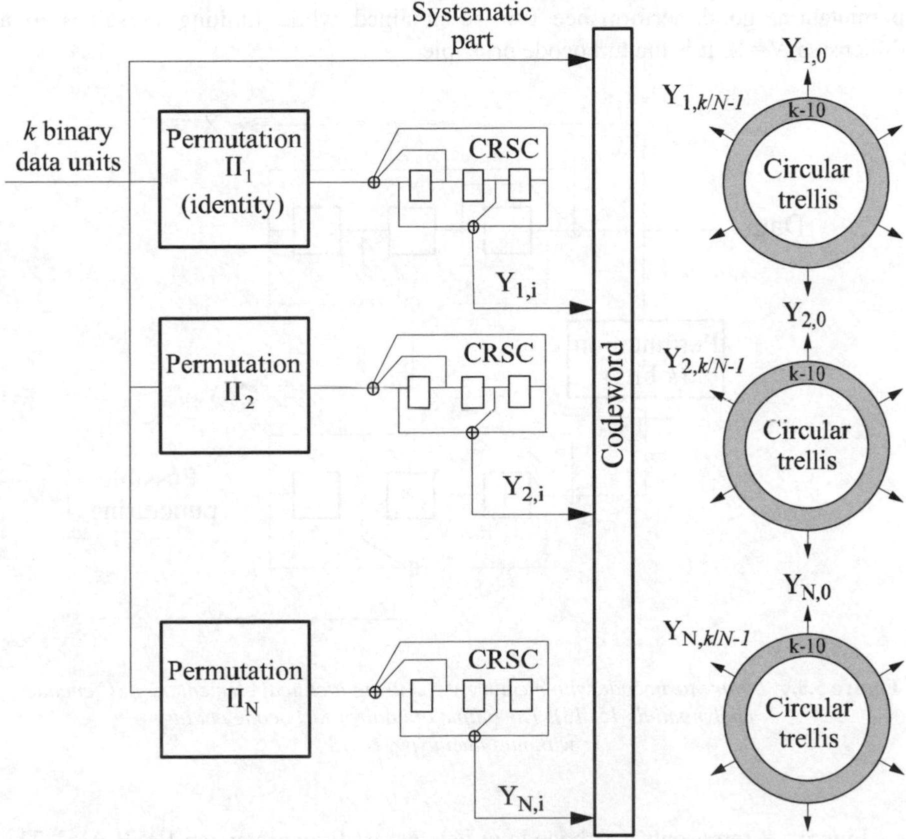

Figure 5.7. *Multiple parallel concatenation of recursive systematic circular convolutional codes (RSCC). Each encoder produces k/N redundancy symbols regularly distributed around each trellis. The total coding output is 1/2*

This value supposes that only one permuted sequence is not RTZ (in the worst case) and that the redundancy *Y* once takes the value "1" on average on out of two times in the corresponding circle. If we take, for example, $N = 8$ and $n = 3$, we obtain $p_8 \approx 10^{-7}$, and for sequences to be coded of length $k = 1024$, we have $d_{min} = 64$, which is quite a sufficient minimum distance.

Extremely fortunately, from the point of view of complexity, it is not necessary to retain such a large dimension N.

In fact, by replacing the random permutation Π_2 by a judiciously elaborate permutation, good performance can be obtained while limiting ourselves to a dimension $N = 2$. It is the turbocode principle.

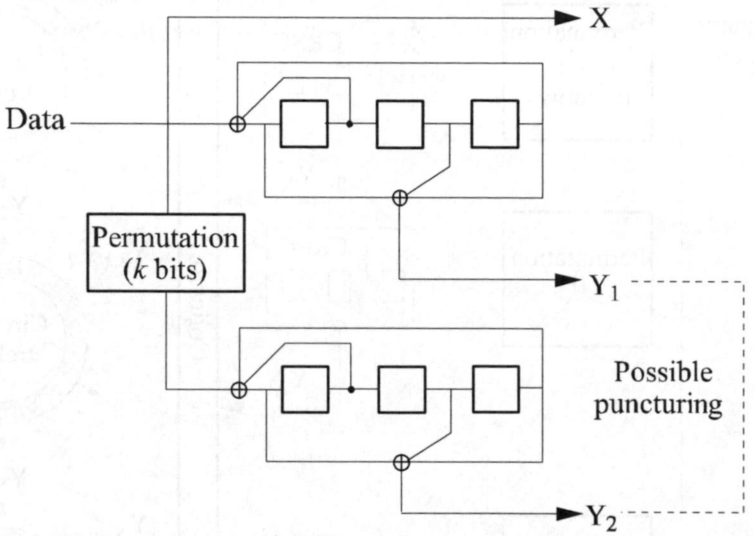

Figure 5.8. *A binary turbocode with memory $\nu = 3$ using identical elementary RSC encoders (polynomials 15, 13). The output of natural turbocode coding, without puncturing, is 1/3*

Figure 5.8 represents a turbocode in its most traditional version [BER 93b]. The input binary message of length k is coded in its natural order and in an upset order by two RSC encoders called C_1 and C_2, which may be circular or not. In this example the two elementary encoders are identical (generating 15 polynomials for recursiveness and 13 for the construction of redundancy), but that is not essential. The output of natural coding, without puncturing, is 1/3. To obtain higher outputs, a puncturing of symbols, generally the redundancy ones, is carried out. Another means of achieving higher outputs is to adopt m-binary codes (see section 5.5).

The permutation function (Π) bearing on a message of finite size *k*, the turbocode is by construction a block code. However, to distinguish it from concatenated algebraic codes decoded in the "turbo" fashion, such as the product codes and the ones later referred to as "block turbocodes", this turbocoding scheme is known as "convolutional" or, more technically, PCCC (Parallel Concatenated Convolutional Code).

The arguments in favor of this coding scheme (some of which have already been introduced in the previous chapters) are as follows:

a) A decoder of convolutional code is vulnerable to the errors occurring in bursts. To code the block twice following two different orders (before and after permutation) is to make somewhat improbable the simultaneous appearance of error bursts at the input of the decoders of C_1 and C_2. If grouped errors occur at the input of the decoder of C_1, the permutation disperses them in time and they become isolated, easily corrigible, errors for the decoder of C_2. The reasoning also holds for error bursts at the input of the second decoder, which before permutation correspond to isolated errors. Thus, two-dimensional coding clearly reduces, in at least one of two dimensions, the vulnerability of convolutional coding with respect to burst disturbances. But, which of the two decoders should be trusted when making the final decision? No criterion makes it possible to grant a greater confidence to one or the other. The answer is provided by the "turbo" algorithm that avoids having to make this choice. This algorithm implements probability exchanges between the two decoders and using these exchanges constrains them to converge towards the same decisions.

b) Parallel concatenation associates two codes with elementary outputs R_1 (C_1 code, with possible puncturing) and R_2 (C_2 code, with possible puncturing) and the total output is

$$R_p = \frac{R_1 R_2}{R_1 + R_2 - R_1 R_2} = \frac{R_1 R_2}{1-(1-R_1)(1-R_2)} \qquad [5.4]$$

This output is higher than the total output of a serially concatenated code ($R_s = R_1 R_2$), for same values of R_1 and R_2, and the difference is all the greater the lower the coding outputs. From that we deduce that for the same correction capacity of the elementary codes, parallel concatenation offers a better coding output, but this advantage is reduced when the outputs considered tend towards 1.

c) Parallel concatenation employs systematic codes. At least one of these codes must be recursive, for a fundamental reason related to the minimum input weight w_{min}, which is only 1 for non-recursive codes but is equal to 2 for recursive codes (see convolutional codes). For confirmation, let us observe Figure 5.9, which represents two non-recursive systematic codes, concatenated in parallel. The input sequence is "all-zero" (sequence of reference) except for one position. This single "1" disturbs the output of the C_1 encoder for a short period of time, equal to the constraint length 4 of the encoder. The redundant information Y_1 is poor, relative to this particular sequence, because it contains only 3 values different from "0". Upon permutation, whatever it is, the sequence is still "all-zero", except for only one position. Again, this "1" disturbs the output of the C_2 encoder during a period of time equal to the constraint length, and the redundancy Y_2 delivered by the second code is as poor as the first. In fact, the minimum distance of this two-dimensional code is not higher than that of a single code with the same output as that of the concatenated code. If we replace at least one of the two non-recursive encoders by a recursive encoder, the "all-zero" sequence except for one position is no longer a *RTZ* sequence for this recursive encoder, and the redundancy that it produces then has a much higher weight.

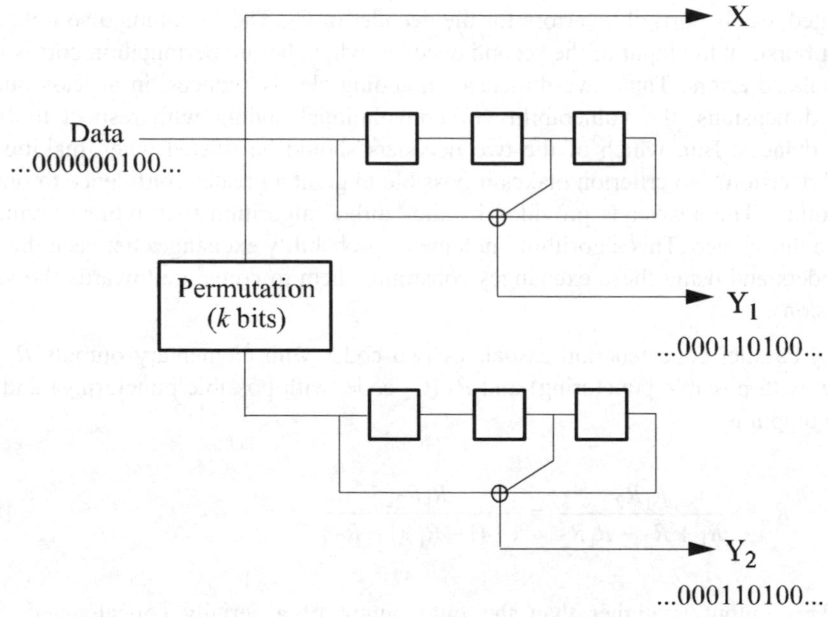

Figure 5.9. *Parallel concatenation of non-recursive systematic codes constitutes a poor code with respect to information sequences of weight 1. In this example, the redundancy symbols Y_1 and Y_2 each contain only 3 values different from "0"*

d) As we saw at the beginning of this chapter, it is possible to increase the dimension of the code using more than two elementary encoders. The result is a significant increase in the minimum distance. Beyond the 4th or 5th dimension with a set of randomly drawn permutations, the turbocode is almost comparable to a random code, with very high minimum distances. Unfortunately, the threshold of convergence of the turbo decoder, i.e. the signal-to-noise ratio, at which this one can start to correct the majority of the errors, is degraded as the dimension grows.

Indeed, the very principle of turbo decoding consists of considering elementary codes one after another, iteratively. Since their redundancy rate decreases as the dimension of the composite code grows, the first stages of decoding are penalized with respect to a dimension 2 code.

This antagonism between increased minimum distance and convergence threshold, which we have already discussed previously, is found in almost all the coding and decoding structures that can be imagined[6]. We can sometimes gain in one of the two behaviors, but it is almost always at the expense of the other.

The parameters defining a particular turbocode are as follows:

a) *m* is the number of bits in the symbols applied to the turboencoder. The applications known to date consider binary ($m = 1$) or double-binary ($m = 2$) symbols (see section 5.5).

b) Each of the two elementary encoders C_1 and C_2 is characterized by:

– its code memory ν;

– its recursiveness and redundancy generating polynomials;

– its output.

The values of ν are in practice lower than or equal to 4. The generating polynomials are generally those used for the traditional convolutional codes and were the subject of numerous works from the 1980s to the 1990s.

c) The way in which we carry out the permutation is important when the target BER is lower than approximately 10^{-5}. Above this value, performance is not very sensitive to the permutation, under the condition, of course, that it at least respects the dispersion principle (that can, for example, be a regular permutation). For low or very low target error rates, performance is dictated by the minimum distance of the code, and the latter strongly depends on the permutation Π (see section 5.4).

d) The puncturing pattern must be as regular as possible, similarly to the normal practice for classical convolutional codes. Apart from this rule, the puncturing pattern is defined in close relationship with the permutation function, when we look for very low error rates.

6. A typical example is the Reed-Solomon code concatenated with a convolutional code. The minimal distance is high, but we can only benefit from it sufficiently far from the theoretical limit.

Puncturing is performed traditionally on the redundancy symbols. In certain cases, it may rather be possible to puncture the information symbols, in order to increase the minimum distance of the code. That is done at the expense of the convergence threshold of the turbo decoder. Indeed, from this point of view, puncturing data shared by the two decoders is more penalizing than puncturing data that is useful only for one of the decoders.

5.3.2. *The termination of constituent codes*

The use of a convolutional code to protect an information block reveals discontinuities during decoding at both ends of this block. Indeed, the decoding of a symbol, whatever it is, must use all of the information, prior and post this symbol. The ends of the block thus cannot simultaneously benefit from the past and the future during the decoding process, and the performance is degraded by a greater vulnerability of information at the beginning and the end of the block[7]. This problem can be circumvented if the decoder knows the initial state and the final state of the encoder. Thus, it is easy to force the initial state of the encoder by a resetting of the register. It is also possible either to transmit the final state of the coding register, or to force it to a value known by the decoder, by the addition of additional bits called *tail bits* to the initial message. We can also adopt the principle of circularity (*tail-biting*) which ensures the continuity of the states at both ends. These various techniques are known under the name of trellis closing.

For a turbocode, the closing of two trellises should be considered and several solutions can be considered:

– Do nothing in particular concerning the final states: the information located at the end of the block, in the natural order as well as in the permutated order, is then less well protected. This leads to a reduction in the asymptotic gain, but this degradation, which depends on the size of the block, can be compatible with certain applications. It should be noted that not-closing the trellis more strongly penalizes the PER (Package Error Rate) that the BER.

– Close the trellises of one or two elementary codes: the CCSDS [CCS 98] and UMTS [3GPP 99] standards use this technique. The bits ensuring the closing of one of the two trellises are not used in the other encoder. These bits are, therefore, not turbocoded which leads, although to a lesser extent, to the same disadvantages as those presented in the previous case. Moreover, the transmission of the closing bits involves a reduction in the coding output and thus spectral effectiveness.

---

7. Neglecting information brought by what precedes or by what follows in the decoding of a symbol protected by a convolutional code leads, on average, to a loss of 3 dB from the coding gain.

– Use an interleaving allowing an automatic closing of the trellis: it is possible to close the trellis of a turbocode automatically, without adding closing bits, by slightly transforming the coding diagram (autoconcatenation) and by using an interleaving complying with certain periodicity rules. This solution described in [BER 96] does not decrease the spectral efficiency but imposes constraints on interleaving, which make it difficult to control the performance for low error rates.

– Adopt a circular coding: a circular convolutional code encoder guarantees that the initial state and the final state of the register are identical. The trellis then takes the shape of a circle, which, from the point of view of the decoder, can be regarded as a trellis of infinite length [BET 98]. This process of closing, already known as *tail-biting* for non-recursive codes, makes it possible to combine spectral efficiency and good performance for strong and low error rates for turbocodes. This technique was retained in the DVB-RCS and DVB-RCT [DVB 00, DVB 01] standards, for example, and is presented in detail in the following section.

5.3.2.1. *Recursive convolutional circular codes*

An example of a recursive systematic convolutional encoder (double-binary code with 3 memory elements) is provided in Figure 5.10. At the moment i the state $\mathbf{S}_i$ of register is a function of the preceding state $\mathbf{S}_{i-1}$ and of the previous input vector $\mathbf{T}_{i-1}$:

$$\mathbf{S}_i = \mathbf{G}\mathbf{S}_{i-1} + \mathbf{T}_{i-1} \qquad [5.5]$$

where $\mathbf{G}$ is the generator matrix of the considered wedged register.

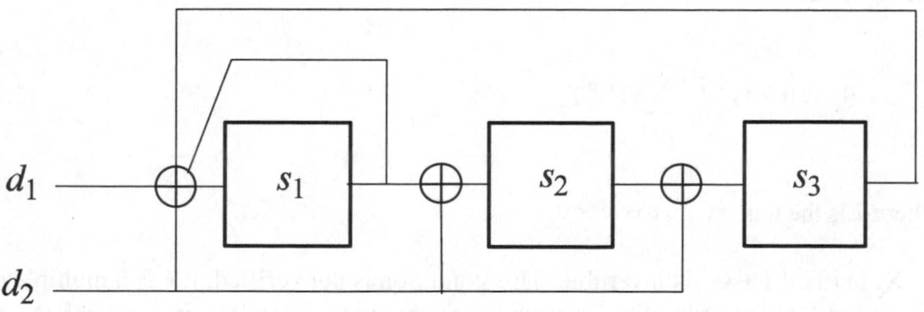

Figure 5.10. *Example of a double-binary recursive convolutional encoder with code memory $v = 3$. The encoder outputs, as well as the temporal index i, are not represented*

The vectors and the matrix in the example in Figure 5.10 are:

$$\mathbf{S}_i = \begin{bmatrix} s_{1,i} \\ s_{2,i} \\ s_{3,i} \end{bmatrix}; \mathbf{T}_i = \begin{bmatrix} d_{1,i} + d_{2,i} \\ d_{2,i} \\ d_{2,i} \end{bmatrix}; \mathbf{G} = \begin{bmatrix} 1 & 0 & 1 \\ 1 & 0 & 0 \\ 0 & 1 & 0 \end{bmatrix}$$

More generally, for a code of memory ν, the vectors **S** and **T** contain ν components and the matrix **G** has the size ν × ν.

From [5.5], $\mathbf{S}_i$ can be expressed according to the initial state $\mathbf{S}_0$ and of the data **T** applied to the encoder:

$$\mathbf{S}_i = \mathbf{G}^i \mathbf{S}_0 + \sum_{p=1}^{i} \mathbf{G}^{i-p} \mathbf{T}_{p-1}$$

If k is the length of the sequence, it is possible to find a state of circulation, noted $\mathbf{S}_c$, such that: $\mathbf{S}_c = \mathbf{S}_0 = \mathbf{S}_k$. We must then have:

$$\mathbf{S}_c = \mathbf{G}^k \mathbf{S}_c + \sum_{p=1}^{k} \mathbf{G}^{k-p} \mathbf{T}_{p-1} \qquad [5.6]$$

which yields:

$$\mathbf{S}_c = \left(\mathbf{I} + \mathbf{G}^k\right)^{-1} \sum_{p=1}^{k} \mathbf{G}^{k-p} \mathbf{T}_{p-1} \qquad [5.7]$$

where **I** is the unitary matrix ν × ν.

$\mathbf{S}_c$ exists if $\mathbf{I} + \mathbf{G}^k$ is invertible. This condition is not verified, if k is a multiple of the period L of the generating sequence of the recursive encoder, since $\mathbf{G}^L = \mathbf{I}$. As an example, the encoder in Figure 5.10 has a period $L = 7$.

If the encoder is initialized at the state $\mathbf{S}_c$ it will return to this state once the given k have been coded. The final state and the initial state can then be fused and the coding trellis can be compared to a circle.

The calculation of $\mathbf{S}_c$ requires pre-processing. First of all, the encoder is initialized to the "all zero" state, then the message to be coded is applied to it for the first time, ignoring the redundancy produced during this stage. Using equation [5.6], the final state, noted $\mathbf{S}_k^0$, is given by:

$$\mathbf{S}_k^0 = \sum_{p=1}^{k} \mathbf{G}^{k-p}\mathbf{T}_{p-1} \qquad [5.8]$$

The circulation state can then be given as follows:

$$\mathbf{S}_c = \left(\mathbf{I} + \mathbf{G}^k\right)^{-1}\mathbf{S}_k^0 \qquad [5.9]$$

In practice, the use of a table makes it possible to determine $\mathbf{S}_c$ on the basis of $\mathbf{S}_k^0$. Finally, the encoder having been initialized to the circulation state, the message to be coded is applied to it again to generate the redundancy sequence.

This elegant and efficient method of transforming a convolutional code into a block code nonetheless has a disadvantage that is the pre-processing stage to find out $\mathbf{S}_k^0$. This introduces latency, which is not, however, a major handicap, because the encoder with a simpler structure than the decoder can function at a higher clock frequency.

The state of circulation not being known by the decoder *a priori*, it must be estimated through a preliminary data processing stage preceding this state. This operation, known as "prolog", relates to a certain amount of data located at the end of the block and the prolog starts by assigning uniform (or metric) probabilities to the initial states of the trellis. The estimate of the circulation state is good as soon as around ten redundancy symbols have been exploited in the prolog.

5.3.3. *Decoding*

The decoding of a turbocode is based on the general diagram in Figure 5.11. The loop makes it possible for each decoder to benefit from all of the available information. The values considered for each node of the set-up are LLRs.

The LLR at the output of a systematic code decoder can be seen as the sum of two terms: intrinsic information, stemming from the transmission channel, and extrinsic information, which this decoder adds to the former to carry out its correction work.

Since intrinsic information is used by the two decoders (at different moments), it is the extrinsic information produced by each decoder that must be transmitted to the other one as new information to ensure joint convergence.

Section 5.3.4 describes the operations performed for the calculation of extrinsic information, by implementation of the MAP algorithm or of its simplified Max-Log-MAP version.

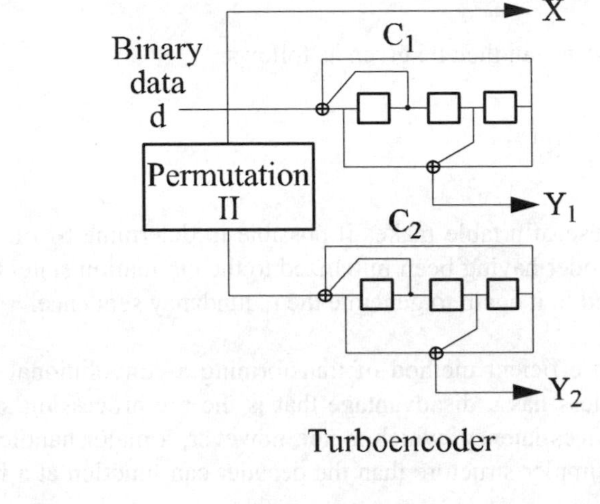

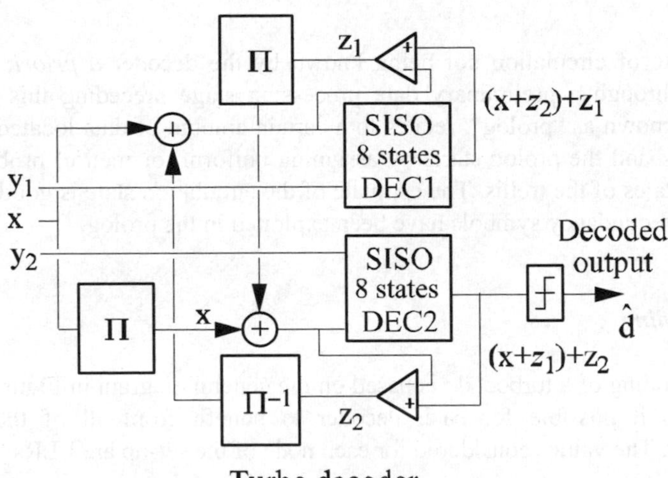

Figure 5.11. *Turboencoder with 8 states and basic structure of the corresponding turbo decoder. The two elementary decoders of the SISO type (soft input/soft output) exchange probabilistic information, known as extrinsic (z)*

The exchange of extrinsic information, in a digital processing circuit, must be implemented through an iterative process: the first decoding by DEC_1 and the memorization of extrinsic information z_1, the second decoding by DEC_2 and the memorization of extrinsic information z_2 (end of the first iteration), new call for DEC_1 and memorization of z_1, etc. Various material architectures with more or less elevated degrees of parallelism are possible to accelerate iterative decoding.

Had we wanted to decode the turbocode using only one decoder, which would take into account all the possible states of the encoder, for each element of the decoded message we would obtain one, and only one, probability of having a binary value equal to "0" or "1". The composite structure in Figure 5.11 in turn employs two decoders working jointly. By analogy with the result that the single decoder would provide, it is necessary for them to converge towards the same decisions with the same probabilities for each unit of data considered. It is the guiding principle of "turbo" processing, which justifies the structure of the decoder, as the following reasoning demonstrates.

The role of a SISO decoder (*soft input/soft output*; see section 5.3.4) is to treat the LLR to try to increase their signal-to-noise ratio, thanks to the energy brought by the redundancy symbols (i.e. y_1 for DEC_1, y_2 for DEC_2). The LLR produced by a binary code decoder with respect to the data unit d can be written simply as

$$LLR_{output}(d) = LLR_{input}(d) + z(d) \qquad [5.10]$$

where $z(d)$ is the extrinsic information with respect to d. The LLR is improved if z is negative, if d is "0", or positive, if d is "1".

After p iterations, the output of DEC1 is

$$LLR_{output,1}^{p}(d) = (x + z_2^{p-1}(d)) + z_1^{p}(d)$$

and the output of DEC2 is

$$LLR_{output,2}^{p}(d) = (x + z_1^{p-1}(d)) + z_2^{p}(d)$$

If the iterative process converges towards a stable solution, $z_1{}^p(d) - z_1{}^{p-1}(d)$ and $z_2{}^p(d) - z_2{}^{p-1}(d)$ tend towards zero when p tends towards infinity. Consequently, the two LLRs with respect to d become identical, which satisfies the fundamental criterion stated higher. As for the proof of convergence in itself, which is not trivial, the reader may refer to [WEI 01, DUA 01], for example.

Apart from the permutation and inverse permutation functions, Figure 5.12 details the operations performed during turbo decoding:

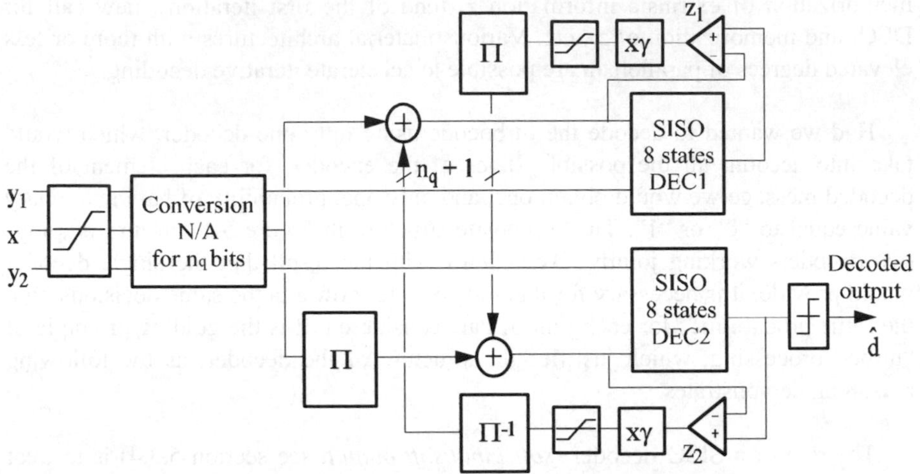

Figure 5.12. *Detailed operations (peaking, quantification, attenuation of extrinsic information) in the turbo decoder in Figure 5.11*

a) Analog-to-digital conversion (A/D) transforms information coming from the demodulator into samples exploitable by the digital decoder. Two parameters are involved in this operation: n_q, the number of quantification bits and Q, the scale factor, i.e. the relation between the average absolute value of the quantified signal and its maximum absolute value. n_q is fixed at a compromise value between the required precision, which depends on the type of modulation, and the decoder complexity: typically 3 or 4 for an PSK4, 5 or 6 for an QAM16, for example. The value of Q depends on the modulation, the coding output and the type of channel. For example, it is larger for a Rayleigh channel than for a Gaussian channel.

b) SISO decoding increases the signal-to-noise ratio equivalent of the LLR, i.e. it provides extrinsic information z_{output} which is more reliable at the output than at the input (z_{input}). The convergence of the iterative process will depend on the transfer function $SNR(z_{output}) = G(SNR(z_{input}))$ of each decoder (see [DIV 01] for example).

When information is not available at the input of a SISO decoder due to puncturing, a neutral value (analog zero) replaces this missing information.

c) When the elementary decoding algorithm is not the optimal algorithm (MAP)[8], but a simplified version, extrinsic information must undergo some transformations before being used by a decoder:

– the multiplication of extrinsic information by the factor γ, less than 1, guarantees the stability of the wedged structure. γ may vary along the iterations, for example of 0.7 at the start of the iterative process to 1 for the last iteration;

– the clipping of extrinsic information simultaneously responds to the need to limit the size of the memories and also to take part in the stability of the process. A typical value of maximal dynamic of extrinsic information is twice the dynamic of the decoder input.

d) Binary decision-making is carried out by a comparison with the analog threshold 0.

The number of iterations required by turbo decoding depends on the size of the block and coding output. The larger the decoded block, the longer the cycles enclosed inside the graph that could be associated to the concatenated code, and convergence is slower. It is the same when coding outputs are low. In practice, we limit the number of iterations to a value ranging between 4 and 10, depending on the constraints of speed, latency and consumption imposed by the material received.

Figure 5.13 gives an example of binary turbocode performance drawn from the UMTS standards [3GPP 99]. We may observe a decrease of the PER, very close to the theoretical limit (given by the method of sphere stacking), but also a rather marked change of slope, due to a minimum distance which is not extraordinary ($d_{min} = 26$) for an output of 1/3.

8. If the MAP algorithm is used, it is preferable that extrinsic information be expressed by a probability and not an LLR, which avoids having to calculate a useless variance [ROB 94].

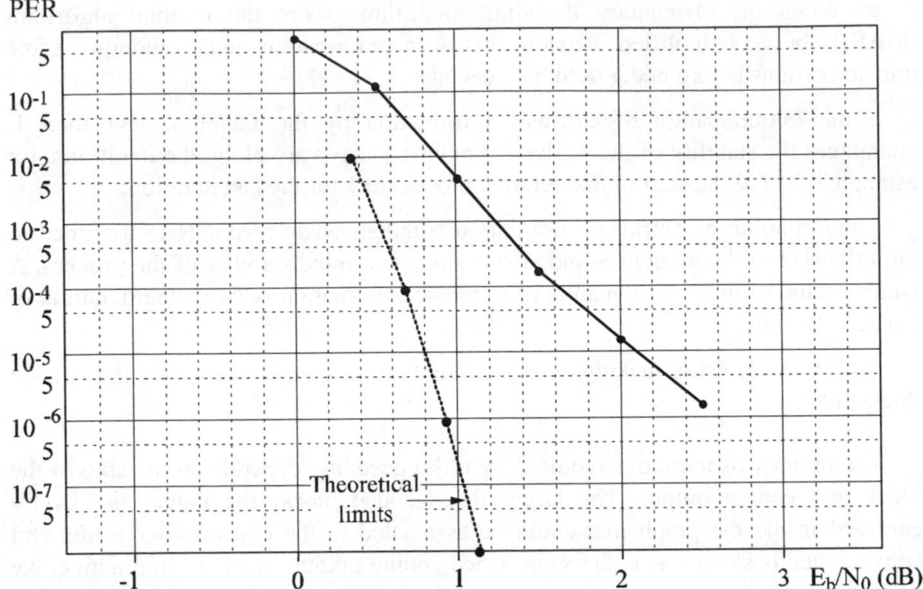

Figure 5.13. *Performance in PER of the turbocode from the UMTS standards for k = 640 and R = 1/3 in a Gaussian channel. Decoding according to the Max-Log-MAP algorithm with 6 iterations*

5.3.4. *SISO decoding and extrinsic information*

Here we develop the processing carried out in practice in a SISO decoder using the MAP algorithm (*Maximum A Posteriori*) [BAH 74] or its simplified version Max-Log-MAP [ROB 97] to decode RSC codes with m binary inputs and to implement iterative decoding.

5.3.4.1. *Notations*

A sequence of data $\mathbf{d}$ is defined by $\mathbf{d} \equiv \mathbf{d}_0^{k-1} = (\mathbf{d}_0 \cdots \mathbf{d}_i \cdots \mathbf{d}_{k-1})$, where $\mathbf{d}_i$ is the vector of m-binary data applied to the encoder input at the moment i: $\mathbf{d}_i = (d_{i,1} \cdots d_{i,l} \cdots d_{i,m})$. The value of $\mathbf{d}_i$ could also be represented by the whole scalar value $j = \sum_{l=1}^{m} 2^{l-1} d_{i,l}$ ranging between 0 and 2^m-1, and we will then write $\mathbf{d}_i \equiv j$.

In the case of a BPSK or QPSK modulation, the coded and modulated sequence $\mathbf{u} \equiv \mathbf{u}_0^{k-1} = (\mathbf{u}_0 \cdots \mathbf{u}_i \cdots \mathbf{u}_{k-1})$ consists of vectors $\mathbf{u}_i$ of the size $m + m'$: $\mathbf{u}_i = (u_{i,1} \cdots u_{i,l} \cdots u_{i,m+m'})$, where $u_{i,l} = \pm 1$ for $l = 1 \cdots m + m'$ and m' is the number of redundancy bits added to the m information bits. The $u_{i,l}$ symbol is representative of a systematic bit for $l \leq m$ and of a redundancy bit for $l > m$.

The sequence observed at the demodulator output is noted $\mathbf{v} \equiv \mathbf{v}_0^{k-1} = (\mathbf{v}_0 \cdots \mathbf{v}_i \cdots \mathbf{v}_{k-1})$, with $\mathbf{v}_i = (v_{i,1} \cdots v_{i,l} \cdots v_{i,m+m'})$. The series of encoder states between moments 0 and k is noted $\mathbf{S} = \mathbf{S}_0^k = (\mathbf{S}_0 \cdots \mathbf{S}_i \cdots \mathbf{S}_k)$. The decoding equations described hereafter are based on the results presented in Chapter 3.

5.3.4.2. *Decoding using the MAP criterion*

At time instant i the soft (probabilistic) estimates provided by the MAP decoder are the 2^m *a posteriori* probabilities (APP) $\Pr(\mathbf{d}_i \equiv j \mid \mathbf{v})$, $j = 0 \cdots 2^m - 1$. The corresponding hard decision $\hat{\mathbf{d}}_i$ is the binary representation of the value j that maximizes the APP.

Each APP can be expressed according to the joint probabilities $p(\mathbf{d}_i = j, \mathbf{v})$:

$$\Pr(\mathbf{d}_i \equiv j \mid \mathbf{v}) = \frac{p(\mathbf{d}_i \equiv j, \mathbf{v})}{p(\mathbf{v})} = \frac{p(\mathbf{d}_i \equiv j, \mathbf{v})}{\sum_{l=0}^{2^m - 1} p(\mathbf{d}_i \equiv l, \mathbf{v})} \qquad [5.11]$$

In practice, we calculate the joint probabilities $p(\mathbf{d}_i \equiv j, \mathbf{v})$ for $j = 0 \cdots 2^m - 1$, then each APP is obtained by normalization.

The trellis representative of a code with memory ν has 2^ν states, taking their scalar value s in $(0, 2^\nu - 1)$. Joint probabilities are calculated on the basis of forward $\alpha_i(s)$ and backward $\beta_i(s)$ probabilities and of branch probabilities $g_i(s', s)$:

$$p(\mathbf{d}_i \equiv j, \mathbf{v}) = \sum_{(s',s)/\mathbf{d}(s',s) \equiv j} \underbrace{p(\mathbf{v}_{i+1}^{k-1} \mid \mathbf{S}_{i+1} = s)}_{\beta_{i+1}(s)} \cdot \underbrace{p(\mathbf{S}_i = s', \mathbf{v}_1^{i-1})}_{\alpha_i(s')} \cdot \underbrace{p(\mathbf{S}_{i+1} = s, \mathbf{v}_i \mid \mathbf{S}_i = s')}_{g_i(s',s)} \qquad [5.12]$$

where $(s', s)/\mathbf{d}(s', s) \equiv j$ designates the set of state to state transitions $s' \to s$ associated to the m-binary information unit j. This unit is, of course, always the same in a trellis that is invariant in time.

The value $g_i(s',s)$ has the expression:

$$g_i(s',s) = \Pr^{a}\left(\mathbf{d}_i \equiv j, \mathbf{d}(s',s) \equiv j\right).p\left(\mathbf{v}_i \middle| \mathbf{u}_i\right) \qquad [5.13]$$

where $\Pr^{a}\left(\mathbf{d}_i \equiv j, \mathbf{d}(s',s) \equiv j\right)$ is the *a priori* probability of transmission of the *m*-tuple of information corresponding to the transition $s' \to s$ of the trellis at time instant i and $\mathbf{u}_i$ is the set of symbols of systematic and redundant information associated to this transition. If the transition $s' \to s$ does not exist for $\mathbf{d}_i \equiv j$, then $\Pr^{a}\left(\mathbf{d}_i \equiv j, \mathbf{d}(s',s) \equiv j\right) = 0$, otherwise it is given by the source statistics, which are generally uniform in practice.

In the case of a Gaussian channel with binary input, the value $p(\mathbf{v}_i \mid \mathbf{u}_i)$ is written:

$$p\left(\mathbf{v}_i \mid \mathbf{u}_i\right) = \prod_{l=1}^{m+m'} \left(\frac{1}{\sigma\sqrt{2\pi}} \exp\left(-\frac{(v_{i,l} - u_{i,l})^2}{2\sigma^2} \right) \right) \qquad [5.14]$$

where σ^2 is the variance of the additive white Gaussian noise.

In practice, we retain only the terms specific to the transition considered and not eliminated by division in the expression [5.11]:

$$p'\left(\mathbf{v}_i \mid \mathbf{u}_i\right) = \exp\left(\frac{\sum_{l=1}^{m+m'} v_{i,l} \cdot u_{i,l}}{\sigma^2} \right) \qquad [5.15]$$

The front and back probabilities are deduced from the following recurrence relations:

$$\alpha_i(s) = \sum_{s'=0}^{2^{\nu}-1} \alpha_{i-1}(s')\, g_{i-1}(s',s) \text{ for } i = 1 \cdots k \qquad [5.16]$$

and

$$\beta_i(s) = \sum_{s'=0}^{2^{\nu}-1} \beta_{i+1}(s') g_i(s,s') \text{ for } i = k-1 \cdots 0 \qquad [5.17]$$

To avoid all the problems of precision or overflow in the representation of these values, it is advisable in practice to standardize them regularly. The initialization of the recursions depends on the availability or absence of knowledge regarding the encoder state at the beginning and the end of coding. If the initial state of the encoder $\mathbf{S}_0$ is known, then $\alpha_0(\mathbf{S}_0) = 1$ and $\alpha_0(s) = 0$ for any other state, otherwise all $\alpha_0(s)$ are initialized to the same value. The same rule applies to the probabilities β_k with respect to the final state $\mathbf{S}_k$. For circular codes, initialization is carried out automatically after the prolog stage, which starts on the basis of identical values for all the trellis states.

In the context of iterative decoding, the composite decoder uses two elementary decoders exchanging *extrinsic probabilities*. Consequently, the building block of decoding described previously must be reconsidered:

a) to take into account an extrinsic probability $\Pr_e^{ext}(\mathbf{d}_i \equiv j | \mathbf{v}')$ at the input in the expression [5.13], calculated by the other elementary decoder of the composite decoder, on the basis of its own input sequence **v'**;

b) to produce its own extrinsic probability $\Pr_s^{ext}(\mathbf{d}_i \equiv j | \mathbf{v})$, which will be used by the other elementary decoder.

In practice, for each value of j $j = 0 \cdots 2^m - 1$:

– a) in expression [5.13], the *a priori* probability $\Pr^a(\mathbf{d}_i \equiv j, \mathbf{d}(s',s) \equiv j)$ is replaced by the modified *a priori* probability $\Pr^{@}(\mathbf{d}_i \equiv j, \mathbf{d}(s',s) \equiv j)$, with the expression, to the nearest standardization factor:

$$\Pr^{@}(\mathbf{d}_i \equiv j, \mathbf{d}(s',s) \equiv j) = \Pr^a(\mathbf{d}_i \equiv j, \mathbf{d}(s',s) \equiv j).\Pr_e^{ext}(\mathbf{d}_i \equiv j | \mathbf{v}') \qquad [5.18]$$

– b) $\Pr_s^{ext}(\mathbf{d}_i \equiv j | \mathbf{v})$ is given by:

$$\Pr_s^{ext}(\mathbf{d}_i \equiv j | \mathbf{v}) = \frac{\sum\limits_{(s',s)/\mathbf{d}(s',s)\equiv j} \beta_{i+1}(s) \alpha_i(s') g_i^*(s',s)}{\sum\limits_{(s',s)} \beta_{i+1}(s) \alpha_i(s') g_i^*(s',s)} \qquad [5.19]$$

The terms $g_i^*(s',s)$ are non-nil if the transition $s' \rightarrow s$ exists in the code trellis. They are then deduced from the expression of $p(\mathbf{v}_i | \mathbf{u}_i)$ by eliminating the terms relating to systematic symbols. In the case of a transmission via a Gaussian channel with binary input, on the basis of the simplified expression [5.15] of $p'(\mathbf{v}_i | \mathbf{u}_i)$, we have:

$$g_i^*(s',s) = \exp\left(\frac{\sum_{l=m+1}^{m+m'} v_{i,l} \cdot u_{i,l}}{\sigma^2} \right) \quad [5.20]$$

5.3.4.3. *The simplified Max-Log-MAP algorithm*

Decoding using the MAP criterion requires a great number of operations, among which calculations of exponentials and multiplications. The rewriting of the decoding algorithm in the logarithmic domain simplifies processing. Balanced estimates provided by the decoder are then values proportional to the logarithms of the APP, known as *Log-APP* and noted L:

$$L_i(j) = -\frac{\sigma^2}{2} \ln \Pr\left(\mathbf{d}_i \equiv j \,|\, \mathbf{v}\right), \; j = 0 \cdots 2^m - 1 \quad [5.21]$$

We define the forward and backward metrics relating to the node s at the moment i, $M_i^\alpha(s)$ and $M_i^\beta(s)$, as well as the branch metric relating to the transition $s' \rightarrow s$ from the trellis at the moment i, $M_i(s',s)$, by:

$$\begin{aligned} M_i^\alpha(s) &= -\sigma^2 \ln \alpha_i(s) \\ M_i^\beta(s) &= -\sigma^2 \ln \beta_i(s) \\ M_i(s',s) &= -\sigma^2 \ln g_i(s',s) \end{aligned} \quad [5.22]$$

Let us introduce the size $A_i(j)$ defined by:

$$A_i(j) = -\sigma^2 \ln \sum_{(s',s)/\mathbf{d}(s',s) \equiv j} \beta_{i+1}(s) \alpha_i(s') g_i(s',s) \quad [5.23]$$

$L_i(j)$ may then be written, with reference to [5.11] and [5.12], in the form:

$$L_i(j) = \frac{1}{2}\left(A_i(j) - \sum_{l=0}^{2^m-1} A_i(l) \right) \qquad [5.24]$$

The expressions [5.23] and [5.24] can be simplified by applying the approximation known as *Max-Log*:

$$\ln(\exp(a) + \exp(b)) \approx \max(a,b) \qquad [5.25]$$

We then obtain for $A_i(j)$

$$A_i(j) \approx \min_{(s',s)/\mathbf{d}(s',s)\equiv j} \left(M_{i+1}^{\beta}(s) + M_i^{\alpha}(s') + M_i(s',s) \right) \qquad [5.26]$$

and for $L_i(j)$

$$L_i(j) = \frac{1}{2}\left(A_i(j) - \min_{l=0\cdots 2^m-1} A_i(l) \right) \qquad [5.27]$$

The hard decision taken by the decoder is the value of j, $j = 0 \cdots 2^m - 1$, which minimizes $A_i(j)$ or, in other words, annuls $L_i(j)$.

Let us introduce the L^{a} values proportional to the logarithms of the *a priori* probabilities $\Pr^{\mathrm{a}}$:

$$L_i^{\mathrm{a}}(j) = -\frac{\sigma^2}{2} \ln \Pr^{\mathrm{a}}(\mathbf{d}_i \equiv j) \qquad [5.28]$$

The branches metrics $M_i(s',s)$ are written, according to [5.13] and [5.22]:

$$M_i(s',s) = 2L_i^{\mathrm{a}}(\mathbf{d}(s',s)) - \sigma^2 \ln p(\mathbf{v}_i \mid \mathbf{u}_i) \qquad [5.29]$$

If the statistics *a priori* transmission of the *m*-tuples $\mathbf{d}_i$ are uniform, the term $2L_i^{\mathrm{a}}(\mathbf{d}(s',s))$ can be removed from the above relation, because the same value then appears in all the branch metrics.

In the case of a transmission via a Gaussian channel with binary input, we have, according to [5.15]:

$$M_i(s',s) = 2L_i^{\mathrm{a}}(\mathbf{d}(s',s)) - \sum_{l=1}^{m+m'} v_{i,l} \cdot u_{i,l} \qquad [5.30]$$

The simplifying *Max-Log* application in the expressions [5.16] and [5.17] leads to the calculation of metrics before and back by the following recurrence relations:

$$M_i^{\alpha}(s) = \min_{s'=0\cdots 2^{\nu}-1}\left(M_{i-1}^{\alpha}(s') - \sum_{l=1}^{m+m'} v_{i-1,l} \cdot u_{i-1,l} + 2L_{i-1}^{\mathrm{a}}(\mathbf{d}(s',s)) \right) \qquad [5.31]$$

$$M_i^{\beta}(s) = \min_{s'=0\cdots 2^{\nu}-1}\left(M_{i+1}^{\beta}(s') - \sum_{l=1}^{m+m'} v_{i,l} \cdot u_{i,l} + 2L_i^{\mathrm{a}}(\mathbf{d}(s,s')) \right) \qquad [5.32]$$

The application of the Max-Log-MAP algorithm in fact amounts to carrying out a double Viterbi decoding, in the forward and backward directions. For that reason it is also called the *dual Viterbi* algorithm.

If the starting state of the encoder $\mathbf{S}_0$ is known, then $M_0^{\alpha}(\mathbf{S}_0) = 0$ and $M_0^{\alpha}(s) = +\infty$ for any other state, otherwise all the $M_0^{\alpha}(s)$ are initialized to the same value. The same rule applies for the initialization of back metrics with respect to the final state $\mathbf{S}_k$. For circular codes all the metrics are initialized to the same value at the start of the prolog.

We will note that the presence of the coefficient σ^2 in the definition [5.21] of $L_i(j)$ makes it possible to dispense with the knowledge of this parameter for the calculation of the metrics and, consequently, for all decoding. It is an important advantage of the Max-Log-MAP algorithm compared to the original MAP algorithm.

In the context of iterative decoding the term $L_i^{\mathrm{a}}(j)$ is modified in order to take into account the extrinsic information input $L_{i,\mathrm{e}}^{\mathrm{ext}}(j)$ coming from the other elementary decoder:

$$L_i^{@}(j) = L_i^{\mathrm{a}}(j) + L_{i,\mathrm{e}}^{\mathrm{ext}}(j) \qquad [5.33]$$

In addition, the extrinsic information produced at output of the decoder is obtained by eliminating the terms containing direct information on $\mathbf{d}_i$ in $L_i(j)$, i.e. intrinsic information and *a priori*:

$$L_{i,s}^{\text{ext}}(j)=\frac{1}{2}\left[\min_{(s',s)/\mathbf{d}(s',s)\equiv j}\left(M_{i+1}^{\beta}(s)+M_i^{\alpha}(s')-\sum_{l=m+1}^{m+m'}v_{i,l}\cdot u_{i,l}\right)-\min_{(s',s)}\left(M_{i+1}^{\beta}(s)+M_i^{\alpha}(s')-\sum_{l=m+1}^{m+m'}v_{i,l}\cdot u_{i,l}\right)\right] \quad [5.34]$$

Let us note j_0 the value of j that minimizes the term $\left(M_{i+1}^{\beta}(s)+M_i^{\alpha}(s')-\sum_{l=m+1}^{m+m'}v_{i,l}\cdot u_{i,l}\right)$, i.e. cancels the extrinsic information $L_{i,s}^{\text{ext}}(j)$.

The expression of $L_i(j)$ can then be reformulated as follows:

$$L_i(j)=L_{i,s}^{\text{ext}}(j)+\frac{1}{2}\sum_{l=1}^{m}v_{i,l}\cdot\left(u_{i,l}\big|_{\mathbf{d}_i\equiv j}-u_{i,l}\big|_{\mathbf{d}_i\equiv j_0}\right)+\left(L_i^{@}(j)-L_i^{@}(j_0)\right) \quad [5.35]$$

This expression shows that in practice extrinsic information $L_{i,s}^{\text{ext}}(j)$ can be extracted from $L_i(j)$ by a simple subtraction. Since the term $\left(u_{i,l}\big|_{\mathbf{d}_i\equiv j}-u_{i,l}\big|_{\mathbf{d}_i\equiv j_0}\right)$ is equal to either 0 or ± 2 in practice, the factor $\frac{1}{2}$ in the definition [5.21] of $L_i(j)$ makes it possible to obtain a soft decision and outgoing extrinsic information on the same scale as the disturbed samples $v_{i,l}$.

5.4. The permutation function

Called interleaving or permutation, the technique that consists of dispersing data in time proves extremely useful in numerical communications. It is used to an advantage, for example, to reduce the effects of the more or less large attenuations in transmissions affected by fading, and, more generally, in situations where noise can deteriorate consecutive symbols. In the case of turbocodes, the permutation also makes it possible to effectively counter the appearance of error packages in at least one of the dimensions of the composite code. However, its role does not stop there: it also determines, in close connection with the properties of the constituent codes, the minimum distance of the concatenated code.

Let us consider the turbocode represented in Figure 5.8. The worst of the permutations that could be used is naturally the identity permutation, which minimizes the diversity of coding (we then have $Y_1 = Y_2$). On the other hand, the best imaginable but probably-non existent [SVI 95] permutation would allow the concatenated code to be equivalent to a sequential machine whose number of irreducible states would be 2^{k+6}. There are indeed $k + 6$ binary memorization characters in the structure: k for the permutation memory and 6 for the two convolutional codes.

Assimilating this sequential machine to a convolutional encoder, and for the usual values of k, the corresponding number of states would be very large; in any case, large enough to guarantee a large minimum distance. For example, a convolutional encoder with a code memory of 60 (10^{18} states!) exhibits a free distance of around 100 (for $R = 1/2$), which is quite sufficient.

Thus, from the worst to the best of permutations, the choice is broad and we still lack a solid and unifying theory on the design of permutations in a turbocode. That said, good permutations could, nevertheless, be defined to prepare standardized turbocoding diagrams.

5.4.1. *The regular permutation*

The starting point in the design of an interleaving is the regular permutation described in Figure 5.14 in two different forms. The first supposes that the block containing k bits can be organized as a table of M rows and N columns.

Interleaving then consists of writing the data to an *ad hoc* memory row by row and reading it column by column (Figure 5.14a). The second form is applied without an assumption regarding the value of k. After writing the data in a linear memory (address i, $0 \leq i \leq k - 1$) the block becomes akin to a circle, the two ends ($i = 0$ and $i = k - 1$) then being contiguous (Figure 5.14b).

Binary data is then extracted so that the jth unit read had been written to the position i, with the value:

$$i = P.j \mod. k \tag{5.36}$$

where P is an integer prime to k. To maximize the spread after permutation, between two consecutive bits in a natural order, whatever they are, and vice versa, P must be close to $\sqrt{2k}$ and such that

$$k \approx \frac{P}{2} \mod. P \qquad [5.37]$$

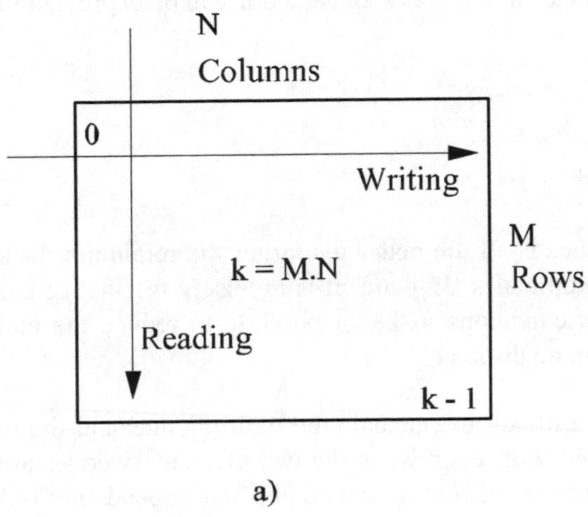

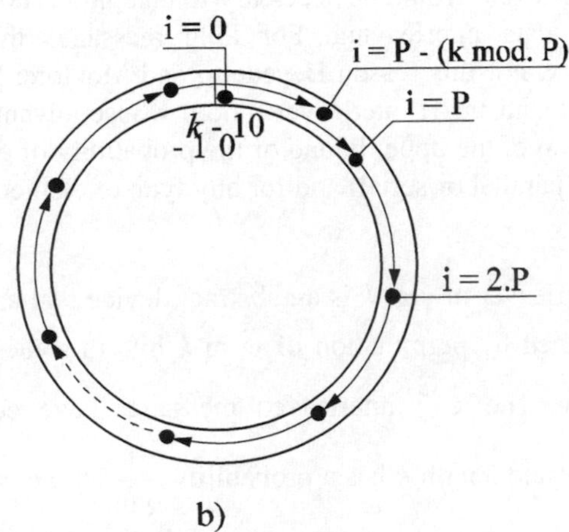

Figure 5.14. *Regular permutation in rectangular or circular form*

5.4.2. *Statistical approach*

The overall performance of a code obtained by parallel concatenation of two codes separated by an interleaver in terms of error probability simultaneously depends on the elementary codes and the interleaver used. Generally speaking, for a Gaussian channel and a binary symbol modulation, the upper bound of the probability of error with respect to the spectral multiplicities or coefficients and of the minimum distance of the concatenated code can be expressed in the form:

$$P_e \leq \sum_{d=d_{\min}}^{\infty} M_d \, e^{-Rd\frac{E_b}{N_0}} \qquad [5.38]$$

The performance is all the better the larger the minimum distance $d_{\min}$ is and the lower the multiplicities M_d are. It is precisely the fact of using an interleaver that under certain conditions makes it possible to reduce the multiplicities and to increase the minimum distance.

Actually, it is difficult to calculate the multiplicities and the minimum distance of the concatenated code, even when the two elementary codes are known, because the redundancy introduced by the second encoder depends not only on the original message, but also on the way in which the data is interleaved before coding. For an interleaver of a given size it would be necessary to take into account exhaustively all the possibilities of data interleaving. For long messages this method quickly becomes too complex. For this reason Benedetto and Montorsi [BEN 96] proposed to use a uniform (or statistical) interleaver model whose advantage lies in making possible the evaluation of the upper bound of the probability of error for any type of code concatenation (parallel or serial) and for any type of elementary code (block or convolutional).

The uniform interleaver of size k is an abstract device that associates one of the C_k^w messages obtained by permutation of w in k bits to a message of k bits and Hamming weight w. The C_k^w interleaved messages have equal probability, a message of weight w and length k has a probability $\dfrac{1}{C_k^w}$ to be coded by the second encoder. This interleaver model provides performances equal to that of an interleaver obtained by taking the average of the performances for all the possible deterministic interleavers of the same size (here equal to k). Thus, there is at least one deterministic interleaver, i.e. whose interleaving rule is fixed, that makes it possible to reach the performances of the uniform interleaver. For low error rates, it is in fact easy to find permutations that lead to much better performances than those obtained with the uniform interleaver.

Let us suppose that without the interleaver there are A_{wd} codewords at the distance d (the "all-zero" reference word implied) generated on the basis of a message of weight w. It is then demonstrated in [BEN 96] that the uniform interleaver associates only $w!k^{1-w}A_{wd}$ messages of weight w and distance d. Thanks to the uniform interleaver, the number of messages of weight w associated to a codeword at the distance d can be reduced, if the factor $w!k^{1-w}$ is less than 1. Consequently, for large interleavers (large k), the reduction factor is the more considerable the larger the weight of the message generating a sequence at the distance d is. For the lowest value of $w = w_{\min}$ the factor $k^{1-w_{\min}}$, called "interleaving gain", makes it possible to evaluate the minimum reduction of A_{wd} obtained using the interleaver.

Let us recall that $w_{\min}$ is the minimum weight of a message generating a codeword with distance d. For block and convolutional non-recursive codes, this parameter $w_{\min}$ being equal to 1, the interleaver does not reduce the multiplicities of code concatenated in parallel and does not bring interleaving gain. On the other hand, the parameter $w_{\min}$ of the recursive convolutional codes is equal to 2 and the interleaving gain is $k^{1-2} = \frac{1}{k}$. The reduction of the multiplicities of the concatenated code is thus proportionate to the size of the interleaver.

In conclusion, the statistical approach to permutation confirms the need to use RSC codes as constituent codes of a turbocode, and makes it possible to observe and evaluate an interleaving gain, as a function of the size of the interleaver and of the minimum weight of a message with distance d.

Coding gain is mainly visible for strong error rates. For low error rates, the minimum distance that should be maximized remains the main parameter. The statistical interleaver does not ensure a maximum minimum distance.

5.4.3. *Real permutations*

The traditional dilemma in the design of good permutations lies in the need to obtain a large minimum distance for two distinct classes of input sequences that require opposite processing. To highlight this problem, let us consider a turbocode with output 1/3, with a regular rectangular permutation (writing along the M rows, reading along the N columns) bearing on blocks of $k = M.N$ bits (Figure 5.15). The elementary encoders are encoders with 8 states whose period is 7 (recursiveness generator 15 in octal).

The first pattern (A) in Figure 5.15 describes a possible information sequence of weight $w = 2$: "10000001" for the C_1 code, which we will also call horizontal code. In fact, it is a minimum RTZ sequence of weight 2 for the encoder considered. The redundancy sequence produced by this encoder has a weight of 6 (exactly: "11001111"). The redundancy sequence produced by the vertical encoder C_2, for which the considered information sequence is also RTZ, is also richer as it is delivered in seven columns. Admitting that Y_2 is equal to "1" on average every other time, the weight of this redundancy sequence is approximately $w(Y_2) \approx \frac{7M}{2}$ leading to a large minimum distance. When we have k tend towards infinity through the values of M and $N (M \approx N \approx \sqrt{k})$, the weight of the redundancy sequence produced by one of the two codes for this type of pattern also tends towards infinity. We then say that the code is *good* for this type of pattern.

The second pattern (b) is that of minimum RTZ sequence of weight 3. There too, the redundancy sequence is poor for the first dimension and has a much higher weight for the second one. The conclusions are the same as previously.

The two other designs (c) represent examples of short RTZ sequences, in each of the two dimensions, combined into composite RTZ patterns with a total weight of 6 and 9. The minimum distances associated to these patterns (30 and 27 respectively for this code with output 1/3) are generally insufficient to ensure a good performance with a low error rate. Moreover, these distances are independent of the size of the block and thus, with respect to the patterns considered, the code is not *good*.

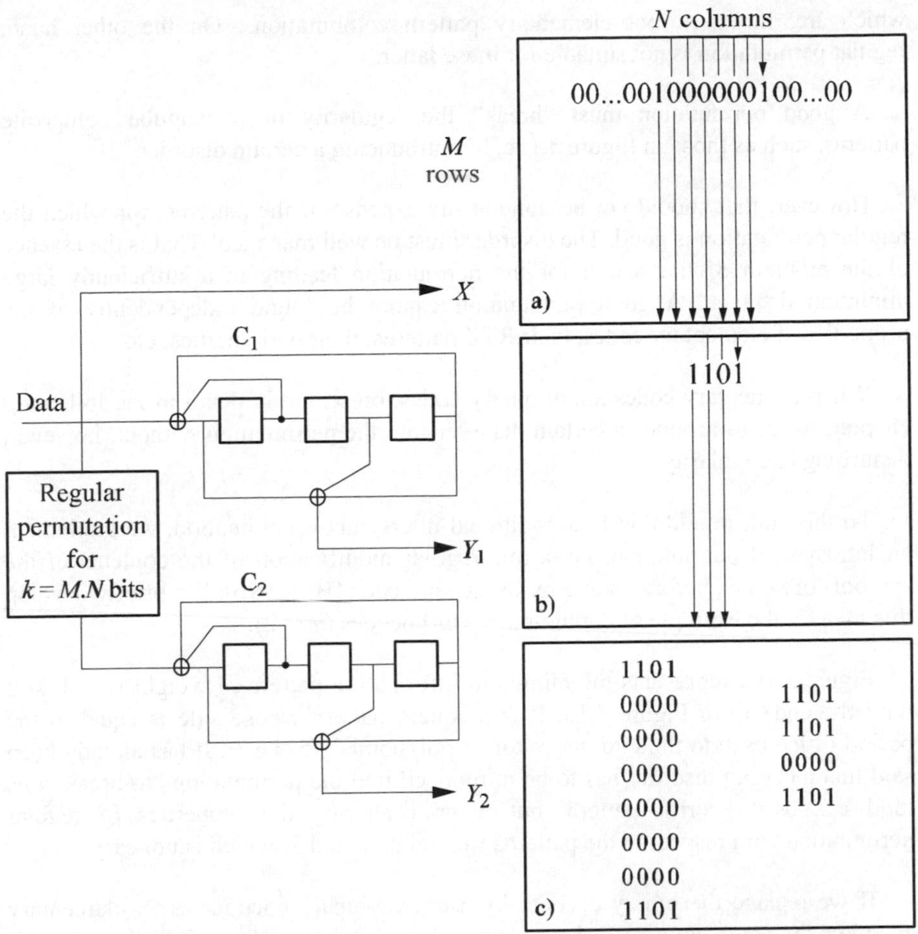

Figure 5.15. *Possible information patterns for weights 2, 3, 6 or 9 with a turbocode whose elementary encoders have a period of 7 and a regular permutation*

As for the sequences that are not RTZ, in at least one dimension, they correspond to sufficiently long redundancy messages so that their weights are not taken into account in the evaluation of the minimum code distance. This is particularly the case if circular codes are used, which are constructed so that any input sequence that is not RTZ influences all of the encoder's redundant output.

Regular permutation is thus a good permutation for the class of RTZ error patterns with a weight of $w \leq 3$, as well as for the patterns with greater weights, which are, however, not elementary pattern combinations. On the other hand, regular permutation is not suitable for these latter.

A good permutation must "break" the regularity of rectangular composite patterns, such as those in Figure 5.15c, by introducing a certain disorder.

However, that should not be done at the expense of the patterns, for which the regular permutation is good. The disorder must be well managed! That is the essence of the problem of the search for the permutation leading to a sufficiently large minimum distance. A good permutation cannot be found independently of the properties of elementary codes, their RTZ patterns, their periodicities, etc.

When elementary codes are m-binary codes, presented in detail in the following chapter, we can introduce a certain disorder into the permutation without, however, disturbing its regularity.

To this end, in addition to a traditional intersymbol permutation, we implement an intrasymbol permutation, i.e. a non-regular modification of the contents of the symbols of m bits, before coding by the second code [BER 99b]. We briefly develop this idea for the example of double-binary turbocodes ($m = 2$).

Figure 5.16a represents the minimum information pattern of weight $w = 4$, still with the code from Figure 5.15. It is a square pattern whose side is equal to the period of the pseudo-random generator of polynomial 15, i.e. 7. It has already been said that a certain disorder has to be introduced into the permutation "to break" this kind of possible error pattern, but without altering the properties of regular permutation with respect to the patterns for weight 2 and 3, which is not easy.

If we replace the binary encoder by a double-binary encoder as an elementary encoder, the error patterns to be considered are no longer formed by bits, but by pairs of bits. Figure 5.16b provides an example of a double-binary encoder, supplied by bit pairs (A, B) and possible error patterns, when the permutation is regular. The (A, B) pairs are numbered from **0** to **3**, according to the following correspondence:

(0,0): **0**; (0,1): **1;** (1,0): **2**; (1,1): **3**.

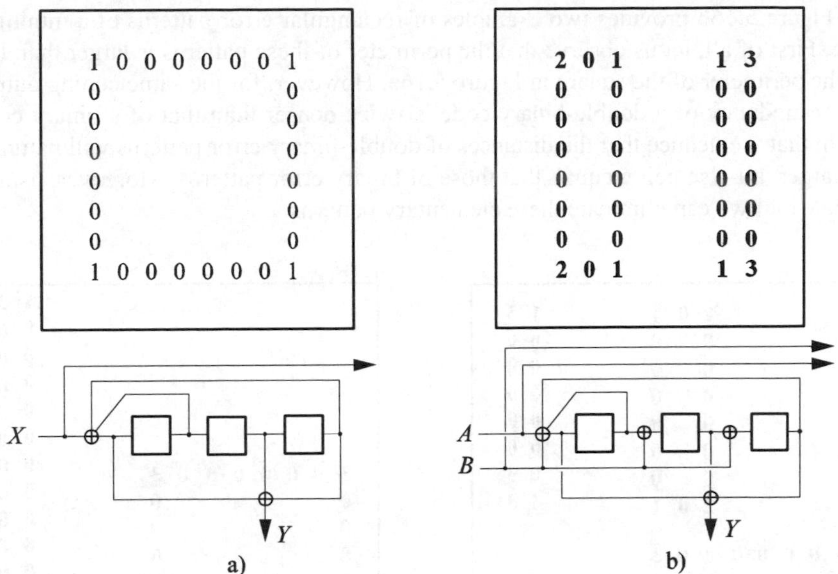

Figure 5.16. *Possible error patterns with low weights, with binary (a) and double-binary (b) turbocodes with 8 states and a regular permutation. The elementary turbocode encoder is represented for each of the two cases*

The periodicities of the double-binary encoder are summarized by the diagram in Figure 5.17. There we find all the combinations of pairs of couples of the *RTZ* type. For example, if the encoder initialized to the 0 state is supplied by the successive pairs **1** and **3**, it immediately enters the 0 state. It is the same for **201**, **2003** or **3000001** sequences, for example.

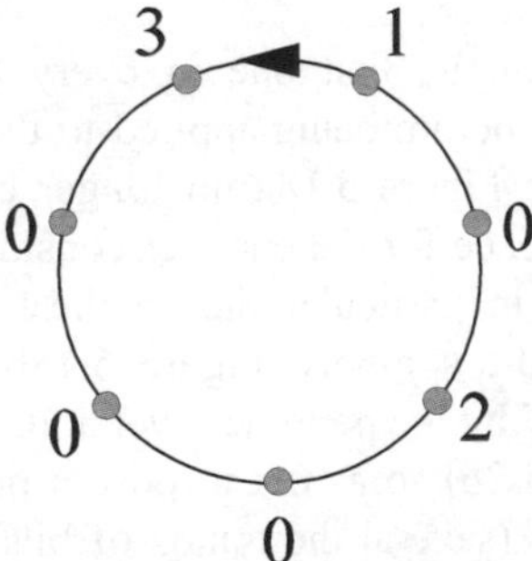

Figure 5.17. *Periodicities of the double-binary encoder from Figure 4.3b. Four input pairs (A, B,) = (0,0), (0,1), (1,0) and (1,1) are noted **0, 1, 2** and **3**, respectively. This diagram provides all the combinations of pairs of couples of the RTZ type*

Figure 5.16b provides two examples of rectangular error patterns of a minimum size. First of all, let us observe that the perimeter of these patterns is larger than half of the perimeter of the square in Figure 5.16a. However, for the same coding output, the redundancy of a double-binary code is twice denser than that of a binary code. From that we deduce that the distances of double-binary error patterns will naturally be larger, all else being equal, that those of binary error patterns. Moreover, using a simple tool we can eliminate these elementary patterns.

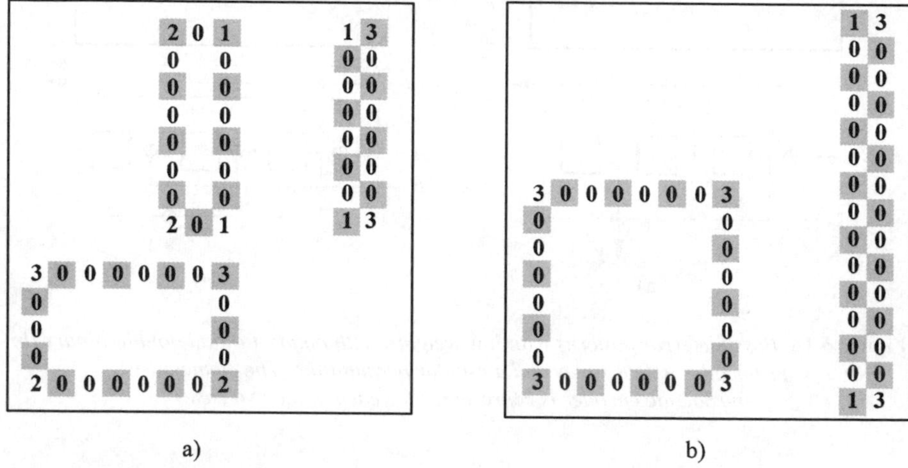

Figure 5.18. *The couples of the gray boxes are inversed before the second coding (vertical).* ***1*** *becomes* ***2***, ***2*** *becomes* ***1***; ***0*** *and* ***3*** *remain unchanged. The patterns in Figure 5.16b, redrawn in (a), are no longer the possible error patterns. Those in (b) still are, with distances 24 and 26, for a coding output of 1/2*

Let us suppose, for example, that one in every two couples is inversed (**1** becomes **2** and reciprocally) before being applied to the vertical encoder. Then the error patterns represented in Figure 5.18a no longer exist; for example, if **30002** does represent an RTZ sequence for the encoder considered, **30001** no longer does. Thus, many error patterns, in particular the smallest ones, disappear due to the disorder introduced inside the symbols. Figure 5.18b provides two examples of patterns that are not "broken" by the periodic inversion. The corresponding distances are sufficiently high (24 and 26) so as not to pose a problem for small or average block sizes. For long blocks (several thousands of bits), an additional intersymbol disorder of low intensity can be added to intrasymbol non-uniformity in order to obtain even greater minimum distances. An example of small "controlled" disorder is provided by the relation [5.36] modified in the following manner:

$$i = P.j + Q \mod. k \quad [5.39]$$

with

$$Q = 0 \quad \text{if } j = 0 \mod. 4$$
$$Q = Q_1 \quad \text{if } j = 1 \mod. 4$$
$$Q = Q_2 \quad \text{if } j = 2 \mod. 4$$
$$Q = Q_3 \quad \text{if } j = 3 \mod. 4$$

where Q_1, Q_2 and Q_3 are small integers, multiples of 4 and, if possible, such that $Q_1 \neq | Q_3 - Q_2 |$. This technique makes it possible to break the error patterns such as those drawn in Figure 5.18b and, more generally, the rectangular patterns whose lengths and widths are both not multiples of 4. The turbocodes retained in the DVB-RCS and DVB-RCT standards [DVB 00, DVB 01] have permutations worked out following this method.

5.5. *m*-binary turbocodes

m-binary turbocodes are constructed on the basis of recursive systematic convolutional codes with *m* binary inputs ($m \geq 2$)[9]. The advantages of this construction compared to the traditional turbocodes diagram ($m = 1$) are varied: better convergence of the iterative process, large minimum distances, reduced sensitivity with respect to the possible puncturing patterns, lower latency, robustness to the sub-optimality of the decoding algorithm, in particular, when the MAP algorithm is simplified in its Max-Log-MAP version.

The $m = 2$ case has already been adopted for the European standards of satellite and ground network feedback channels: DVB-RCS and DVB-RCT [DVB 00, DVB 01]. Combined with the circular trellises technique these double-binary turbocodes with 8 states offer good average performances and a great flexibility of adaptation to different block sizes and different outputs, while keeping a reasonable decoding complexity.

9. There are at least two ways to construct an m-binary convolutional code: either on the basis of the Galois body $GF(2^m)$ or of the Cartesian product $(GF(2))^m$. Here we only consider the latter, more convenient, construction method. Indeed, a code worked out in $GF(2^m)$, with a memory depth v, has 2^{vm} possible states, whereas the number of states for the code defined in $(GF(2))^m$, for the same depth, can be limited to 2^v.

5.5.1. *m-binary RSC encoders*

Figure 5.19 represents the general structure of an m-binary RSC encoder.

It uses a pseudo-random generator with code memory ν and a generator matrix of a wedged register $\mathbf{G}$ (sized n × n).

The input vector **d** with m components is connected to the various possible sockets via a grid of interconnections whose binary matrix, sized n × m, is noted $\mathbf{C}$.

The vector **T** applied to ν possible takes of the register at the moment i is given by:

$$\mathbf{T}_i = \mathbf{C}.\mathbf{d}_i \qquad [5.40]$$

with $\mathbf{d}_i = \left(d_{1,i} \ldots d_{m,i}\right)^{\mathrm{T}}$.

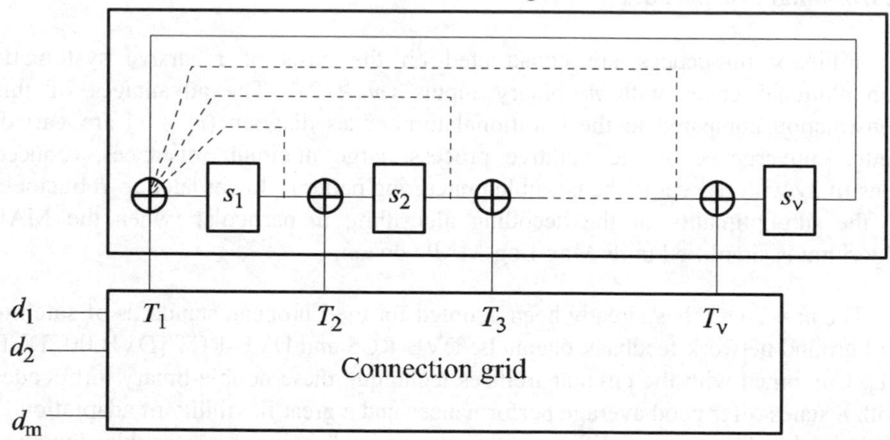

Figure 5.19. *General structure of an m-binary RSC encoder with code memory ν. Neither the temporal index nor the encoder output are represented here*

If we wish to avoid parallel transitions in the trellis code, the condition $m \leq \nu$ must be observed.

Except for very particular cases, this encoder is not equivalent to an encoder with a single input where we would successively present d_1, d_2, ... d_m. An m-binary encoder is thus generally not decomposable.

The redundant machine output (not represented in the figure) is calculated at the moment i by the expression:

$$y_i = \sum_{j=1...m} d_{j,i} + \mathbf{R}^{\mathrm{T}}\mathbf{S}_i \qquad [5.41]$$

where $\mathbf{S}_i = (s_{1,i}, s_{2,i}, ..., s_{\nu,i})^{\mathrm{T}}$ is the state vector at the moment i and $\mathbf{R}^{\mathrm{T}}$ is the transposed redundancy vector. The pth component of $\mathbf{R}$ equals "1", if the pth component of $\mathbf{S}_i$ is used in the construction of yi, and equals "0" otherwise. We can demonstrate that y_i can also be expressed in the form:

$$y_i = \sum_{j=1...m} d_{j,i} + \mathbf{R}^{\mathrm{T}}\mathbf{G}^{-1}\mathbf{S}_{i+1} \qquad [5.42]$$

provided that:

$$\mathbf{R}^{\mathrm{T}}\mathbf{G}^{-1}\mathbf{C} \equiv \mathbf{0} \qquad [5.43]$$

The expression [5.41] ensures, on the one hand, that the Hamming weight of the vector $(d_{1,i}, d_{2,i}, ... d_{m,i}, y_i)$ is at least equal to two, when we deviate from the reference path ("all-zero" path), in the trellis. Indeed, inversing a single component of $\mathbf{d}_i$ modifies the value of y_i.

In addition, expression [5.42] indicates that the Hamming weight of the same vector is also at least equal to two, when the reference path has been retaken.

In conclusion, relations [5.41] and [5.42] together guarantee that the free distance of the code, whose output is $R = m/(m + 1)$, is at least equal to 4, regardless of m.

Since the minimum distance of a concatenated code is much larger than that of each constituent elementary code, we can imagine being able to obtain large minimum distances, for the low as well as for strong outputs.

The choice of large values for m could, of course, imply a great complexity of decoding, since the trellis representing the code has 2^m paths per node.

However, recent work on the decoding of *m*-binary convolutional codes by means of the dual code has shown that the complexity of decoding can be reduced to that of binary codes [BER 98].

5.5.2. *m-binary turbocodes*

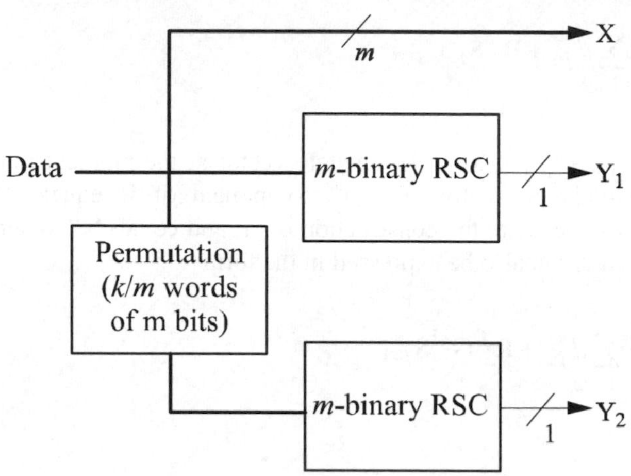

Figure 5.20. *M-binary turboencoder*

We consider a parallel concatenation of two RSC m-binary encoders associated to an interleaving function of words with *m* bits (Figure 5.20). The blocks of *k* bits (*k* being a multiple of *m*) are coded twice by this two-dimensional code, whose output is $m/(m + 2)$. The principle of circular trellises is adopted to enable the coding of blocks without termination sequences and edge effects.

The advantages of this construction with respect to the traditional turbocodes are the following:

– Better convergence. This point was first observed in [BER 97] and commented on in [BER 99c]. A better convergence in a two-dimensional iterative process is explained by a lower density of erroneous paths in each dimension, thus reducing the correlation effects between the constituent decoders. Figure 5.21 compares, for the same block size (*k* bits of information), a possible situation after a certain number of iterations, for a binary turbo decoder and a double-binary turbo decoder. To simplify matters, each block is presented as a square and a regular permutation is used. The lines in each square symbolize the places where the elementary decoders, for each of the two dimensions, made mistakes (erroneous paths in the trellises). Figure 5.21a depicts a particular case of severe error locking, schematized by the four dashes forming a rectangle. This kind of situation is typical of the difficulties faced by the turbo decoder in its exchange of probabilities; the correlation between the noises over extrinsic information in this short cycle is then a serious barrier to convergence.

The length of error patterns in the double-binary case is on average divided by 2, since the density of redundancy symbols in the corresponding trellis is twice larger than that of the binary code. The ratio between the sides of the two squares is only $\sqrt{2}$, which explains this low density of erroneous paths for each of the two dimensions. The gain for each density is exactly $\frac{2}{\sqrt{2}}$. We can note, for example, that the rectangle initially present in Figure 5.21a has disappeared.

The advantage of the reduction of density of errors is pronounced when we replace binary codes with double-binary codes, but the additional gain is less considerable for $m > 2$.

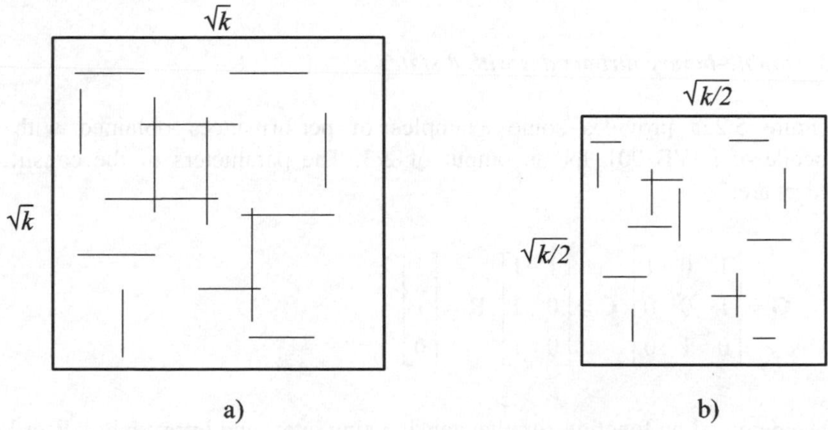

Figure 5.21. *Examples of erroneous paths in the elementary decoders of a turbocode. The density of errors is lower in the case of a double-binary turbocode (b) than in the case of a traditional binary turbocode (a)*

– Larger minimum distances. In addition to the argument developed previously for the *m*-binary convolutional codes and their minimum distance at least equal to 4 regardless of the output, the composite *m*-binary code adds another degree to the construction of permutations: the intrasymbol permutation. This point is developed in section 5.4.3.

– Reduced sensitivity with respect to punctured sequences. To obtain coding outputs higher than m/(m + 1) on the basis of the encoder in Figure 5.19, it is not necessary to remove as many redundancy symbols as with a binary encoder. It is the same for m-binary turbocodes.

– Reduced latency. From the point of view of coding as well as of decoding, latency is divided by m, since the data is treated by groups of m bits.

– Robustness of the decoder. For binary turbocodes the difference in performance between the MAP algorithm and its simplified versions, or between the MAP and the SOVA algorithms, varies from 0.2 to 0.6 dB, according to the size of the blocks and the coding outputs. This difference is divided by two when we use double-binary turbocodes and can be even smaller for m > 2. This favorable (and slightly surprising) property can be explained in the following manner: for a block of a given size (k bits), the smaller the number of stages in the trellis is, the closer is the decoder, regardless of the algorithm on which it is based, to the Maximum Probability (MP) decoder. In extreme cases, a trellis reduced to a single stage and thus containing all the possible codewords is equivalent to an MP decoder.

5.5.3. *Double-binary turbocodes with 8 states*

Figure 5.22a provides some examples of performances obtained with the turbocode of [DVB 00], for an output of 2/3. The parameters of the constituent encoders are:

$$\mathbf{G} = \begin{bmatrix} 1 & 0 & 1 \\ 1 & 0 & 0 \\ 0 & 1 & 0 \end{bmatrix}; \mathbf{C} = \begin{bmatrix} 1 & 1 \\ 0 & 1 \\ 0 & 1 \end{bmatrix}; \mathbf{R} = \begin{bmatrix} 1 \\ 1 \\ 0 \end{bmatrix}$$

The permutation function simultaneously using inter- and intrasymbol disorder is described in [DVB 00]. In particular, we may observe:

– good average performances for this code, whose decoding complexity remains very reasonable (approximately 18,000 gates per iteration + memory);

– a certain coherence with respect to performance variation with the size of the blocks (according to [DOL 98], for example). The same coherence could also be observed for performance variation with coding output;

– quasi-optimality of decoding for low error rates. The theoretical asymptotic curve for 188 bytes was calculated using only the knowledge of the minimum code distance (13 in this case) and not using the total spectrum of distances. Despite that, the difference between the asymptotic curve and the curve obtained by simulation is merely 0.2 dB for a PER of 10-7.

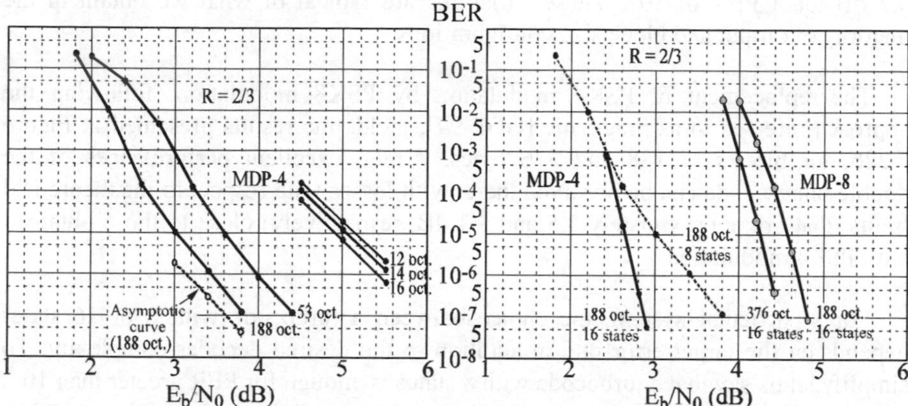

Figure 5.22. *(a) Performance, expressed in PER, of a double-binary turbocode with 8 states for blocks of 12, 14, 16, 53 and 188 bytes. PSK4, AWGN noise and output 2/3. Max-Log-MAP decoding with input samples of 4 bits and 8 iterations. (b) Performance, expressed in PER, of a double-binary turbocode with 16 states for blocks of 188 bytes (PSK4 and PSK8) and 376 bytes (PSK8). AWGN noise and output 2/3. Max-Log-MAP decoding with input samples of 4 bits (PSK4) or 5 bits (PSK8) and 8 iterations*

5.5.4. *Double-binary turbocodes with 16 states*

The extension of the preceding diagram to elementary encoders with 16 states clearly makes it possible to increase minimum distances. For example, we may choose:

$$\mathbf{G} = \begin{bmatrix} 0 & 0 & 1 & 1 \\ 1 & 0 & 0 & 0 \\ 0 & 1 & 0 & 0 \\ 0 & 0 & 1 & 0 \end{bmatrix}; \mathbf{C} = \begin{bmatrix} 1 & 1 \\ 0 & 1 \\ 0 & 0 \\ 0 & 1 \end{bmatrix}; \mathbf{R} = \begin{bmatrix} 1 \\ 1 \\ 1 \\ 0 \end{bmatrix}$$

For the turbocode of output 2/3, still with blocks of 188 bytes, the minimum distance obtained is equal to 18 instead of 13 for the code with 8 states. Figure 5.22b shows the gain obtained for a low error rate: approximately 1 dB for a PER of 10^{-7}, and 1.4 dB asymptotically considering the respective minimum distances. We may note that the convergence threshold is approximately the same for decoders with 8 and 16 states, the curves being practically identical for PER superior to 10^{-4}. The theoretical limits (TL) for $R = 2/3$, block size of 188 bytes and target PER of 10^{-4} and 10^{-7} are 1.9 and 2.2 dB respectively. The performances of the decoder with 16 states, in this example, are therefore: TL plus 0.6 dB for a PER of 10^{-4} and TL plus 0.7 dB for a PER of 10^{-7}. These variations are typical of what we obtain in the majority of output and block size configurations.

The replacement of PSK4 modulation by PSK8 modulation, following the approach referred to as pragmatic [GOF 94], yields the results presented in Figure 5.22b, for blocks of 188 and 376 bytes. Again, excellent performances of the double-binary code can be observed there, with losses with respect to the theoretical limits (that are approximately 3.5 and 3.3 dB, respectively) close to those obtained with PSK4 modulation.

For a particular system, the choice between a turbocode with 8 or 16 states depends on the target error rate in addition to the desired decoder complexity. To simplify let us say that a turbocode with 8 states is enough for PER greater than 10^{-4}. It is generally the case for transmissions with possibility of repetition (ARQ: *Automatic Repeat reQuest*). For lower PER, typical for diffusion or mass memory applications, the code with 16 states is largely preferable.

5.6. Bibliography

[3GPP 99] 3GPP Technical Specification Group, *Multiplexing and Channel Coding (FDD)*, TS 25.212 v2.0.0, June 1999.

[ALE 99] ALEXANDER P. D., REED M. C., ASENSTORFER J. A., SCHLEGEL C. B., "Iterative Multi-User Interference Reduction: Turbo CDMA", *IEEE Trans. Comm.*, vol. 47, no. 7, p. 1008-1014, July 1999.

[BAH 74] BAHL L. R., COCKE J., JELINEK F., RAVIV J., "Optimal Decoding of Linear Codes for Minimizing Symbol Error Rate", *IEEE Trans. Inform. Theory*, IT-20, p. 248-287, March 1974.

[BAT 87] BATTAIL G., "Pondération des symboles décodés par l'algorithme de Viterbi", *Ann. Télécommun.*, Fr., 42, no. 1-2, p. 31-38, January 1987.

[BAT 89] BATTAIL G., "Coding for the Gaussian Channel: the Promise of Weighted-Output Decoding", *International Journal of Satellite Communications*, vol. 7, p. 183-192, 1989.

[BEN 96] BENEDETTO S., MONTORSI G., "Design of Parallel Concatenated Convolutional Codes", *IEEE Trans. Comm.*, vol. 44, no. 5, p. 591-600, May 1996.

[BER 93a] BERROU C., ADDE P., ANGUI E., FAUDEIL S., "A Low Complexity Soft-Output Viterbi Decoder Architecture", *Proc. of ICC'93*, p. 737-740, Geneva, May 1993.

[BER 93b] BERROU C., GLAVIEUX A., THITIMAJSHIMA P., "Near Shannon Limit Error-Correcting Coding and Decoding: Turbo Codes", *Proc. of IEEE ICC'93*, p. 1064-1070, Geneva, May 1993.

[BER 96] BERROU C., JÉZÉQUEL M., "Frame-Oriented Convolutional Turbo Codes", *Electronics Letters*, vol. 32, no. 15, p. 1362-1364, July 1996.

[BER 97] BERROU C., "Some Clinical Aspects of Turbo Codes", *Int'l Symposium on Turbo Codes et Related Topics*, p. 26-31, Brest, France, September 1997.

[BER 98] BERKMANN J., "On Turbo Decoding of Nonbinary Codes", *IEEE Comm. Letters*, vol. 2, no. 4, p. 94-96, April 1998.

[BER 99a] BERROU C., DOUILLARD C., JÉZÉQUEL M., "Multiple Parallel Concatenation of Circular Recursive Convolutional (CRSC) Codes", *Ann. Télécomm.*, vol. 54, no. 3-4, p. 166-172, March-April 1999.

[BER 99b] BERROU C., DOUILLARD C., JÉZÉQUEL M., "Designing Turbo Codes for Low Error Rates", IEEE colloquium, *Turbo Codes in Digital Broadcasting – Could it Double Capacity?*, p. 1-7, London, November 1999.

[BER 99c] BERROU C., JÉZÉQUEL M., "Non-Binary Convolutional Codes for Turbo Coding", *Elect. Letters*, vol. 35, no. 1, p. 39-40, January 1999.

[BET 98] BETTSTETTER C., Turbo Decoding with Tail-Biting Trellises, Diplomarbeit, Technischen Universität München, July 1998.

[CCS 98] Consultative Committee for Space Data Systems, "Recommendations for Space Data Systems. Telemetry Channel Coding", *BLUE BOOK*, May 1998.

[DIV 01] DIVSALAR D., DOLINAR S., POLLARA F., "Iterative Turbo Decoder Analysis Based on Density Evolution", *IEEE Journal on Selected Areas in Comm.*, vol. 19, no. 5, p. 891-907, May 2001.

[DOL 98] DOLINAR S., DIVSALAR D., POLLARA F., "Code Performance as a Function of Block Size", *TMO progress report 42-133*, JPL, NASA, May 1998.

[DOU 95] DOUILLARD C., PICART A., DIDIER P., JÉZÉQUEL M., BERROU C., GLAVIEUX A., "Iterative Correction of Intersymbol Interference: Turbo-Equalization", *European Trans. on Telecomm.*, vol. 6, no. 5, p. 507-511, September/October 1995.

[DUA 01] DUAN L., RIMOLDI B., "The Iterative Turbo Decoding Algorithm has Fixed Points", *IEEE Trans. Inform. Theory*, vol. 47, no. 7, p. 2993-2995, November 2001.

[DVB 00] DVB, "Interaction Channel for Satellite Distribution Systems", ETSI EN 301 790, V1.2.2, p. 21-24, December 2000.

[DVB 01] DVB, "Interaction Channel for Digital Terrestrial Television", ETSI EN 301 958, V1.1.1, p. 28-30, August 2001.

[FOR 66] FORNEY G. D. Jr., *Concatenated Codes*, MIT Press, Cambridge, USA, 1966.

[GAL 62] GALLAGER R. G., "Low-Density Parity-Check Codes", *IRE Trans. Inform. Theory*, vol. IT-8, p. 21-28, January 1962.

[GLA 97] GLAVIEUX A., LAOT C., LABAT J., "Turbo Equalization Over a Frequency Selective Channel", in *Proc. of the First Symposium on Turbo Codes and Related Topics*, p. 96-102, Brest, France, September 1997.

[GOF 94] LE GOFF S., GLAVIEUX A., BERROU C., "Turbo Codes and High Spectral Efficiency Modulation", *Proc. of IEEE ICC'94*, p. 645-649, New Orleans, May 1994.

[HAG 89a] HAGENAUER J., HOEHER P., "A Viterbi Algorithm with Soft-Decision Outputs and its Applications", *Proc. of Globecom'89*, Dallas, Texas, p. 47.11-47-17, November 1989.

[HAG 89b] HAGENAUER J., HOEHER P., "Concatenated Viterbi-Decoding", *Proc. Int. Workshop on Inf. Theory*, Gotland, Sweden, August/September 1989.

[LEK 00] LEK S., "Turbo Space-Time Processing to Improve Wireless Channel Capacity", *IEEE Trans. Comm.*, vol. 48, no. 8, p. 1347-1359, August 2000.

[LOD 93] LODGE J., YOUNG R., HOEHER P., HAGENAUER J., "Separable MAP 'FILTERS' for the Decoding of Product and Concatenated Codes", *Proc. of ICC'93*, Geneva, p. 1740-1745, May 1993.

[LOE 98] LOELIGER H.-A., LUSTENBERGER F., HELFENSTEIN M., TARKÖY F., "Probability Propagation and Decoding in Analog VLSI", *Proc. of ISIT'98*, p. 146, Cambridge, MA, August 1998.

[NAR 99] NARAYANAN K. R., STÜBER G. L., "A Serial Concatenation Approach to Iterative Demodulation and Decoding", *IEEE Trans. Comm.*, vol. 47, p. 956-961, July 1999.

[POD 95] PODEMSKI R., HOLUBOWICZ W., BERROU C., BATTAIL G., "Hamming Distance Spectra of Turbo Codes", *Ann. Télécomm.*, vol. 50, no. 9-10, p. 790-797, September-October 1995.

[ROB 94] ROBERTSON P., "Illuminating the Structure of Parallel Concatenated Recursive Systematic (Turbo) Codes", *Proc. of Globecom'94*, San Francisco, p. 1298-1303, November 1994.

[ROB 97] ROBERTSON P., HOEHER P., VILLEBRUN E., "Optimal and Suboptimal Maximum A Posteriori Algorithms Suitable for Turbo Decoding", *European Trans. Telecommun.*, vol. 8, p. 119-125, March-April 1997.

[SHA 48] SHANNON C. E., "A Mathematical Theory of Communication", *Bell System Technical Journal*, vol. 27, October 1948.

[SVI 95] SVIRID Y. V., "Weight Distributions and Bounds for Turbo Codes", *European Trans. on Telecomm.*, vol. 6, no. 5, p. 543-55, September-October 1995.

[TAN 81] TANNER R. M., "A Recursive Approach to Low Complexity Codes", *IEEE Trans. Inform. Theory*, vol. IT-27, p. 533-547, September 1981.

[THI 93] THITIMAJSHIMA P., "Les codes convolutifs récursifs systématiques et leur application à la concaténation parallèle", Thesis no. 284, University of Bretagne Occidentale, Brest, France, December 1993.

[WEI 01] WEISS Y., FREEMAN W. T., "On the Optimality of Solutions of the Max-Product Belief-Propagation Algorithm in Arbitrary Graphs", *IEEE Trans. Inform. Theory*, vol. 47, no. 2, p. 736-744, February 2001.

Chapter 6

Block Turbocodes

6.1. Introduction

The turbocode principle was introduced by C. Berrou during the ICC Geneva congress in 1993 [BER 93] where, for the first time, an error correcting code operating within less than 0,5 dB of the Shannon limit [SHA 48] was announced. These results have initially surprised all the specialists in the field who were persuaded that it was not possible to reach this level of performance with a reasonable complexity. Very quickly many researchers, such as Hagenauer, Benedetto, Divsalar [HAG 96, BEN 96, DIV 95, ROB 94, WIB 95] and a number of others confirmed the results of Berrou and within a few years the turbocode became essential in the field of the error corrector coding as the 21st century solution.

The principle described by Berrou consists of carrying out iterative decoding of two CRS (convolutional recursive systematic) codes concatenated in parallel through a random or non-uniform interleaver. This iterative processing is based on SISO (soft input soft output) decoding and on the optimal transfer of the decoding information from one decoder to the next. To that end he has introduced the concept of extrinsic information, which plays a fundamental part in the operation of the convolutional turbocode (CTC).

Chapter written by Ramesh PYNDIAH and Patrick ADDE.

In view of the first results of Berrou it was obvious that it became possible to obtain performances comparable to block codes. To get to that point several problems had to be solved:

– which type of concatenation should be adopted?

– how should a SISO decoder of reasonable complexity for block codes be produced?

– how should information be transmitted in an optimal way from one decoder to the next?

The first results for the BTC were presented during the San Francisco Globecom conference in 1994 [PYN 94], that is, 18 months after Berrou's publication. This chapter presents the various concepts used in BTC.

After a study of the various types of concatenation for block codes, we will successively approach SISO decoding, iterative decoding used for the BTC and the performances of the BTC for a Gaussian and Rayleigh channel in the case of MDP4 or MAQ modulations with *M* states.

6.2. Concatenation of block codes

The general principle of coding retained for the turbocode consists of associating (or concatenating) two or more elementary codes in order to build a more powerful code than the elementary codes used. In the case of BTC the concatenated code is constructed on the basis of elementary codes [MAC 78] of the BCH (Bose-Chandhuri-Hocquenghen), RS (Reed-Solomon) or other types.

In practice we distinguish between two types of concatenation, which are respectively parallel and serial concatenation of elementary codes. The principles of parallel and serial concatenation are illustrated by Figures 6.1 and 6.2 respectively, where Π indicates the interlacing function.

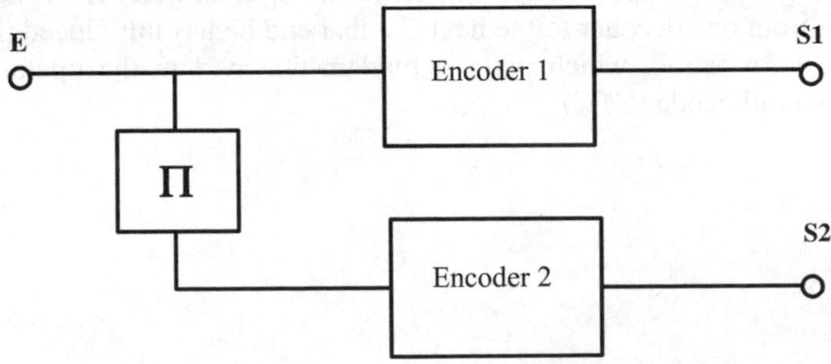

Figure 6.1. *General diagram of parallel concatenation of two codes*

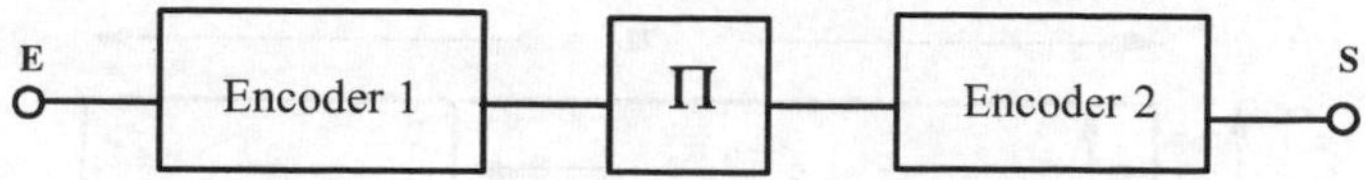

Figure 6.2. *General diagram of serial concatenation of two codes*

In addition, the concatenated code also depends on the nature of the interleaver, noted Π in Figures 6.1 and 6.2, which can be uniform or pseudo-random.

Thus, there are four different ways to construct a concatenated block code, according to the type of concatenation (parallel or serial) and to the nature of the interleaver used (uniform or pseudo-random).

We will examine these various possibilities and provide some results that will enable us to justify the choice of the concatenated code used for the BTC. To simplify this discourse, we will limit ourselves to the case of concatenation of two BCH codes.

Moreover, we will only consider the case of codes in systematic form. This restriction makes it possible to simplify the placement of the decoder and in practice we almost exclusively use systematic block codes.

6.2.1. *Parallel concatenation of block codes*

First of all, we will consider the parallel concatenation of two BCH codes using uniform interlacing.

The term uniform interlacing refers to an interleaver where the data is written into a matrix by rows and then read by columns. Let us consider the concatenation of two BCH codes noted $\boldsymbol{C}^1$ and $\boldsymbol{C}^2$ with the parameters (n_1, k_1, δ_1) and (n_2, k_2, δ_2), where n_i is the length, k_i is the dimension and δ_i is the minimum distance (Hamming) of the code $\boldsymbol{C}^i$. Initially, the data is placed in a matrix of size $k_1 \times k_2$, which is the size of the concatenated code, noted K_p. The columns of the matrix are coded by the first code $\boldsymbol{C}^1$ and we obtain the matrix (see Figure 6.3) of size $n_1 \times k_2$.

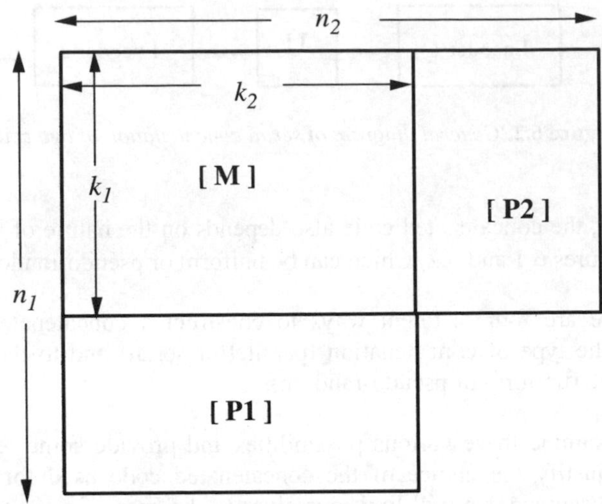

Figure 6.3. *Example of a coded matrix in the case of parallel concatenation*

The k_1 rows of this matrix are then coded by the second code C^2 and we obtain the concatenated code illustrated in Figure 6.3. This code has a length $N_p = ((n_1 \times k_2) + k_1(n_2 - k_2))$ and from it we deduce its output, which is given by:

$$R_p = \left(\frac{K_p}{N_p}\right) = \left(\frac{R_1 R_2}{R_1 + R_2 - R_1 R_2}\right) \qquad [6.1]$$

where $R_i = k_i / n_i$ is the output of the code C^i. Let us consider the case where the two codes have the same output. Table 6.1 shows the evolution of R_p following the output of the elementary codes. We note that the ratio between the output of the elementary code and that of the concatenated code decreases and tends towards one when the former is increased. We will now consider the third parameter of the concatenated code, which is the minimum (Hamming) distance noted Δ_p. In the case of block codes concatenated in parallel this minimum distance is given [PYN 97] by the relation:

$$\Delta_p = (\delta_1 + \delta_2 - 1) \qquad [6.2]$$

This result is obtained very simply by using the linearity and the weight spectrum of the C^1 and C^2 codes. Since C^1 and C^2 are linear, the code concatenated in parallel is also linear.

The minimum distance of the concatenated code is given by the weight of the codeword with a minimum weight different from zero [MAC 78]. Let us consider the weight of the various codewords of the code concatenated in parallel, which we classify according to the growing weight of K_p binary information symbols contained in the matrix $k_1 \times k_2$ in Figure 6.3, which we will note [**M**]. For a matrix of weight **P**([**M**]) = 1, the codeword with the lowest weight is such that (see Figure 6.4) the column of the coded matrix containing the binary information symbol at "one" has a weight of δ_1 and the row of the coded matrix containing the binary symbol of information at "one" has a weight of δ_2.

Indeed, the k_2 columns of the coded matrix must verify the coding equation $\boldsymbol{C}^1$ and the k_1 rows of the coded matrix must verify that of $\boldsymbol{C}^2$. The weight of this codeword is provided by relation [6.2]. For **P**([**M**]) = 2, the codeword with the lowest weight is such that two columns of the coded matrix containing binary symbols at "one" have a weight of δ_1 and the row of the coded matrix containing the two binary information symbols at "one" has a weight of δ_2, if $\delta_1 < \delta_2$. The weight of this codeword is given by:

$$(\delta_2 + 2\delta_1 - 2) = \Delta_p + (\delta_1 - 1) \geq \Delta_p \qquad [6.3]$$

Considering the weight of the codewords with **P**([**M**]) binary information symbols at one with **P**([**M**]) > 1, we show that these weights are always ≥ Δ_p. The codeword with a weight of Δ_p exists because the majority of block codes contain at least one codeword with a minimum weight associated with a message with the weight of one.

If we consider the case of two Hamming codes concatenated in parallel, the minimum distance is $\Delta_p = (3+3-1) = 5$. In the case of two codes concatenated in parallel with the same minimum distance, the ratio between the minimum distance of the concatenated code and that of the elementary codes tends towards two when the latter tends towards infinity.

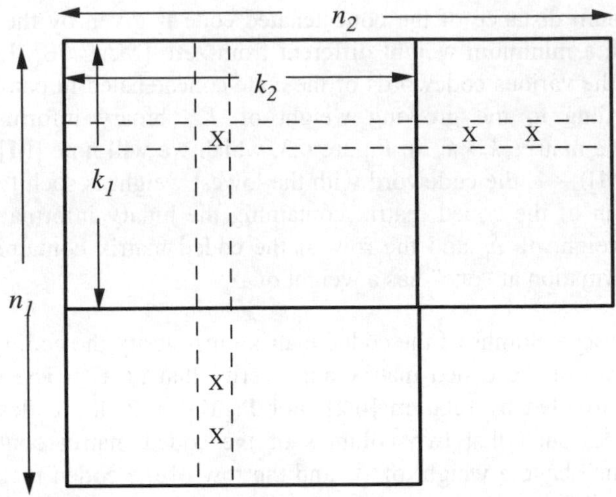

Figure 6.4. *Example of codeword with a minimum weight of a code concatenated in parallel with uniform interlacing for* $\delta_1 = \delta_2 = 3$ *(X: indicates the position of binary symbols at "one")*

Now let us consider the parallel concatenation of two BCH codes with random pseudo interlacing. The data is arranged in a $k_1 \times k_2$ size matrix that we note [**M**]. The k_2 columns of the matrix are coded by the code C^1. The data is then interlaced in a pseudo-random fashion (hence the term: pseudo-random interlacing) and arranged in a matrix of size $k_1 \times k_2$ noted [**M'**]. The matrix [**M'**] contains the same data as [**M**] but it is arranged in a different order. The k_1 rows of the matrix are coded by the code C^2. The parameters *N* and *K* of this concatenated code are the same as previously. Consequently, the coding output *R* is not modified. On the other hand, the minimum Hamming distance is no longer given by the relation of equality [6.2] which is transformed into a lower limit:

$$\Delta_p \geq (\delta_1 + \delta_2 - 1) \qquad [6.4]$$

In practice, for the majority of known codes (BCH or RS) this lower limit is reached and, thus, in the case of BTC it does not make sense to use pseudo-random interlacing.

This is explained by the fact that the majority of block codes have a codeword with a minimum weight associated with a message with a weight of one.

A comparison of the performances of BTC with the two types of interleavings carried out by Hagenauer [HAG 96] shows that they are nearly identical.

However, the conclusion should not be drawn that the limit is always reached, because this issue has not been fully explored. The discovery of block codes with all the codewords with minimum weights associated with messages of weight higher than one could call the above conclusions into question.

6.2.2. *Serial concatenation of block codes*

Serial concatenation is distinguished from parallel concatenation by the fact that the second code $\boldsymbol{C}^2$ is also applied to the binary parity symbols generated by the first code $\boldsymbol{C}^1$ (see Figure 6.2). As previously, let us first consider the case of uniform interlacing.

The data is placed in the matrix [**M**] of size $k_1 \times k_2$. The k_2 columns of the matrix are coded by code $\boldsymbol{C}^1$ and then the n_1 rows of the $n_1 \times k_2$ matrix are coded by code $\boldsymbol{C}^2$ (see Figure 6.5). The size of the concatenated code is $K_s = k_1 \times k_2$, its length is $N_s = n_1 \times n_2$ and its coding output is $R_s = R_1 \times R_2$. For identical elementary codes $R_s < R_p$. On the other hand, as the output of elementary codes increases, the ratio $R_s / R_p \leq 1$ tends towards 1.

To prove the inequality it suffices to show that $(R_s / R_p - 1) \leq 0$. For that we use the following relation:

$$\left(R_s / R_p - 1\right) = \left(1 - R_1\right)\left(R_2 - 1\right) \leq 0 \qquad [6.5]$$

Thus, it is enough that at least one of the two terms (R_1 or R_2) tends towards 1 for the ratio to become $R_s / R_p \rightarrow 1$.

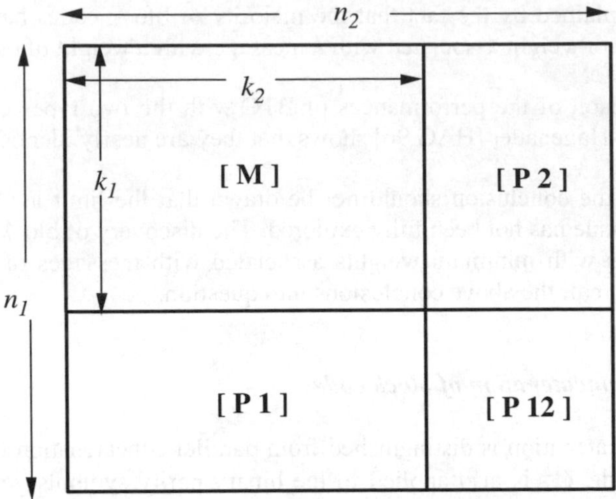

Figure 6.5. *Example of a code matrix concatenated serially*

Table 6.1 illustrates this evolution for $(R_1 = R_2)$.

$R_1 = R_2$	1/2	2/3	3/4	4/5	5/6	6/7
R_p	1/3	2/4	3/5	4/6	5/7	6/8
R_s	1/4	4/9	9/16	16/25	25/36	36/49
R_s/R_p	3/4	8/9	15/16	24/25	35/36	48/49

Table 6.1. *Evolution of the concatenated code output according to the elementary codes output*

We observe that for $R_1 = R_2 > 2/3$ the output is $R_s \geq 0.9R_p$ and that there is no significant advantage in using parallel concatenation from the point of view of coding output.

We will now demonstrate that the minimum distance of a serially concatenated code with uniform interlacing is equal to the product of the minimum distance of the two codes: $\Delta_s = \delta_1 \times \delta_2$. To establish this result we must first show that the n_2 columns of the coded matrix respect the coding equation C^1 and the n_1 rows respect C^2. The second point is verified very easily using the fact that the n_1 rows of the coded matrix have been generated by code C^2. In addition, the k_2 columns of the coded matrix verify the coding equation C^1 because they have been generated by code C^1. It suffices to show that the last $(n_2 - k_2)$ columns of the coded matrix verify the coding equation C^1.

Let [G^i] be the generator matrix, of size $k_i \times n_i$, of the C^i code. In the case of a systematic code, this matrix has the form:

$$\left[\mathbf{G}^i\right] = \left[\mathbf{I}_{k_i \times k_i} \middle| \mathbf{Q}^i\right] \quad [6.6]$$

where the k_i first columns constitute the identity matrix of size k_i and [$\mathbf{Q}^i$] is the sub-matrix of size $(k_i \times (n_i - k_i))$, generating the binary parity symbols. The parity sub-matrices [$\mathbf{P}^1$] [$\mathbf{P}^2$] and [$\mathbf{P}^{12}$] of the coded matrix are provided by the following equations:

$$\mathbf{P}^1 = \left[\left[\mathbf{M}\right]^T \left[\mathbf{Q}^1\right]\right]^T = \left[\mathbf{Q}^1\right]^T [\mathbf{M}] \quad [6.7]$$

$$\mathbf{P}^2 = \left[\mathbf{M}\right]\left[\mathbf{Q}^2\right] \quad [6.8]$$

$$\mathbf{P}^{12} = \left[\mathbf{P}^1\right]\left[\mathbf{Q}^2\right] = \left[\mathbf{Q}^1\right]^T [\mathbf{M}]\left[\mathbf{Q}^2\right] \quad [6.9]$$

We will now verify that the last $(n_2 - k_2)$ columns of the coded matrix verify the coding equation C^1. To this end it suffices to demonstrate that the sub-matrix obtained by encoding the sub-matrix [$\mathbf{P}^2$] using the C^1 code is equal to [$\mathbf{P}^{12}$].

$$\left[\left[\mathbf{P}^2\right]^T \left[\mathbf{Q}^1\right]\right]^T = \left[\left[[\mathbf{M}]\left[\mathbf{Q}^2\right]\right]^T \left[\mathbf{Q}^1\right]\right]^T = \left[\mathbf{Q}^1\right]^T [\mathbf{M}]\left[\mathbf{Q}^2\right] = \left[\mathbf{P}^{12}\right] \quad [6.10]$$

The relation [6.10] shows that the sub-matrix [$\mathbf{P}^{12}$] associated with the sub-matrix [$\mathbf{P}^2$] verifies the C^1 coding constraint.

In the case of serial concatenation of two linear and systematic block codes (BCH, RS or other) using uniform interlacing, the n_2 columns respect the coding equation C^1 and the n_1 rows respect C^2. Thus, we can define a codeword as a $n_1 \times n_2$ matrix, such that the n_2 columns verify the coding equation C^1 and the n_1 rows verify C^2. Moreover, we can permute the coding order (rows followed by columns instead of columns followed by rows) without modifying the result. This property will be exploited hereafter together with iterative decoding.

To show that, in the case of serial concatenation with uniform interlacing, the minimum distance of the code is given by $\Delta_s = \delta_1 \times \delta_2$, we will use the linearity properties of the concatenated code and the weight of the codewords of $\boldsymbol{C}^1$ and $\boldsymbol{C}^2$. The operations used to carry out concatenated serial coding being linear operations (linear coding and permutation), we deduce from it that the concatenated code is a linear code.

To determine its minimum distance it is enough to know the codeword with a non-zero minimum weight. We will consider the binary matrices of size $n_1 \times n_2$ noted [**A**], which we will classify according to the weight of the Hamming matrix **P**([**A**]). It is easy to demonstrate that there is no matrix with weight lower than $(\delta_1 \times \delta_2)$ (other that that of zero weight), such that the n_2 columns verify the coding equation $\boldsymbol{C}^1$ and the n_1 rows verify $\boldsymbol{C}^2$.

For **P**([**A**]) $=(\delta_1 \times \delta_2)$ there is at least one matrix [**A**] that verifies the coding equation of the concatenated code (see Figure 6.6), which is where the announced result stems from. This codeword of minimum (Hamming) weight contains δ_1 "one" along the columns and δ_2 "one" along the rows (see example in Figure 6.6).

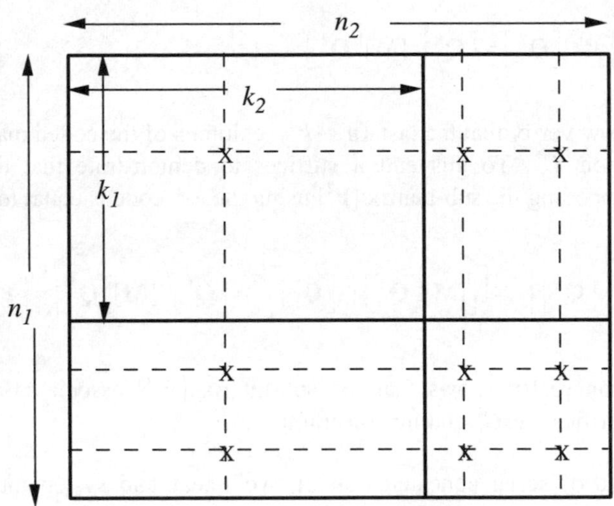

Figure 6.6. *Illustration of a minimum weight word of a serially concatenated code with uniform interlacing for* $\delta_1 = \delta_2 = 3$

Thus, the minimum (Hamming) distance for a serially concatenated code with uniform interlacing makes it possible to obtain a greater minimum distance than that of a parallel concatenation, using the same elementary codes. For example, let us consider the concatenation of two Hamming codes ($\delta_1 = \delta_2 = 3$), $\Delta_p = 5$ and $\Delta_s = 9$. On the other hand, parallel concatenation makes it possible to obtain a larger coding output than that obtained with the serial concatenation. In order to compare these two types of concatenation we will use the product of the coding output with the minimum distance of the concatenated code, which yields an upper limit of the asymptotic coding gain:

$$(G_a)_i \leq 10\log(R_i \times \Delta_i) \qquad [6.11]$$

with $i = s$ or p. Table 6.2 makes it possible to compare serial and parallel concatenation of the four BCH codes. We note that serial concatenation offers a better asymptotic coding gain for the various cases considered and the variation lies between 2 and 4 decibel. It is thus preferable to use serial concatenation when we are looking to maximize asymptotic gain. In the general case, relation [6.11] is the upper limit for asymptotic coding gain. The serial concatenated code, which has been initially proposed by Elias in 1954 [ELI 54], is also called a product code or iterative code and will be studied hereafter.

	Parallel			**Serial**		
$C^1 = C^2$	R_p	Δ_p	$(G_a)_p$	R_s	Δ_s	$(G_a)_s$
(15,11,3)	0.579	5	≤4.6 dB	0.538	9	≤6.8 dB
(63,57,3)	0.826	5	≤6.2 dB	0.818	9	≤8.7 dB
(15,7,5)	0.304	9	≤4.4 dB	0.218	25	≤7.4 dB
(63,51,5)	0.680	9	≤7.9 dB	0.655	25	≤12.1 dB

Table 6.2. *Comparison of serial and parallel concatenation*

Lastly, we consider the case of a serial concatenation of two block codes with pseudo-random interlacing. In this case the data of the matrix obtained after coding by $\boldsymbol{C}^1$ undergoes pseudo-random interlacing. The n_1 rows of the matrix of interlaced data are then coded by the code $\boldsymbol{C}^2$. The size of the code $K_s = k_1 \times k_2$, its length $N_s = n_1 \times n_2$ and its coding output $R_s = R_1 \times R_2$ are unchanged with respect to uniform interlacing. On the other hand, its minimum distance depends on interlacing, as for the convolutional turbocode [BEN 96]. In this case, it has not been shown that the last $(n_2 - k_2)$ columns of the coded matrix respect the coding equation $\boldsymbol{C}^1$. In the worst case $\Delta_s = \sup(\delta_1, \delta_2)$ and this situation occurs when the interlacing groups some δ_1 "one" of a codeword in a column of weight δ_1 in the same line and when this row generates a codeword of weight δ_2. The optimization of interlacing is very complex and has been barely studied until now. In addition, the fact that the $(n_2 - k_2)$ last columns of the coded matrix do not respect the coding equation $\boldsymbol{C}^1$ will have negative consequences on the operation of iterative decoding thereafter. Lastly, on a practical level, the implementation of pseudo-random interlacing can, in certain cases (large-sized blocks), lead to prohibitive complexity.

We realize that serial concatenation of block codes with uniform interlacing (or product codes) constitutes the best concatenated code for the BTC.

6.2.3. *Properties of product codes and theoretical performances*

The product code was introduced by Elias [ELI 54] in 1954 for the case of Hamming codes. The process of coding described in section 6.2.2 can be applied to any systematic linear block codes. The systematic nature of the elementary code is necessary in order to show that for any word of a product code $n_1 \times n_2$, the n_2 columns verify the coding equation $\boldsymbol{C}^1$ and the n_1 lines verify $\boldsymbol{C}^2$. Thus, this property is true for any systematic linear block code (BCH, RS or others). This result remains true when the number of elementary codes is higher than two. As a result, the parameters of a product code noted (N, K, Δ) are expressed as the product of the parameters of the elementary codes.

In the remainder of this section we will limit ourselves to the case of BCH codes in order to simplify the discussion. Generally, primitive BCH codes have a length of $n = 2^m - 1$, with a whole positive m. The minimum distance of the code is odd and its correction capacity is given by:

$$t = \left\lfloor \frac{\delta - 1}{2} \right\rfloor \qquad [6.12]$$

where $\lfloor \; \rfloor$ indicates the whole part. In practice we will limit ourselves to BCH codes with correction power $t \leq 2$, which make it possible to obtain a good compromise between complexity and performances. Let us consider the case of product codes formed by two identical elementary codes. Thus, for t = 1 and 2, the minimum distance of the product code is Δ = 9 and 25 (see Table 6.2) respectively. We can increase the minimum distance of the product code significantly by adding a binary parity symbol to the elementary codes, following the principle known as code extension.

Let us consider a code ***C*** with parameters (n,k,δ). This code has 2^k codewords $(c_1,c_2,....,c_n)$ of length *n*. The extension of the code consists of adding a binary symbol given by:

$$c_{n+1} = \sum_{i=1}^{n} c_i \qquad [6.13]$$

This binary symbol equals "one" if the word contains an odd number of "one" and equals "zero" otherwise. The weights of all the words with odd weights are incremented by "one". Thus, the minimum distance from the extended code is incremented by a unit and all the codewords have even weights. The parameters of the extended code constructed on the basis of the primitive BCH code ***C*** are $(n+1,k,\delta+1)$. The extension of the code makes it possible to increase its minimum distance at the price of a slight reduction in coding output. If we replace the elementary code by its extended code, the minimum distance of the product code increases from 9 to 16 for a code with parameter $t = 1$ and from 25 to 36 for a code with parameter $t = 2$ (see Table 6.3).

In addition, we note that the upper limit of the asymptotic coding gain increases considerably (> 2 dB for t = 1 and > 1 dB for t = 2), when we pass from the primitive code to the extended code. The impact on the complexity of the decoder will be discussed later on.

We will now consider the theoretical performances of product codes. For a transmission by phase shift keying (PSK) with two or four states in a channel with additive white Gaussian noise with optimal decoding (MAP: maximum *a posteriori* probability), the upper limit of the probability of error per block [PROA1] is given by:

$$(P_e)_{block} \leq \left(\frac{1}{2}\right) \sum_{m=\Delta}^{N} w_m \times \operatorname{erfc}\left(\sqrt{\frac{RmE_b}{N_0}}\right) \qquad [6.14]$$

where w_m is the number of codewords of weight *m*, E_b is the energy received by the binary symbol and N_0 is the one-sided spectral density of the noise.

	Primitive BCH code			Extended BCH code		
Code	R	Δ	G_a	R	Δ	G_a
(31,26,3)	0.70	9	≤ 8.0 dB	0.66	16	≤ 10.2 dB
(31,21,5)	0.46	25	≤ 10.6 dB	0.43	36	≤ 11.9 dB

Table 6.3. *Comparison of product codes constructed on the basis of primitive BCH and extended BCH codes*

Let us first consider the product codes constructed on the basis of two Hamming codes with identical parameters (n, k, 3). The minimum distance of this product code is $\Delta = \delta^2 = 9$. Taking into account the constraint imposed by the product code (all the rows and columns are codewords of the Hamming code) we can verify that there is not a product codeword of weight m with $\delta^2 < m < \delta \times (\delta + 1)$, that is, $9 < m < 12$ in our case. We may also verify that there is at least one product codeword of weight $m = 12$ that corresponds to a rectangular pattern in the coding matrix (see Figure 6.7). Thus, $w_m = 0$ for $9 < m < 12$. We may also verify that $w_{13} = 0$, as well as a certain number of other terms. Thus, for a high signal-to-noise ratio ($RE_b >> N_0$) we can limit ourselves to the contribution of the first non-nil term of relation [6.14]. We then obtain a lower limit of the probability of error per block:

$$(P_e)_{block} \geq \left(\frac{w_\Delta}{2}\right) \operatorname{erfc}\left(\sqrt{\frac{R\Delta E_b}{N_0}}\right) \qquad [6.15]$$

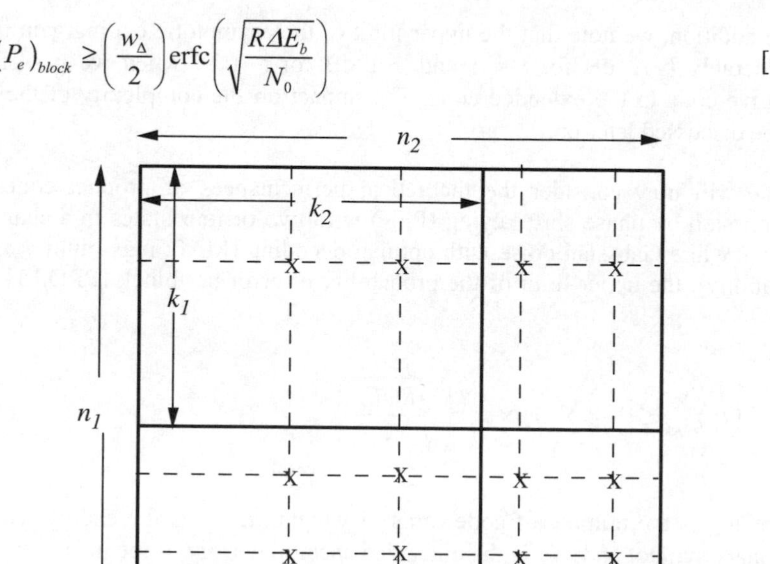

Figure 6.7. *Example of a codeword of weight 12 of a product code constructed on the basis of two Hamming codes*

This limit becomes finer as the signal-to-noise ratio increases. For a given signal-to-noise ratio, it will be finer as the variation between δ^2 and $\delta(\delta+1)$ grows and the $w_{\delta(\delta+1)}/w_{\delta^2}$ ratio becomes smaller. For a Hamming code $\delta(\delta+1)-\delta^2=\delta=3$.

We will now calculate the first two terms $w_m \neq 0$ to verify if the low $w_{\delta(\delta+1)}/w_{\delta^2}$ condition is met. These words of product code correspond to rectangular or square patterns of "one" (see Figure 6.7), formed by words of the Hamming code with a weight of $(\delta+i)$ for the rows and $(\delta+j)$ for the columns with $0<i,j<n$ and $m=(\delta+i)\times(\delta+j)$. To simplify the notation we will indicate by $w_{(\delta+i)\times(\delta+j)}$ the number of codewords of the product code with a weight of $m=(\delta+i)\times(\delta+j)$ and $w_{(\delta+i)}$ the number of codewords of the Hamming code with a weight of $(\delta+i)$. The number of codewords of the product code is then given by:

$$w_{(\delta+i)\times(\delta+j)} = w_{(\delta+i)} \times w_{(\delta+j)} \qquad [6.16]$$

When the code comprises few codewords, we can determine $w_{(\delta+i)}$ in an exhaustive manner. In the opposite case we can estimate $w_{(\delta+i)}$ by using the following relation [MAC 78]:

$$w_{(\delta+i)} = \frac{\binom{\delta+i}{n}}{2^{n-k}} \qquad [6.17]$$

Let us consider two examples, for instance, the Hamming codes (15,11,3) and (31, 26, 3). Using relation [6.17] we obtain $w_{3\times4} = w_{4\times3} \approx 3w_{3\times3}$ for the first code and $w_{3\times4} = w_{4\times3} \approx 7w_{3\times3}$ for the second. Thus, the contribution of the second non-zero term in relation [6.14] very quickly becomes negligible compared to the first term and the approximation [6.15] is justified.

With a high signal-to-noise ratio $RE_b > N_0$, the probability of error per block is very low and it is given by relation [6.15]. When there are decoding errors, there is a very high probability that the decoded word will be a codeword with minimum (Hamming) distance from the transmitted word. From this we deduce a lower limit for the probability of error per binary symbol equal to:

$$P_{eb} \geq \left(\frac{\Delta}{N}\right) \times \left(\frac{w_\Delta}{2}\right) \times \operatorname{erfc}\left(\sqrt{\frac{R\Delta E_b}{N_0}}\right) \qquad [6.18]$$

Now let us consider the case of product codes elaborated on the basis of two extended Hamming codes with the parameters $(n+1,k,3+1)$. The minimum distance of this product code is $\Delta=(3+1)^2=16$. By taking into account the constraint imposed by the product code (all the lines and columns are words of the extended Hamming code), there is at least one codeword with a rectangular pattern with a weight of $m=4\times6$. We can verify that there is no word of the product code of weight m with $16<m<24$ (the number of "one" per row and column is even and always higher than $(\delta+1)=4$). We notice that the difference between the indices of the first two terms $w_m\neq0$ is greater $(24-16=8)$ for an extended code.

The number of codewords of the product code in the case of the extended elementary code has the form:

$$w_{(\delta+1+i)\times(\delta+1+j)}=w_{(\delta+1+i)}\times w_{(\delta+1+j)} \qquad [6.19]$$

with i and j being positive and even integers, and:

$$w_{(\delta+1+i)}=\frac{\binom{\delta+i}{n}}{2^{n-k}}+\frac{\binom{\delta+1+i}{n}}{2^{n-k}} \qquad [6.20]$$

since the codewords with an even weight in the form of $4\leq 2i\leq n+1$ of the extended Hamming code are given by the sum of the codewords of weight $(2i-1)$ and the codewords of weight $2i$.

For a Hamming code (16,11,4), $w_{4\times6}=w_{6\times4}\approx4w_{4\times4}$ and for a code (32,26,4), $w_{4\times6}=w_{6\times4}\approx25w_{4\times4}$.

From this we deduce that the contribution of the codewords of weight 24 will be negligible in relation [6.14] as soon as $RE_b>N_0$. Thus, the lower limit of the probability of error per block will be finer in the case of product codes constructed on the basis of extended codes.

Product codes elaborated on the basis of extended codes provide a higher limit of asymptotic gain and this limit is reached quicker. It is thus advisable to use extended codes to construct product codes.

We can also increase the minimum (Hamming) distance of a BCH primitive code by using the version known as expurgated. In this case we remove the codewords of odd weight among the 2^k codewords and the parameters of the expurgated code are $(n,k-1,\delta+1)$. A binary information symbol is replaced by a binary parity symbol. The properties of product codes constructed on the basis of expurgated codes have the same nature as those of extended codes. They therefore constitute an interesting alternative.

6.3. Soft decoding of block codes

The decoding of product codes can be carried out iteratively. Indeed, let us take the case of a product code obtained on the basis of the serial concatenation of two BCH codes. Let $\boldsymbol{C}^1$ be the code applied along the columns and $\boldsymbol{C}^2$ along the rows. Iterative decoding consists of decoding the columns (using the decoder of the code $\boldsymbol{C}^1$) followed by a decoding of the rows (using $\boldsymbol{C}^2$) and then reiterating the process. The decoding of columns possibly leads to codewords $\boldsymbol{C}^1$ (along the columns) but the rows are not necessarily codewords $\boldsymbol{C}^2$. The decoding of rows may lead to codewords $\boldsymbol{C}^2$ (along the rows), but the columns are not necessarily codewords $\boldsymbol{C}^1$. By reiterating the process we converge towards a codeword of the product code, such that all the columns are codewords $\boldsymbol{C}^1$ and all the rows are codewords $\boldsymbol{C}^2$. The problem consists of finding the good criterion and the associated algorithms to carry out this iterative decoding. This section is devoted to the decoding of block codes $\boldsymbol{C}^1$ or $\boldsymbol{C}^2$.

Optimal decoding of block codes can be carried out using two criteria according to the nature of observations presented at the decoder. In the case of binary decoder input, optimal decoding consists of finding the codeword with the minimum observation Hamming distance. This type of decoder is also called binary decoding or hard decoding.

Much work has been carried out on binary decoding of block codes in order to reduce its complexity. We may cite the contributions of Berlekamp [BER 68, MAC 78] in the case of cyclic codes. These decoders are of relatively low complexity but their coding gain is lower than that brought by decoding known as soft or flexible decoding.

Soft decoding is often used in the case of a transmission of the codewords by linear modulation of the PSK or QAM (quadrature amplitude modulation of two carriers) type in a channel with additive white Gaussian noise (Gaussian channel). Let us consider the case of a transmission by BPSK in a Gaussian channel associated with a coherent receiver. The observations at the output of the optimal receiver have the form:

$$r_i = e_i + b_i \qquad [6.21]$$

where e_i is the transmitted binary symbol taking its values in $\{-1,+1\}$ and b_i is the sample of noise with a standard deviation σ. It is demonstrated that optimal decoding of the observations consists of determining the codeword with the minimum Euclidean distance (this quantity will be developed hereafter) of observations. This decoder makes it possible to significantly improve the coding gain (between 1.5 dB and 2.5 dB), compared to binary decoding. On the other hand, its implementation proves to be more complex, as we will see further on in the chapter.

In the 1970s Reddy and Robinson [RED 70, RED 72] have studied the iterative decoding of product codes using binary decoders. This process proves to be sub-optimal for soft data at the decoder input. Indeed, when the data input has real values, the decoder carries out a thresholding of the data to transform it into binary before carrying out decoding. We then have a loss of information because the transformation of real into binary is equivalent to a quantification of the data for a binary symbol, which simply indicates the sign of real observation. As an indication, the first binary decoding of the rows led to a loss ranging between 1.0 and 2.0 dB, as we will see hereafter. It is thus necessary to use soft decoding of the elementary codes. The following section deals with the soft decoding of block codes, which provides a binary decision for soft data at the input.

6.3.1. *Soft decoding of block codes*

Let us consider the transmission of words of a binary block code C with the parameters (n,k,δ) by BPSK modulation in a Gaussian channel. The transmitted codeword $E=(e_1,e_2,\ldots,e_n)$ is one of the codewords $C^i=(c_1^i,c_2^i,\ldots,c_n^i)$ with $1 \le i \le 2^k$. The operation of coding is defined for the binary symbols {0, 1} and the modulation carries out the following transformation: $\{0 \leftrightarrow -1, 1 \leftrightarrow +1\}$. In order to simplify notations, we will consider that the binary symbols of the codewords have a value of $\{-1,+1\}$.

The transmission of the e_i elements is governed by relation [6.21] and at the decoder input we have an observation vector $R=(r_1,r_2,\ldots,r_n)$ associated with the transmitted codeword E. The system is supposed to be ideal and perfectly synchronized. On the basis of observation R the decoder must work out an optimal decision in the sense of a criterion to be defined. The most natural criterion is that of minimization of the probability of error per binary information symbol $(Pe)_{bit}$. This criterion is relatively complex to implement and will be treated in section 6.4. In practice, we prefer to use the criterion of minimization of the probability of error per codeword $(Pe)_{block}$ (or block), which is simpler to realize. These two criteria lead to almost identical results asymptotically (with low probability of error). Thus, we will consider the decoding of R for the minimization of $(Pe)_{block}$. We demonstrate that the minimization of $(Pe)_{block}$ is achieved with MAP stated as follows:

$$D = \arg\max_{C^i \in C} \left\{ \Pr\left\{ E = C^i / R \right\} \right\} \quad [6.22]$$

where $D = (d_1, d_2, ..., d_n)$ and $\mathrm{P}\{X\}$ indicates the probability of X. This decision rule is very general and is therefore not limited to the Gaussian channel. Using the relation of Bayes we can express this rule in the following form:

$$D = \arg\max_{C^i \in C} \left\{ \frac{\mathrm{P}\left\{ R / E = C^i \right\} \Pr\left\{ E = C^i \right\}}{\mathrm{P}\{R\}} \right\} \quad [6.23]$$

where $\mathrm{P}\{X\}$ indicates the probability density of X. Supposing that the transmitted data is mutually independent and has equal probabilities, the binary blocks of k information symbols are independent and have the same probability of 2^{-k}. Coding is a bijective application between the messages and the codewords, the latter are also independent and of probability 2^{-k}. As the probability density of R is independent of the codewords, the decision rule can be also written:

$$D = \arg\max_{C^i \in C} \left\{ \mathrm{P}\left\{ R / E = C^i \right\} \right\} \quad [6.24]$$

In the case of a Gaussian channel, the probability density of R conditionally to the transmission of a codeword C^l is given by:

$$\mathrm{P}\left\{ R / E = C^i \right\} = \prod_{l=1}^{n} \left(\frac{1}{\sqrt{2\pi}\sigma} \exp\left\{ \frac{-\left(r_l - c_l^i \right)^2}{2\sigma^2} \right\} \right) = \left(\frac{1}{\sqrt{2\pi}\sigma} \right)^n \exp\left\{ \frac{\sum_{l=1}^{n} -\left(r_l - c_l^i \right)^2}{2\sigma^2} \right\} \quad [6.25]$$

By transferring relation [6.24] to [6.25] we show very easily that the optimal decision is given by:

$$D = \arg\min_{C^i \in C} \left\{ \sum_{l=1}^{n} \left(r_l - c_l^i \right)^2 \right\} = \arg\min_{C^i \in C} \left\{ \left(\mathrm{d_E}\left(R, C^i \right) \right)^2 \right\} \quad [6.26]$$

where $d_E(X,Y)$ indicates the Euclidean distance between two variables X and Y. The quantity to be minimized is a measurement of the distortion introduced by the transmission channel, called square of the Euclidean distance, between the observation and the codeword C^i. This decision rule makes it possible to minimize the probability of error per word of decoded code. It is also known under the term soft decoding. In the case of a BPSK (or QPSK) modulation we may easily show that the minimization of the Euclidean distance amounts to finding the codeword which maximizes the correlation:

$$D = \arg\min_{C^i \in C} \left\{ \sum_{l=1}^{n} (r_l)^2 + \sum_{l=1}^{n} (c_l^i)^2 - 2\sum_{l=1}^{n} (r_l \times c_l^i) \right\} = \arg\max_{C^i \in C} \left\{ \sum_{l=1}^{n} (r_l \times c_l^i) \right\} \quad [6.27]$$

This function is simpler to implement and is the one used in practice. It corresponds to the codeword having maximum correlation with the observation. If we apply an exhaustive search for the most probable word of the code, the complexity of decoding is given by the number of codewords, that is 2^k. The complexity of decoding remains reasonable for codes with a small size, i.e. ($k \leq 8 \Leftrightarrow 2^k \leq 256$). For example, the extended Hamming code (16,11,4) contains 2,048 codewords and an exhaustive search is relatively complex.

Block codes used are often large in order to obtain high coding outputs. Exhaustive search for the most probable word of the code is not possible for these codes and the first applications of block codes have primarily used binary decoding.

6.3.2. *Soft decoding of block codes (Chase algorithm)*

In 1972, Chase [CHA 72] proposed a slightly sub-optimal algorithm of reduced complexity to carry out the soft decoding of block codes. This algorithm is based on the fact that the required codeword is the one with the minimum Euclidean distance from the observation, and one that can restrict the search to the codewords that are the closest to the observation in the sense of Euclidean distance. It is thus necessary to generate a subset of codewords with a short Euclidean distance of observation and this subset must contain the word with the minimum distance with a probability close to one.

The vector of observation R can be regarded as a point in the space of real numbers of dimension n. Each component of R is associated with a dimension of space and the value of this component indicates the projection of the point on the axis associated with this dimension in space. The codewords C^i are vectors with n components where each component has a value in $\{-1,+1\}$. They are also points in the space with the restriction that the components can only take two possible values

$\{-1,+1\}$. There are thus 2^k points in space associated with the codewords. Let us note that all the vectors of size n with a value in $\{-1,+1\}$ are not codewords and there are $(2^n >> 2^k)$ of them. We can show that the Euclidean distance between two binary vectors with value in $\{-1,+1\}$ is related to the Hamming distance by relation:

$$d_E\left(C^i,C^j\right)=\sqrt{4\times d_H\left(C^i,C^j\right)} \qquad [6.28]$$

On the basis of this representation of the problem we can define the zone containing the codewords closest to the observation in the sense of Euclidean distance. This zone is defined by a sphere in the space of size n. The coordinates of the center of the sphere are given by the vector $Y=\left(y_1,y_2,...,y_n\right)$ with $y_i=\operatorname{sgn}\left(r_i\right)$ and its radius is equal to $\sqrt{4(\delta-1)}$. The vector Y corresponds to the binary vector with a minimum Euclidean distance of R because it presents the maximum correlation with R. If the Hamming distance between E and Y is less than $(\delta-1)$, then E belongs to the sphere. When $d_H\left(E,Y\right)>(\delta-1)$, E is outside the sphere but we then show that the codeword with the minimum Euclidean distance of R is different from E. In this case, exhaustive search and the Chase algorithm yield erroneous resulted. Thus, the Chase algorithm makes it possible to correct $(\delta-1)$ errors at the most.

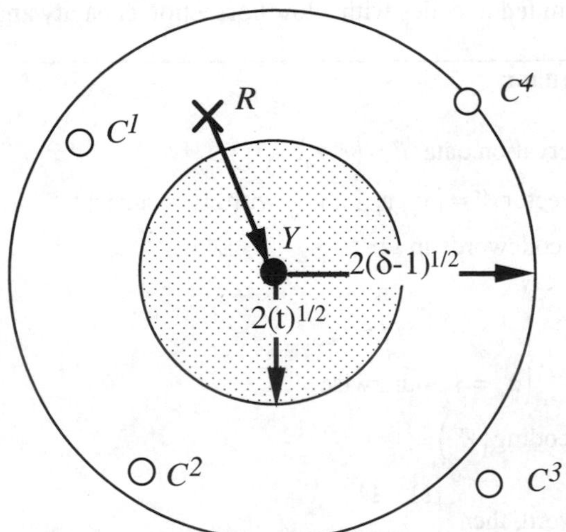

Figure 6.8. *Principle of decoding by the Chase algorithm*

We now will tackle the construction of the subset of codewords Ω contained in the sphere with a radius of $\sqrt{4(\delta-1)}$ centered in Y. For that we use a binary decoder (with binary input) and we start by the binary decoding of the vector Y. The decoder behaves as a codeword detector, which scans the sphere of radius $\sqrt{4t}$ centered in Y. We demonstrate that there is at most only one codeword in this sphere and the decoder provides this codeword if it exists. To find all of the codewords contained in the sphere of radius $\sqrt{4(\delta-1)}$ centered in Y, it is enough to present the set of the binary vectors contained in the sphere of radius $\sqrt{4t}$ centered in Y to the binary decoder. This operation makes it possible to scan a sphere of radius $\sqrt{(4t)+(4t)} = \sqrt{4(2t)} = \sqrt{4(\delta-1)}$ (see Figure 6.8). Let us note that the square of the radius of the scanned sphere is given by the sum of the squares of the radii of the two spheres because the radii are not co-linear, contrary to what Figure 6.8 may lead to believe. Indeed, this figure is merely a projection onto a plane of a volume defined in a space with n dimensions. The number of binary vectors contained in the sphere of radius $\sqrt{4t}$ centered in Y determines the extent of binary decodings that need be performed and is given by:

$$N_{dec} = 1 + \sum_{i=1}^{t} \binom{i}{n} \qquad [6.29]$$

The number of decodings to be carried out is in the order of ($n^t/t!$) and, thus, this algorithm is limited to codes with a low correction capacity and low length.

First Chase algorithm:
Start
Loading the observation data $R = (r_1, r_2, \ldots, r_n)$.
Calculating the vector $Y = (y_1, y_2, \ldots, y_n)$ with $y_i = \text{sgn}(r_i)$.
Determining the codewords in Ω:
For $I = 0$ with $i < N_{dec}$

$$Z^i = (z_1^i, z_2^i, \ldots, z_n^i): \begin{cases} z_j^i = -y_j & \text{if } j = i \\ z_j^i = y_j & \text{otherwise} \end{cases}$$

X^i = Binary decoding(Z^i)

If X^i is a code word, then $\begin{cases} \Omega = \Omega \cup X^i \\ met(i) = \mathrm{d_E}(R, X^i) \end{cases}$

$D = (d_1, d_2, \ldots, d_n)$: codeword in Ω associated with the smallest Euclidean distance.
End.

This algorithm is called the first Chase algorithm and above we provide a description of this algorithm in the case of a Hamming code ($t = 1$). If we consider the case of the code (63,57,3), the exhaustive method carries out a search among $2^{57} \approx 1.5\times10^{17}$ codewords, whereas the first Chase algorithm considers only 63. In the case of a Hamming code the complexity of the first Chase algorithm grows linearly with the length of code and it is applicable only to codes with a small length ($n \leq 15$).

To reduce the complexity of decoding, Chase proposed an algorithm with a reduced search zone in order to minimize the number of binary decodings. Among the binary vectors contained in the sphere of radius $\sqrt{4t}$ centered in Y we use only a subset of them to construct the subset of codewords Ω. This subset of binary vectors uses the measurement of reliability of binary data y_j which is defined on the basis of the log of probability λ_j associated with the decision y_j and using the rule of Bayes and relation [6.21] in the case of BPSK (QPSK) in a Gaussian channel we demonstrate that:

$$\lambda_j = \ln\left(\frac{\Pr\{e_j = +1/r_j\}}{\Pr\{e_j = -1/r_j\}}\right) = \left(\frac{2}{\sigma^2}\right)r_j \qquad [6.30]$$

Let us write $P^+ = \Pr\{e_j = +1/r_j\}$ and $P^- = \Pr\{e_j = -1/r_j\}$. Let us note that $0 \leq P^+, P^- \leq 1$, $P^+ + P^- = 1$ and that $\lambda_j \propto r_j$. When $\lambda_j > 0$, $P^+ > P^-$ and the decision is $e_j = +1$. When $\lambda_j < 0$, $P^+ < P^-$ and the decision is $e_j = -1$. Thus, the sign λ_j makes it possible to make a decision on the value of e_j. When $|\lambda_j| = 0$, $P^+ = P^- = 1/2$, the two values of e_j have equal probability and the probability of error when we base the decision on the sign of λ_j is 1/2. On the other hand, when $\lambda_j \to +\infty$ (or $-\infty$), $P^+ >> P^-$ (or $P^+ << P^-$) the probability of error tends towards zero and the decision is made with greater reliability. $|\lambda_j|$ is a measurement of the reliability of the decision y_i. We can thus classify the components of Y by ascending order of reliability and we note $(i_1, i_2, ..., i_q)$ the position of the q least reliable binary symbols in Y.

For the simplified algorithm, also called second Chase algorithm [CHA 72], among the binary vectors contained in the sphere of radius $\sqrt{4t}$ centered in Y we preserve only those obtained by permuting the binary symbols taken from the q least reliable binary symbols. When we reduce the search zone by decreasing q, the probability that the transmitted word is outside the search zone increases and we degrade the performance of the decoder. It is therefore necessary to find a compromise between the reduction of the search zone and degrading the performance of the decoder. Chase proposes to use the following empirical relation:

$$q = \left\lfloor \frac{\delta}{2} \right\rfloor \qquad [6.31]$$

to determine q. The number of binary vectors used to build Ω and thus the number of binary decodings are then given by:

$$N_{dec} = (2)^q \qquad [6.32]$$

In the case of a Hamming code, $q = 1$ and the number of binary decodings passes from n to two using only the least reliable component of the decision vector Y. The complexity of this algorithm no longer depends on n and thus there are no more restrictions on the length of code used which then grows exponentially with q (see relation [6.32]). Let us note that this algorithm requires the search for the least reliable components of Y, but the complexity of this search is negligible compared to the considerable reduction in the number of binary decodings that need to be carried out. In addition, the degradation of performance is relatively low. A thorough study of the impact of q on the performances of the decoder has been carried out by S. Jacq [JAC 95]. Figure 6.9 shows the evolution of the binary error rate (BER) according to E_b/N_0 for three various values of the number of binary vectors used in the second Chase algorithm applied to the code (64,57,4). We note that the coding gain increases with N_{dec} but that the increase in gain with each time that we double N_{dec} decreases and tends towards zero.

A description of the second Chase algorithm is given below in the case of a Hamming code ($t = 1$).

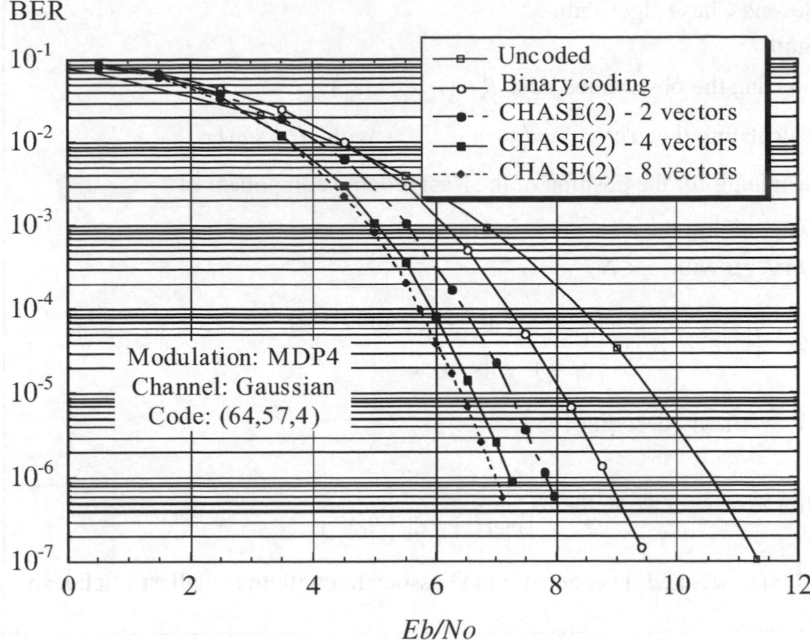

Figure 6.9. *Evolution of the BER according to E_b/N_0 according to the number of binary vectors used in the simplified Chase algorithm*

In the case of product codes, we have shown in section 6.2 that it was more advantageous to use as an elementary code the extended code obtained by adding a binary parity symbol to the word of primitive code (the parity of the word of primitive code). It is thus important to evaluate the impact of this binary parity symbol on the complexity of the Chase algorithm. Looking again at relation [6.31] we note that the value of q increases by one. Thus, the number of binary decodings to be carried out is multiplied by two. In addition, binary decoding is carried out in two stages. We start by considering only the data of the primitive code to which we apply:

– thresholding of the data R;

– the search for the q least reliable components of Y;

– the binary decoding of binary vectors $Z^i \Rightarrow X^i$;

– the construction of the subset of codewords Ω.

Second Chase algorithm:
Start
Loading the observation data $R=(r_1,r_2,...,r_n)$.
Calculating the vector $Y=(y_1,y_2,...,y_n)$ with $y_i=\operatorname{sgn}(r_i)$.
Searching for the position of the least reliable component of Y: i_1
Determining the codewords in Ω:
For $I=0$ with $i<N_{dec}$

$$Z^i=\left(z_1^i,z_2^i,...,z_n^i\right):\begin{cases} z_j^i=-y_j \text{ if } (j=i_1 \text{ and } i\neq 0) \\ z_j^i=y_j \text{ otherwise}\end{cases}$$

$$X^i = \text{Binary decoding}\left(Z^i\right)$$

$$\text{If } X^i \text{ is a codeword, then } \begin{cases} \Omega=\Omega\cup X^i \\ met(i)=\mathrm{d_E}\left(R,X^i\right)\end{cases}$$

$D=(d_1,d_2,...,d_n)$: codeword in Ω associated with the smallest Euclidean distance.
End.

Ω then contains codewords belonging to the primitive code of length n. Afterwards, we introduce the contribution of the binary parity symbol in the following manner:

– for each codeword of Ω we calculate the parity of the word that is concatenated, then use the latter to form the extended codeword of length n+1;

– we calculate the Euclidean distance between the extended codewords and R;

– we search for the word of extended code associated with the smallest Euclidean distance.

This last modification of the algorithm induces a negligible increase in complexity.

In short, the complexity of decoding of the extended codes is twice greater than that of the primitive code. However, it should be noted that the complexity of the (second) Chase algorithm is very low in the case of codes with a short minimum distance (Hamming for example). Indeed, the complexity of decoding changes from two binary decodings to four, regardless of the length of the code.

Figure 6.10 represents the BER according to the E_b/N_0 ratio for a transmission by BPSK in a Gaussian channel for two extended BCH codes of length $n = 64$. We note that Chase decoding applied to the code (64,57,4) yields better performances than binary decoding applied to the code (64,51,6). The difference, in terms of coding gain, is approximately 1 dB for a BER of 10^{-6} and the coding output is higher by more than 10% (0.89 instead of 0.80). In addition, we note that Chase decoding applied to the code (64,51,6) makes it possible to improve its coding gain by approximately 1.7 dB, with a BER of 10^{-6}.

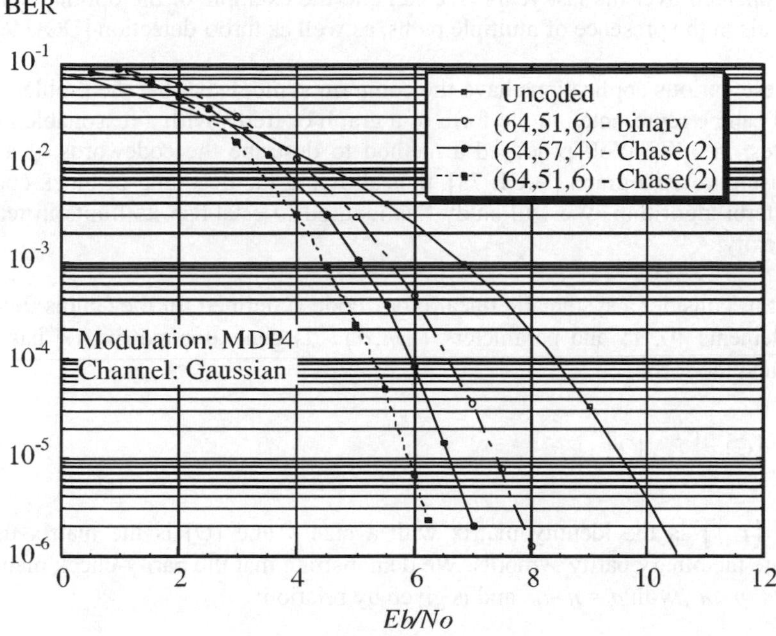

Figure 6.10. *Evolution of the BER according to E_b/N_0 for different BCH codes of length 64*

There are other algorithms for the soft decoding of block codes. We will discuss them further on briefly because the Chase algorithm is the one offering the best complexity/performance compromise. The latter will be the one used as starting point to build the SISO decoder, which is a fundamental function in the decoding of the BTC.

6.3.3. *Decoding of block codes by the Viterbi algorithm*

The Viterbi algorithm [SAW 77], also known as the Djikstra algorithm, makes it possible to considerably reduce the complexity of the search of the minimum (or maximum) cost path in a graph. This algorithm is very effective when the number of states of the graph is reasonable (< 100) and its complexity increases linearly with the length of the graph. The most widespread application of this algorithm, in the field of the digital communications, relates to optimal decoding (ML) of convolutional codes. Other applications of this algorithm in digital communications have appeared over the last years. We can cite the example of the optimal detection of signals in the presence of multiple paths, as well as turbo-detection [DOU 95].

These various applications have the common symbolistic that the problem being solved can be represented in the form of a graph or trellis with a reasonable number of states. In 1978 Wolf proposed a method to describe the codewords of a block code using a trellis graph [WOL 78], thus allowing the decoding of block codes by the Viterbi algorithm. We will study this method to establish its limits in terms of complexity.

Let us consider a systematic linear block code ***C*** defined for the Galois field with two elements {0, 1} and parameters (n,k,δ). Its generator matrix [*G*] has a size $k\times n$ and the form:

$$[G]=\left[I_{k\times k}\,|Q\right] \qquad [6.33]$$

where $\left[I_{k\times k}\right]$ is the identity matrix with a size *k* and [*Q*] is the matrix used to generate the binary parity symbols. We demonstrate that the parity-check matrix has the size $p\times n$, with $p = n - k$, and is given by relation:

$$[H]=\left[Q^T\,|I_{p\times p}\right] \qquad [6.34]$$

This matrix makes it possible to verify if a binary vector $U=(u_1,u_2,....,u_n)$ verifies the coding equation of the code ***C***. For that it suffices to calculate the syndrome of *U*:

$$S=U\times\left[H^T\right] \qquad [6.35]$$

The syndrome S is a binary vector with p components. We demonstrate that S is zero, if and only if U is a codeword. Each component of S makes it possible to verify one of the p parity equations of the code. If the calculation of the syndrome is developed, the following relations are obtained:

$$s_j = \sum_{i=1}^{n} u_i \times h_{ji} (\text{mod}2) \tag{6.36}$$

with $1 \leq j \leq p$ and h_{ji} is an element of $[H]$ located at the intersection of the row j and the column i of the matrix. By adopting a condensed notation, we can write the above relation in the form:

$$S = \sum_{i=1}^{n} u_i \times \vec{h}_i (\text{mod}2) \tag{6.37}$$

where $\vec{h}_i$ is the vector of size p obtained by transposing the ith column of the matrix $[H]$. The sum modulo-2 is applied to each component of the vectors as indicated by relation [6.36].

We will now use the equations of the syndrome to define the graph associated with the code **C**. For that it is necessary to define the states of the graph and the transitions between states. The states of the graph are defined by the various values taken by the syndrome. The number of states of the graph associated with the code is thus equal to $2^p = 2^{n-k}$. In order to define the transitions between the states of the graph we introduce the syndrome of the L first binary symbols of U by:

$$S(l) = \sum_{i=1}^{l} u_i \times \vec{h}_i (\text{mod}\, 2) \tag{6.38}$$

The syndrome for $l = 0$ is initialized to zero. Then the following binary symbol can take two values: $u_{l+1} = 0$ or 1. The transitions are then given by relation:

$$S(l+1) = \begin{cases} S(l) \text{ if } u_{l+1} = 0 \\ S(l) \oplus \vec{h}_{l+1} \;(\text{mod}2) \text{ if } u_{l+1} = 1 \end{cases} \tag{6.39}$$

It suffices then to apply the constraint $S(n) = (0, 0, .., 0)$ and all the paths in the graph starting at the zero state in $l = 0$ and arriving at the zero state in $l = n$ verify the equations of coding of $\boldsymbol{C}$.

The binary sequences associated with the paths of the graph are codewords.

To illustrate the principle, let us consider the example of the Hamming code (7,4,3). The matrices [G] and [H] of this code are provided below:

$$[G] = \begin{bmatrix} 1 & 0 & 0 & 0 & 0 & 1 & 1 \\ 0 & 1 & 0 & 0 & 1 & 0 & 1 \\ 0 & 0 & 1 & 0 & 1 & 1 & 0 \\ 0 & 0 & 0 & 1 & 1 & 1 & 1 \end{bmatrix} \qquad [6.40]$$

$$[H] = \begin{bmatrix} 0 & 1 & 1 & 1 & 1 & 0 & 0 \\ 1 & 0 & 1 & 1 & 0 & 1 & 0 \\ 1 & 1 & 0 & 1 & 0 & 0 & 1 \end{bmatrix} \qquad [6.41]$$

The number of syndromes is $2^p = 2^{n-k} = 8$ which we number from 0 to 7 by transforming the binary triplet of the syndrome into an integer.

By adopting the method described above we obtain the graph in Figure 6.11 for the code (7,4,3). The data $u_l = 0$ are indicated by branches in dotted lines and $u_l = 1$ by branches in solid lines.

The paths making it possible to pass from $S(0) = 0$ to $S(n) = 0$ are shown in bold lines. We verify that there are $2^k = 16$ paths verifying $S(0) = S(n) = 0$. It suffices then to apply the Viterbi algorithm to the received data to find the codeword with the minimum Euclidean distance.

We can thus reduce the complexity of decoding compared to an exhaustive search. Indeed, the Viterbi algorithm does only take into account at most 2^{n-k} codewords closest to the observation instead of considering 2^k codewords. In the case of the code (7,4,3) this gain is relatively low, as it is two.

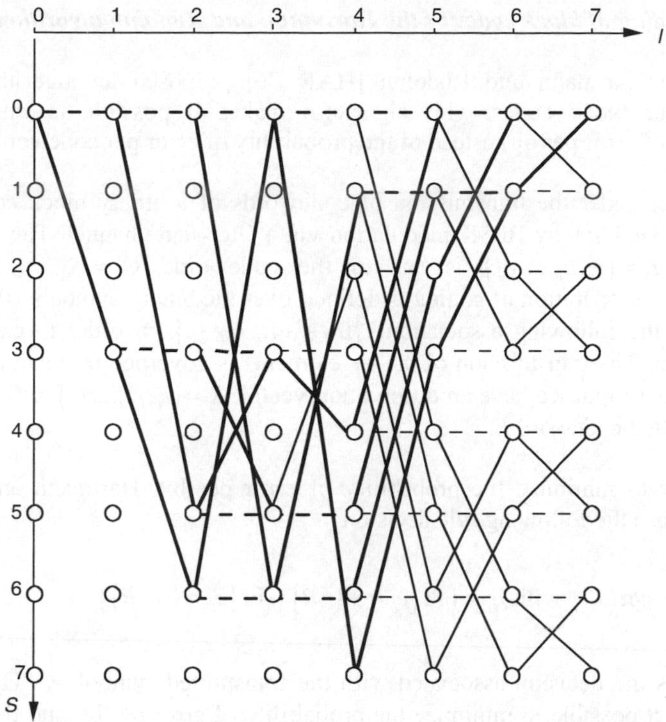

Figure 6.11. *Graph describing the codewords of the Hamming code (7,4,3)*

If we consider the Hamming codes, the number of states of the graph grows to 2^m and the gain in complexity grows to $2^{\left(2^m-1-2m\right)}$ and increases very quickly. In the case of a BCH code (15,11,3) the increase in complexity is 128 for a number of states of the graph that is 16. On the other hand, for codes with a correction capacity higher than 1 the number of states of the graph increases very quickly. In the case of the BCH (15,7,5) the number of states of the graph is 128 and the complexity of decoding by the Viterbi algorithm becomes prohibitive.

Finally, let us consider the case of extended Hamming codes used to construct the product codes. The introduction of an additional binary parity symbol multiplies the number of states of the graph by two. For iterative decoding of product codes by the Viterbi algorithm, we must limit ourselves to Hamming codes with a number of states lower than 64.

6.3.4. *Decoding of block codes by the Hartmann and Rudolph algorithm*

In 1976 Hartmann and Rudolph [HAR 76] proposed an algorithm of soft decoding for block codes. This algorithm makes it possible to minimize the probability of error per bit instead of the probability of error per codeword.

Let us consider the transmission of codewords of a binary block code **C** with parameters (n,k,δ) by BPSK modulation via a Gaussian channel. The transmitted codeword $E=(e_1,e_2,...,e_n)$ is one of the codewords $C^i=(c_1^i,c_2^i,...,c_n^i)$ with $1\le i\le 2^k$. The operation of coding is defined over the binary symbols $\{0,1\}$ and we apply then the following association: $\{0\leftrightarrow +1, 1\leftrightarrow -1\}$ in order to carry out the transmission. The transmission of the e_i elements is governed by relation [6.1] and at the decoder input we have an observation vector $R=(r_1,r_2,...,r_n)$ associated with the transmitted codeword *E*.

In order to minimize the probability of error per bit, Hartmann and Rudolph propose to use the following rule decision:

$$d_i=\operatorname{sgn}(\mu_i)\text{ with }\mu_i=\left(\Pr\{e_i=+1/R\}-\Pr\{e_i=-1/R\}\right) \qquad [6.42]$$

where d_i is the decision associated with the transmitted symbol e_i. This decision rule makes it possible to minimize the probability of error per bit and the difficulty consists of evaluating the quantity μ_1. It is shown (see [HAR 76]) that μ_1 is given by the following relation:

$$\mu_i=\left(\sum_{j=1}^{2^{n-k}}\prod_{l=1}^{n}\left(\frac{1-\varphi_l}{1+\varphi_l}\right)^{u_l^j\oplus\delta_{i,l}}\right) \qquad [6.43]$$

where $U^j=(u_1^j,u_2^j,...,u_n^j)$ is the codeword number *j* of the dual code of *C* with $1\le j\le 2^{n-k}$. Let us recall that the dual code of *C* is obtained using the generator matrix [*H*] of the code *C*. This code has a length *n* and size *n-k*. $\delta_{i,l}$ is the Kronecker symbol, which is equal to "one" for $l=i$ and zero otherwise. The term φ_l is given by:

$$\varphi_l=\left(\frac{\Pr\{r_l/c_l=1\}}{\Pr\{r_l/c_l=0\}}\right)=\left(\frac{\Pr\{r_l/e_l=-1\}}{\Pr\{r_l/e_l=+1\}}\right)=\exp\left\{\frac{\left((r_l-1)^2-(r_l+1)^2\right)}{2\sigma^2}\right\} \qquad [6.44]$$

in the case of a BPSK transmission via a Gaussian channel. The calculation of the quantity μ_1 is relatively complex because we have to sum 2^{n-k} terms formed by the product of n terms. In practice, this complexity can be slightly reduced because a certain number of terms that form part of the product are neutral elements for it, i.e. "one" (case where $u_l^j \oplus \delta_{i,l} = 0$). To illustrate the principle let us consider the case of the Hamming code (7,4,3). The matrices [G] and [H] of this code are given in section 6.3.3. Let us write:

$$\rho_l = \left(\frac{(1-\varphi_l)}{(1+\varphi_l)} \right) = \left(\frac{\Pr\{r_l / e_l = +1\} - \Pr\{r_l / e_l = -1\}}{\Pr\{r_l / e_l = +1\} + \Pr\{r_l / e_l = -1\}} \right) \qquad [6.45]$$

and let us consider the decoding of the first binary symbol given by the sign of μ_1. The eight codewords of the dual code are enumerated in Table 6.4.

If we develop the expression of μ_1 using the codewords of the dual code, we find the expression:

$$\mu_1 = \begin{cases} (\rho_1\rho_2\rho_3\rho_4\rho_5 + \rho_3\rho_4\rho_6 + \rho_2\rho_4\rho_7 + \rho_2\rho_5\rho_6) + \\ (\rho_3\rho_5\rho_7 + \rho_1\rho_2\rho_3\rho_6\rho_7 + \rho_1\rho_4\rho_5\rho_6\rho_7 + \rho_1) \end{cases} \qquad [6.46]$$

To decode the following binary symbol we use the cyclic property of the Hamming codes. It is enough to apply a circular shift of order one to the received data and to use relation [6.46] in order to decode the second binary symbol. We reiterate the process to decode the other binary symbols. We note that the number of terms involved in the product is lower than n, which reduces the complexity of the decoder.

Generally, the complexity of this algorithm is given by 2^{n-k} products with more than n terms and the sum of 2^{n-k} terms per decoded binary symbol. If we consider the extended version of the primitive code, complexity is multiplied by two. As for the Viterbi algorithm, it is limited mainly to the extended Hamming codes with a length less or equal to 32.

l	**1**	**2**	**3**	**4**	**5**	**6**	**7**
$\delta_{0,l}$	1	0	0	0	0	0	0
U^1	0	1	1	1	1	0	0
U^2	1	0	1	1	0	1	0
U^3	1	1	0	1	0	0	1
U^4	1	1	0	0	1	1	0
U^5	1	0	1	0	1	0	1
U^6	0	1	1	0	0	1	1
U^7	0	0	0	1	1	1	1
U^8	0	0	0	0	0	0	0

Table 6.4. *Words of the dual code of the code (7,4,3)*

6.4. Iterative decoding of product codes

The principle of the BTC rests on the iterative decoding of concatenated block codes. In section 6.2 we have demonstrated that serial concatenation of block codes was more advantageous in terms of asymptotic coding gain than parallel concatenation and that uniform interlacing was preferable to non-uniform interlacing. Similarly, we have shown that it was more interesting to use the extended (or expurgated) version of the primitive codes in order to increase the asymptotic coding gain. The code retained for the BTC is, finally, obtained by serial concatenation of extended block codes with uniform interlacing. This code called product code was proposed by Elias in 1954 [ELI 54]. It lends itself ideally to iterative decoding, which consists of sequentially decoding the rows and the columns of the matrix and reiterating the process.

In addition, in section 6.3 we have demonstrated that it is preferable to use a soft decoding algorithm that increases the asymptotic coding gain (between 1.5 and 2.0 dB) compared to a binary decoding. The second Chase algorithm makes it possible to solve the problem of soft decoding with a very good performance/complexity compromise. However, the Chase algorithm provides a binary decision at output. In a process of iterative decoding the following decoder will benefit from binary input, which is certainly more reliable, but will not have the soft data. Thus, it will not be able to exploit soft decoding and that applies to the following decodings of the iterative process.

So that iterative decoding can benefit from soft decoding at each iteration, it is essential to associate a weighting to the decisions provided by the Chase algorithm. The modification of the Chase algorithm to carry out decoding with SISO is the subject of the following section.

6.4.1. *SISO decoding of a block code*

To simplify, we will consider the decoding of a systematic block code with parameters (n,k,δ). The codewords defined over $\{0,1\}$ are transmitted via Gaussian channel by an amplitude modulation with two states $\{-1,+1\}$ using the following association: $0 \rightarrow -1$ and $1 \rightarrow +1$. In order to simplify the notation, we will suppose that the codewords are directly defined in $\{-1,+1\}$. The observation $R=(r_1,...,r_n)$ at the channel output for a transmitted codeword $E=(e_1,....,e_n)$ is given by:

$$R = E + B \qquad [6.47]$$

where $B=(b_1,....,b_n)$ is a vector of additive white Gaussian noise. SISO decoding of the data R can be performed using the logarithm likelihood ratio (LLR) of the transmitted symbol e_j [PYN 94, PYN 98]:

$$\lambda_j = \ln\left(\frac{\Pr\{e_j = +1/R\}}{\Pr\{e_j = -1/R\}}\right) \qquad [6.48]$$

The sign λ_j provides the optimal decision in the sense of minimization of the probability of error per binary symbol and its absolute value yields a weighting of this decision according to its reliability. The numerator of the LLR can be written in the following way:

$$\Pr\{e_j = +1/R\} = \sum_{1\le i\le N} \Pr\{e_j = +1, E = C^i / R\} \qquad [6.49]$$

where $C^i = \left(c_1^i,....,c_n^i\right)$ is the codeword number i with $c_j^i \in \{-1,+1\}$ and $N = 2^k$ is the number of the codewords. Using the Bayes rule we can put [6.49] in the form:

$$\Pr\{e_j = +1/R\} = \sum_{1 \le i \le N} \Pr\{e_j = +1/E = C^i, R\} \times \Pr\{E = C^i / R\} \quad [6.50]$$

Using the fact that:

$$\Pr\{e_j = +1/E = C^i, R\} = \Pr\{e_j = +1/E = C^i\} = \begin{cases} 1 \text{ if } c_j^i = +1 \\ 0 \text{ if } c_j^i = -1 \end{cases} \quad [6.51]$$

we can rewrite [6.50] in the form:

$$\Pr\{e_j = +1/R\} = \sum_{C^i \in S^{+1(j)}} \Pr\{E = C^i / R\} \quad [6.52]$$

where $S^{+1(j)}$ is the set of codewords with a +1 in position j. Similarly, we show that the denominator of [6.48] is equal to:

$$\Pr\{e_j = -1/R\} = \sum_{C^i \in S^{-1(j)}} \Pr\{E = C^i / R\} \quad [6.53]$$

where $S^{-1(j)}$ is the set of codewords having a –1 in position j. Using the Bayes rule again, we show that:

$$\Pr\{E = C^i / R\} = \frac{\mathrm{P}\{R / E = C^i\} \times \Pr\{E = C^i\}}{\mathrm{P}\{R\}} \quad [6.54]$$

where P{•} indicates a probability density. Supposing that the binary symbols of the message are mutually independent and identically distributed (i.i.d), we have:

$$\Pr\{E = C^i\} = \frac{1}{N} \quad [6.55]$$

Taking again relation [6.48] and the equations above, we show that the LLR is given by the following equation:

$$\lambda_j = \ln\left(\frac{\sum\limits_{C^i \in S^{+1(j)}} \mathrm{P}\left\{R/E = C^i\right\}}{\sum\limits_{C^i \in S^{-1(j)}} \mathrm{P}\left\{R/E = C^i\right\}}\right) \qquad [6.56]$$

where:

$$\mathrm{P}\left\{R/E = C^i\right\} = \prod_{l=1}^{n}\left(\mathrm{P}\left\{r_l/e_l = c_l^i\right\}\right) = \prod_{l=1}^{n}\left(\frac{1}{\sqrt{2\pi}\sigma}\exp\left(-\frac{\left(r_l - c_l^i\right)^2}{2\sigma^2}\right)\right) \qquad [6.57]$$

is the probability density of R for an transmitted word $E = C^i$ and σ represents the standard deviation of the noise. By transferring [6.57] into [6.56] we obtain:

$$\lambda_j = \ln\left(\frac{\sum\limits_{C^i \in S^{+1(j)}} \exp\left(-\frac{\left\|R - C^i\right\|^2}{2\sigma^2}\right)}{\sum\limits_{C^i \in S^{-1(j)}} \exp\left(-\frac{\left\|R - C^i\right\|^2}{2\sigma^2}\right)}\right) \qquad [6.58]$$

where:

$$\left\|R - C^i\right\|^2 = \sum_{l=1}^{n}\left(r_l - c_l^i\right)^2 \qquad [6.59]$$

represents the square of the Euclidean distance between R and C^i. Let $C^{+1(j)}$ be the codeword belonging to $S^{+1(j)}$ with a minimum Euclidean distance of R and $C^{-1(j)}$ the codeword belonging to $S^{-1(j)}$ with a minimum Euclidean distance of R. We demonstrate that the LLR can be expressed in the form:

$$\lambda_j = \frac{1}{2\sigma^2}\left(\left\|R - C^{-1(j)}\right\|^2 - \left\|R - C^{+1(j)}\right\|^2\right) + \ln\left(\frac{\sum\limits_i A_i}{\sum\limits_i B_i}\right) \qquad [6.60]$$

where:

$$A_i = \exp\left(\frac{\left\| R - C^{+1(j)} \right\|^2 - \left\| R - C^i \right\|^2}{2\sigma^2} \right) \leq 1; \text{ with } C^i \in S^{+1(j)} \quad [6.61]$$

and:

$$B_i = \exp\left(\frac{\left\| R - C^{-1(j)} \right\|^2 - \left\| R - C^i \right\|^2}{2\sigma^2} \right) \leq 1; \text{ with } C^i \in S^{-1(j)}$$

[6.62]

Supposing that the codewords are distributed uniformly in the space of the codewords, we show that $\sum_i A_i \approx \sum_i B_i$ and that the ratio of the two terms tends towards one. Thus, the second term of [6.60] becomes negligible compared to the first and the LLR is approximated well by:

$$\lambda_j = \frac{1}{2\sigma^2} \left(\left\| R - C^{-1(j)} \right\|^2 - \left\| R - C^{+1(j)} \right\|^2 \right) \quad [6.63]$$

Developing [6.63] we obtain:

$$\lambda_j = \frac{2}{\sigma^2} \left(r_j + \sum_{l=1; l \neq j}^{n} r_l c_l^{+1(j)} p_l \right) \quad [6.64]$$

with:

$$p_l = \begin{cases} 0 \text{ if } c_l^{+1(j)} = c_l^{-1(j)} \\ 1 \text{ if } c_l^{+1(j)} \neq c_l^{-1(j)} \end{cases} \quad [6.65]$$

In the case of a Gaussian channel, σ is constant. By normalizing relation [6.64] with respect to $2/\sigma^2$ we obtain the following equation:

$$r'_j = \left(\frac{\sigma^2}{2}\right)\lambda_j = r_j + w_j \qquad [6.66]$$

with:

$$w_j = \sum_{l=1;l\neq j}^{n} \eta c_l^{+1(j)} p_l \qquad [6.67]$$

The quantity r'_j represents the soft output of the SISO decoder. Relation [6.66] shows that the soft output is expressed as the sum of the soft input and a quantity w_j, which we call extrinsic information given by relation [6.67]. This extrinsic information plays a fundamental part in the operation of a turbo decoder.

We note that w_j is a linear combination of the soft input data of the decoder. It depends on the two codewords with a minimum Euclidean distance of R with a +1 and a – 1 in position j respectively. In addition, w_j does not depend on r_j.

The term w_j in [6.66] is the contribution of the SISO decoder to the soft decision r'_j. Since it does not depend on r_j, it is also independent of the latter. It contains information on the sign of the binary symbol e_j contained in the other binary symbols of the codeword and which stems from the correlation between these binary symbols introduced by coding.

In a turbo decoder, extrinsic information w_j makes it possible to transmit to the decoder following the information on the sign of the e_j bit contained in the other binary symbols of the decoded codeword.

6.4.2. *Implementation of the weighting algorithm*

We will now consider the implementation of the SISO decoder on the basis of the Chase algorithm. The principle consists of generating a set Ω of codewords with minimum Euclidean distance from the observation R. To simplify, we write that R is the LLR vector (normalized) of the binary symbols of the codeword transmitted at the decoder input. Among this set of codewords, the decision provided by the Chase algorithm is D.

From this decision we will calculate the weighting of the components d_j of this decision. To calculate this weighting on the basis of [6.63] we need the two codewords $C^{+1(j)}$ and $C^{-1(j)}$. It is obvious that *D* is one of these two codewords. Let *C* be the codeword in Ω (if it exists) with a minimum Euclidean distance of *R* and verifying $c_j \neq d_j$. We easily deduce from it that *C* is the second codeword among the two cited above. We can then write the equation of normalized SISO decoder output in the following way:

$$r'_j = \left(\frac{\|R-C\|^2 - \|R-D\|^2}{4} \right) \times d_j \qquad [6.68]$$

We note that the sign of the soft output is given by d_j, which is the decision provided by Chase and that the amplitude of the output depends on the difference between the two metrics $\|R-C\|^2$ and $\|R-D\|^2$. This difference is always positive according to the definition of *D* and *C*: $\|R-C\|^2 \geq \|R-D\|^2$. When $\|R-C\|^2 \to \|R-D\|^2$, the two codewords *C* and *D* have equal probability and the probability of error of d_j tends towards 0.5. This results then in an amplitude of the soft output, which tends towards zero according to [6.68]. We then say that the reliability of the decision is zero. As the difference of the two metrics increases, the probability of *D a posteriori* tends towards one and that of *C* towards zero. The decision d_j becomes increasingly reliable and that results in an amplitude of the soft output, which increases and tends towards infinity. For the moment we suppose that there exists a codeword *C* in Ω called codeword concurrent with *D*. But for this assumption to be always verified it would be necessary to generate a set of words of a very large Ω code. We would then face prohibitive complexity. A simple and relatively effective solution consists of generating a set Ω on the basis of a reasonable number of test sequences (8, 16 or 32). For the components of *D* without a concurrent word code we apply the empirical relation:

$$r'_j = \beta \times d_j \qquad [6.69]$$

This empirical solution is justified by the fact that we know the optimal decision (sign of d_j) but not its reliability, which depends on *C*. When *C* is not in the set Ω, that implies that it is outside the search zone defined in the Chase algorithm by the test sequences. Consequently, the concurrent *C* is relatively distant from *R* (in the sense of Euclidean distance) and the decision is relatively reliable, hence we get a relatively high value of *β*. However, it is necessary to be careful and not assign too high a reliability to erroneous decisions, which can occur during the first iterations.

This reliability is thus according to the probability of errors at the output of the SISO decoder. One can determine the value of β using the following relation [PYN 98]:

$$\beta \propto \ln\left(\frac{\Pr\{d_j = e_j\}}{\Pr\{d_j \neq e_j\}}\right) \quad [6.70]$$

This modification of the Chase algorithm makes it possible to carry out SISO decoding of block codes with relatively low complexity if we limit ourselves to codes with a Hamming distance lower or equal to 8. We will now approach the iterative decoding of product codes.

6.4.3. *Iterative decoding of product codes*

Let us consider the iterative decoding of product codes formed by serial concatenation of two extended and systematic BCH codes. Each column of the coded matrix is a word of the code $\boldsymbol{C}^1$ and each row is a word of the code $\boldsymbol{C}^2$. Iterative decoding consists of carrying out a decoding of the rows followed by a decoding of the columns (or vice versa) using the corresponding SISO decoder described above and reiterating this procedure several times. With each iteration the decoder exploits the observation stemming from the transmission channel, as well as the extrinsic information provided by the preceding decoder. It delivers the corresponding extrinsic information to the current decoding. Nevertheless, it is necessary to take some precautions concerning the information exchanged between the decoders. Indeed, it is necessary to avoid the propagation of errors that can induce an increase in the number of errors during iterations and force the algorithm to diverge.

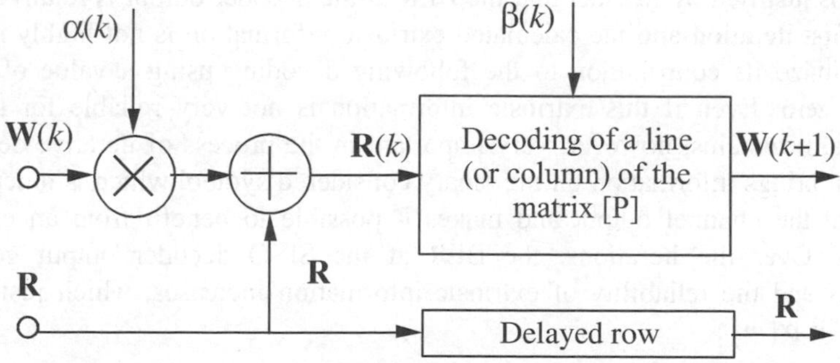

Figure 6.12. *Synoptic of the elementary decoder*

The synoptic of the elementary decoder used to carry out the iterative decoding of a product code is illustrated in Figure 6.12. This decoder has two inputs and two outputs. The first input receives the extrinsic information matrix $[W(m)]$ from the previous decoding. During the first decoding this matrix is initialized to zero. The second input receives the matrix of normalized LLR $[R]$ calculated on the basis of the observations at the output of the transmission channel. The decoder carries out SISO decoding of the rows or the columns of the matrix and delivers a new matrix with extrinsic information $[W(m+1)]$ obtained by the difference between the LLR at the output and the data at the input of the calculation unit. The second output delivers a copy of the matrix $[R]$.

We note (see Figure 6.12) that the input of the calculation unit that carries out SISO decoding of the row and the columns is:

$$\left[R(m)\right]=\left[R\right]+\alpha(m)\left[W(m)\right] \qquad [6.71]$$

The coefficient $\alpha(m)$ takes into account the fact that the components of $[R]$ and $[W(m)]$ are random variables with the same law but with different parameters. Indeed, if we consider the case of a phase modulation with two states in a Gaussian channel, we demonstrate that the observations at the output of the transmission channel follow a Gaussian law of average ± 1 with the same standard deviation σ as the additive noise. On the other hand, the components of the matrix $[W(m)]$ are linear combinations of the observations at the decoder input [6.67]. They are also random variables according to a Gaussian law, but with a different average and standard deviation. This point will be illustrated by an example later on.

In practice, the coefficient $\alpha(m)$ is optimized by successive approximation and it moves from one iteration to the next. For the first iterations it takes a value close to zero and grows with the iterations to reach a value close to one with the last. This strategy is justified by the fact that the BER at the decoder output is relatively high for the first iteration and the calculated extrinsic information is not highly reliable. We minimize its contribution to the following decoding using a value of $\alpha(m)$ close to zero. Even if this extrinsic information is not very reliable for the first iterations, it remains, however, very important in the process of iterative decoding. Indeed, it brings information on the binary considered symbol which is independent of that at the channel output and makes it possible to benefit from an effect of diversity. Over the iterations, the BER at the SISO decoder output generally decreases and the reliability of extrinsic information increases, which justifies an increase in $\alpha(m)$.

The coefficient $\beta(m)$ follows an evolution similar to that of $\alpha(m)$. Indeed, $\beta(m)$ is a function of the BER at the SISO decoder output [6.70]. When the BER is high, $\beta(m)$ takes a value close to zero and grows along the iterations. The optimization of the two coefficients is relatively complicated because they are not independent and the solution is not unique. In addition, they also depend on the codes used. A manner of reducing the dependence between the two coefficients consists of normalizing to one the average of the absolute value of the components of $[W(m)]$ calculated from [6.66] [PYN 98].

6.4.4. *Comparison of the performances of BTC*

We will now consider the performances of different block turbocodes constructed on the basis of extended BCH codes. We will limit ourselves to product codes constructed on the basis of two identical codes with a correction capacity ≤ 2. Table 6.5 provides the parameters of the various product codes considered. The first column indicates the parameters of two BCH codes used and the following columns indicate the length, size, minimum distance, coding output and asymptotic coding gain of the product code.

We note that for a BCH code with a given minimum (Hamming) distance, the coding output and the coding gain increase with the length of the code. We can thus construct product codes with very high output with considerable coding gain on the basis of Hamming codes of length 256 or 512. In addition, the minimum distance of the product code does not depend on the dimension of the product code K. Thus, we can obtain a relatively high minimum distance even for product codes of small sizes.

Product code	N	K	Δ	R	Gain (dB)
(16,11,4) × (16,11,4)	256	121	16	0.473	8.80
(16,7,6) × (16,7,6)	256	49	36	0.191	8.40
(32,26,4) × (32,26,4)	1,024	676	16	0.660	10.25
(32,21,6) × (32,21,6)	1,024	441	36	0.431	11.90
(64,57,4) × (64,57,4)	4,096	3,249	16	0.793	11.05
(64,51,6) × (64,51,6)	4,096	2,601	36	0.635	13.60
(128,120,4) × (128,120,4)	16,384	14,440	16	0.879	11.50
(128,113,6) × (128,113,6)	16,384	12,769	36	0.779	14.50
(256,247,4) × (256,247,4)	65,536	61,009	16	0.931	11.75
(512,502,4) × (512,502,4)	262,144	252,004	16	0.961	11.87

Table 6.5. *Parameters of product BCH codes*

In order to objectively compare the performances of various BTC, we use the same set of coefficients for all simulations. The number of test sequences is fixed at 16 and they are built on the basis of the four least reliable bits. These test sequences are obtained by commutating all the combinations of one, two, three and four binary symbols among the four least reliable binary symbols in the input. The average of the absolute value of the components of $[W(m)]$, calculated on the basis of [6.66], is normalized to one after decoding (row or column) of the matrix. The number of iterations (an iteration corresponds to the decoding of the rows then columns or vice versa) is more than six. The values used for α and β in simulations are as follows:

$$\alpha(m) = [0.0,\ 0.2,\ 0.3,\ 0.5,\ 0.7,\ 0.9,\ 1.0,\ 1.0,\ 1.0,\ 1.0];$$
$$\beta(m) = [0.2,\ 0.4,\ 0.6,\ 0.8,\ 1.0,\ 1.0,\ 1.0,\ 1.0,\ 1.0,\ 1.0] \qquad [6.72]$$

We now will consider the performances of the BTC in various cases of figures.

a) Performance of BTC in a Gaussian channel with QPSK modulation

Figure 6.13 makes it possible to compare the evolution of the BER according to the E_b/N_0 ratio for the first six iterations of the product code (64,51,6) × (64,51,6). We note that the BER decreases after each iteration. For a E_b/N_0 ratio of 2.7 dB, the BER is 3×10^{-2} after the first iteration, 9×10^{-3} after the second, 5×10^{-4} after the third, 3×10^{-5} after the fourth and 5×10^{-7} after the sixth. Each additional iteration makes it possible to decrease the BER thanks to the extrinsic information taken at the decoder output and reintroduced at the input. It is the effect known as of turbodecoding. In practice, for a given BER we note that the gain brought by each additional iteration decreases and tends towards zero and that we can limit ourselves to four or maybe five iterations.

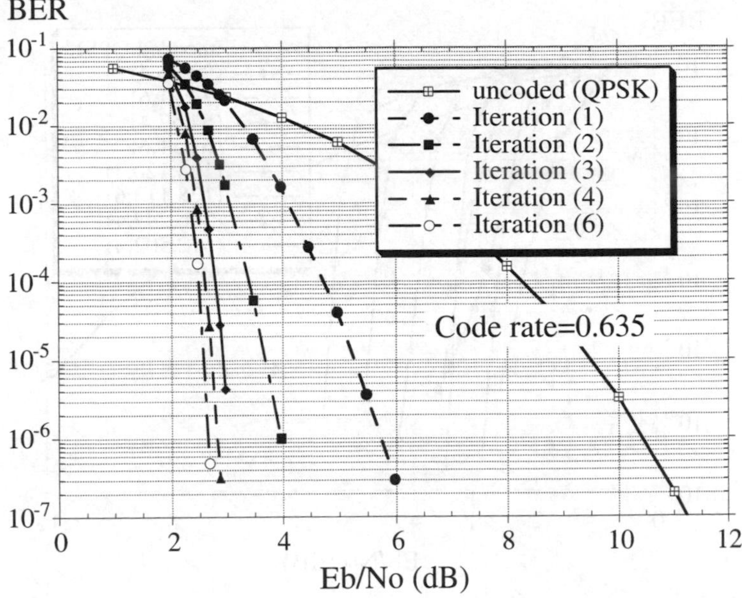

Figure 6.13. *Evolution of the BER according to E_b/N_0 for various iterations in the case of the product code (64,51,6) × (64,51,6)*

Figure 6.14 shows the evolution of the BER according to E_b/N_0 to the fourth iteration, for various product codes. We note that the decrease of the BER according to E_b/N_0 increases with the length of code, but also with its minimum distance. In addition, we observe that the fast decrease of the BER occurs beyond a (threshold) value of E_b/N_0. In the lower part of the threshold the turbo effect does not kick in. In this area, the relationship between the BER in the input and output of the SISO decoder is not sufficient to initiate the avalanche phenomenon leading to turbo decoding.

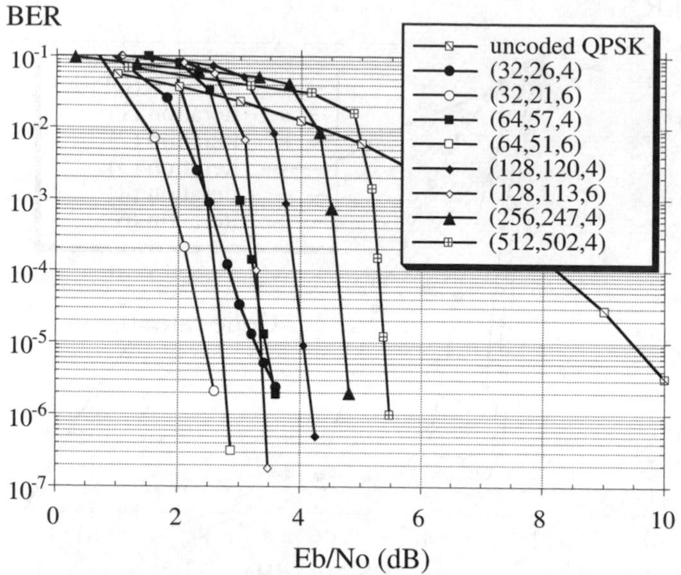

Figure 6.14. *Evolution of the BER according to* E_b/N_0 *after the fourth iteration of the product codes of Table 6.5*

Figure 6.15 makes it possible to compare the performances of different turbocodes with the Shannon limit in the case of a QPSK modulation in a Gaussian channel and supposing the use of a code of infinite size. The curve represented in Figure 6.15 indicates the E_b/N_0 ratio necessary to ensure a zero probability of error for a given coding output. The rhombuses and the triangles indicate the E_b/N_0 ratio necessary to obtain a BER of 10^{-5} after four iterations for various BTC. We note that the performances of the BTC with high output are very close to the Shannon limit (# 0.7 dB). For BTC of poor output the variation compared to the Shannon limit is explained by the fact that the size K of the codes decreases.

Figure 6.16 makes it possible to compare the probability distribution of the observations at the output of the transmission channel with that of extrinsic information for various iterations. As indicated previously, the average of the absolute value of extrinsic information calculated according to [6.68] has been normalized to one. We observe that extrinsic information follows a Gaussian distribution and that its variance decreases with iterations. For the first iteration we notice a higher dispersion of samples around the average compared to the samples coming from the transmission channel. That justifies the use of the coefficient α in [6.71]. Along the iterations, the dispersion of the samples of extrinsic information around the average value decreases and becomes increasingly reliable. We are then brought to increase the value of α.

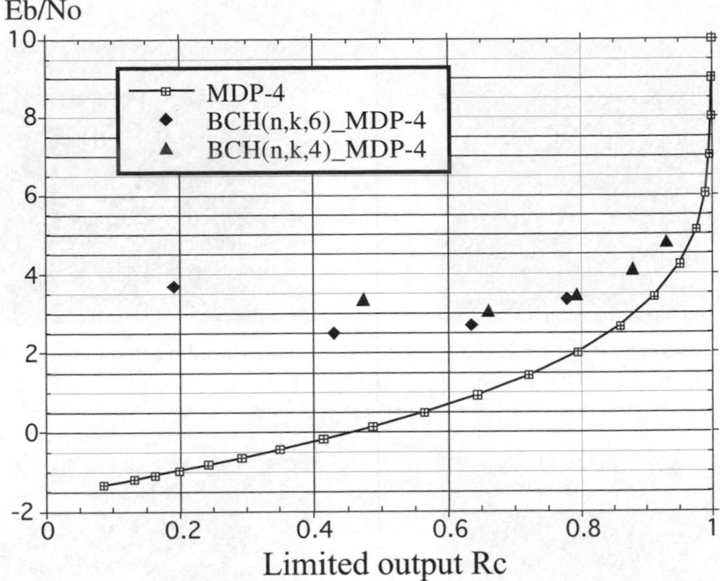

Figure 6.15. *Comparison of E_b/N_0 with a BER of 10^{-5} with the Shannon limit for the product codes of Table 6.5*

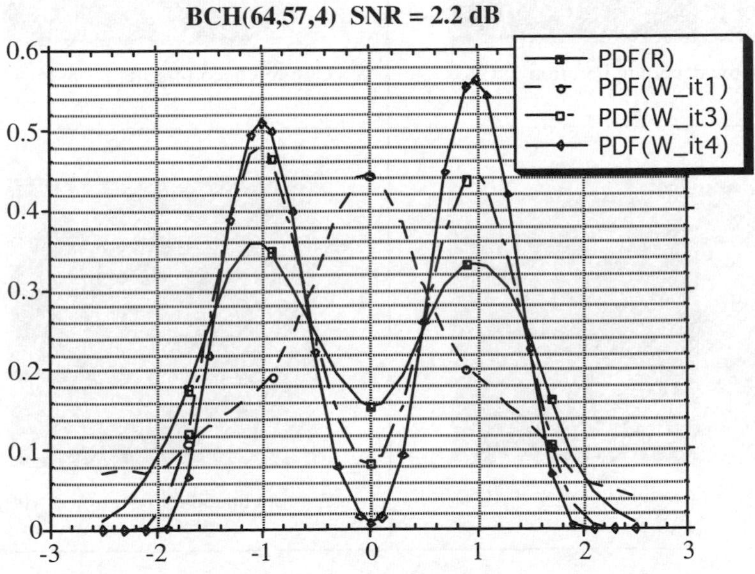

Figure 6.16. *Distribution of samples in the matrices [R], [W(1)], [W(2)] and [W(3)] in the case of a product code $(64,57,4) \times (64,57,4)$*

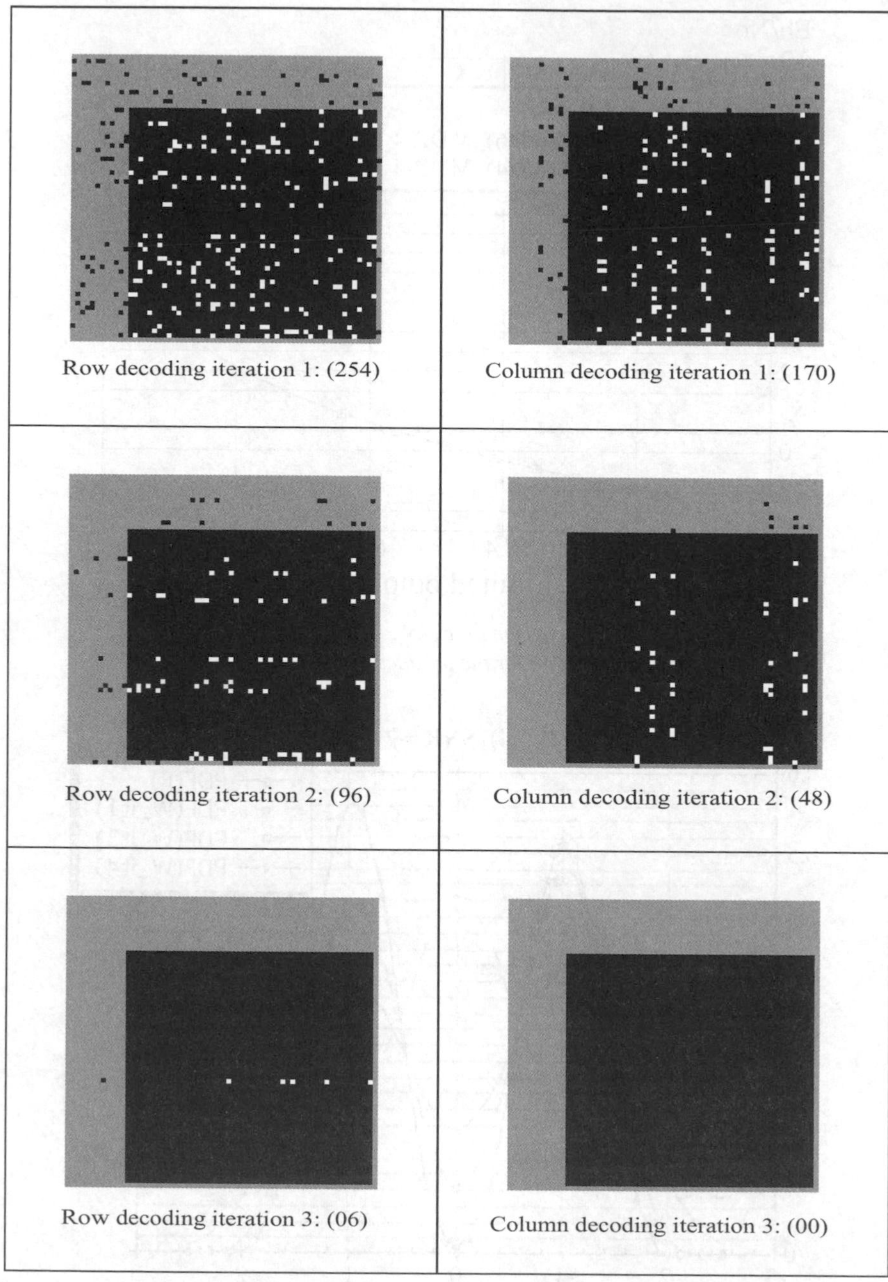

Figure 6.17. *Example of the evolution of errors during iterative decoding of the matrix (64,51,6) × (64,51,6)*

Figure 6.17 makes it possible to visualize the evolution of errors in a coded matrix during iterations in the case of the code (64,51,6) × (64,51,6). The black background indicates the zone containing binary information symbols and the gray background contains binary parity symbols. The light points against the black background and the black spots against the gray background indicate the position of errors.

In this example the error count at the output of the transmission channel is 263 for a total of 4,096 binary symbols, which leads to a BER of 6.4×10^{-2}. During the first iteration, the error count passes successively to 254 then 170. The efficiency of the decoder in the first iterations is very low.

At the second iteration, the error count passes to 96 and then 48. The improvement is more sensitive because the error count has been divided by three and five respectively. Finally, at the third iteration, the error count passes to 6 and then 0.

It is important to note that, as opposed to traditional codes, in certain cases turbocodes are able to correct error counts much higher than its minimum distance.

In the above example the decoder made it possible to correct a pattern with 263 errors, whereas its minimum distance is only 36. This property of the turbo decoder stems directly from the process of iterative decoding. Indeed, during the first iteration, this latter corrects only certain errors, making it possible to initialize the process of correction which develops during the following iterations and produces an "avalanche" effect.

We can also observe that after the decoding of rows (or columns), the errors form patterns along the rows (or columns) containing an even number of errors higher than or equal to 6. Indeed, the SISO decoder delivers codewords along the rows (or columns) of the matrix. Thus, all the rows (or columns) after decoding of rows (or columns) are codewords.

However, if the columns and the rows of the matrix are codewords, then the latter is a codeword of the product code. We can use this property to detect the convergence of the algorithm after each iteration. It suffices to calculate the syndrome of the columns after the decoding of rows (or vice versa).

If the syndromes are all zero, the decoding algorithm has converged and we can stop the iterative process. If the syndromes are non-nil at the last iteration, the algorithm has not converged and we detect the presence of errors (see [PYN 97]).

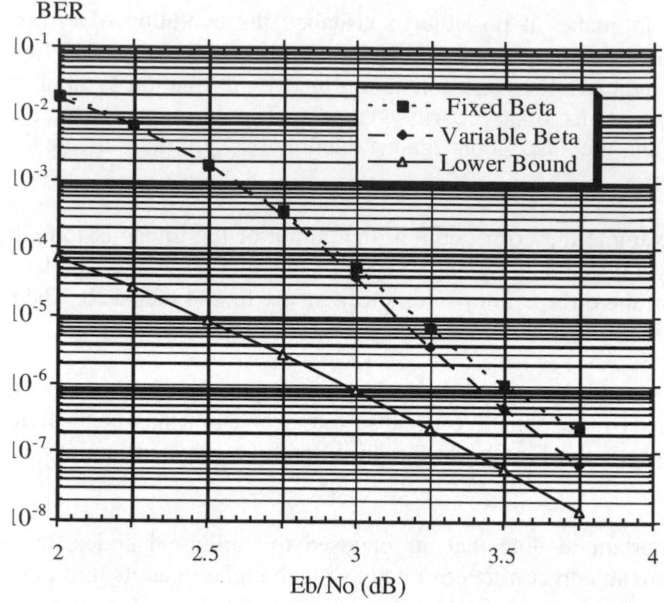

Figure 6.18. *Comparison between the performances of a simulation and the theoretical performances in the case of the product code (32,26,4)*^{2}

Figures 6.18 and 6.19 compare the results of a simulation of a BTC with its theoretical performances. We notice that the results of simulation are very close to the theoretical performances established in section 6.2.3 [6.18].

The gap is lower than 0.5 dB for the code (32,26,4)^2 and 0.7 dB for the code (32,21,6)^2. In order to improve the performances of the iterative decoder with low BER we can use the mean of the absolute value of the least reliable samples at the decoder input for the variable β, and this can be done for each row or column of the matrix [ADD 00].

This average value is multiplied by a scale factor in order to be optimized. In Figures 6.18 and 6.19 we notice that the results with the β variable improve the performances of the decoder and reduce the variation to 0.25 dB and 0.12 dB respectively. The improvement is even more spectacular in the case of the code (32,21,6)^2.

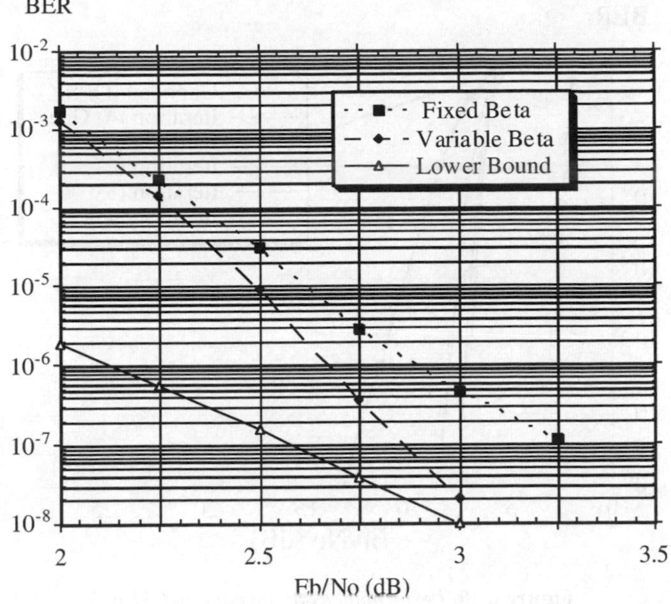

Figure 6.19. *Comparison between the performances of a simulation and the theoretical limit in the case of the product code (32,21,6)*^{2}

b) Performance of BTC associated with the QPSK in a Rayleigh channel

The performances of BTC associated with QPSK modulation in a Rayleigh channel have been evaluated using the following model for the channel:

$$r_i = \rho_i \times e_i + b_i \tag{6.73}$$

where $e_i \in \{-1,+1\}$ is the symbol transmitted via the phase or quadrature path, ρ_i is the attenuation brought by the channel according to a Rayleigh law and b_i is the additive white Gaussian noise (AWGN). We demonstrate that optimal decoding consists of using the same algorithm as for the Gaussian channel by replacing r_i by $\rho_i \times r_i$ insofar as we use the correlation instead of the Euclidean distance as a metric (see [6.27]). Finally, we can use the same decoder to treat the case of the Gaussian and Rayleigh channels.

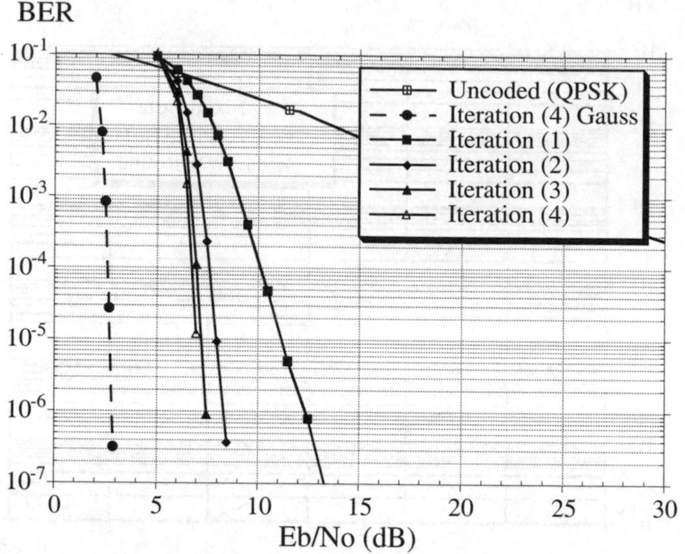

Figure 6.20. *Performances of the code $(64,51,6)^2$ associated with QPSK modulation in a Rayleigh channel*

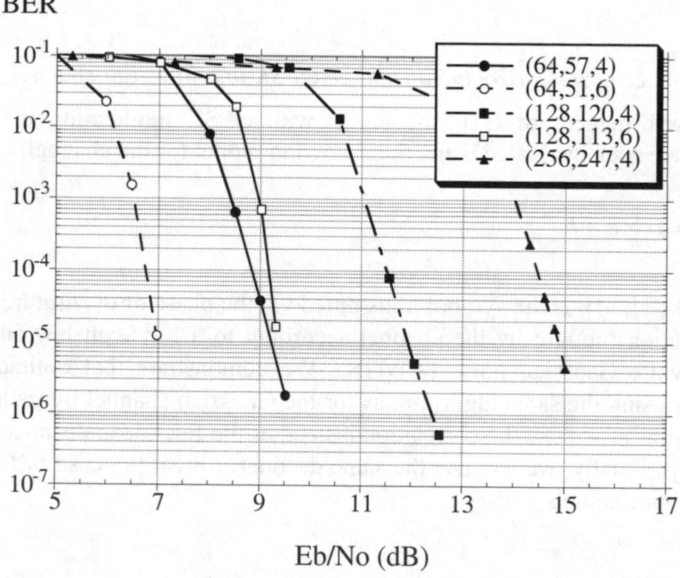

Figure 6.21. *Performances at the fourth iteration of the different product codes associated with QPSK modulation in a Rayleigh channel*

Figure 6.20 represents BER according to E_b/N_0 for the product code $(64,51,6)^2$ associated with a QPSK modulation in a Rayleigh channel. We observe that the performances improve with each new iteration and that the improvement becomes negligible after 4 iterations.

For comparison we have also indicated in Figure 6.20 the curve (dotted) of the BER according to E_b/N_0 for a Gaussian channel to the fourth iteration. We note a degradation of performances in the Rayleigh channel but the curves are parallel to the low BER ($<10^{-2}$). Figure 6.21 shows the curves of the BER according to E_b/N_0 to the fourth iteration for various product codes associated with the QPSK in a Rayleigh channel.

c) Performance of BTC associated with QAM modulations in a Gaussian channel

In general, other works [UNG 82] recommend using the coded modulation when we associate a corrector coding to modulations with a large number of states. In the particular case of turbocodes, the best performance/complexity compromise is provided by the approach known as pragmatic Viterbi [SAW 89] and that is the approach, which we will consider.

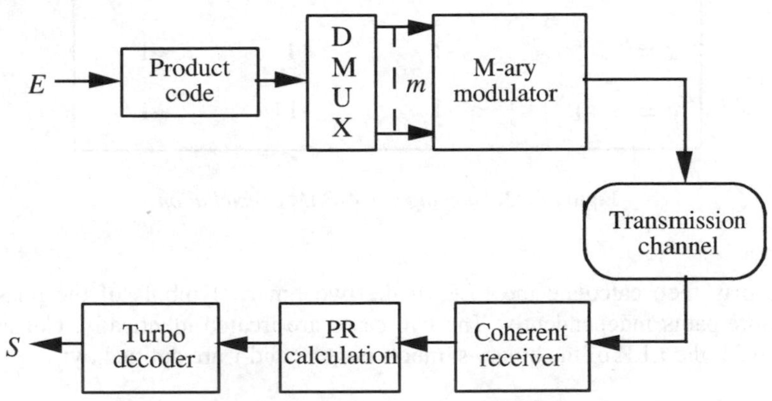

Figure 6.22. *Synoptic of the pragmatic transmission system*

The pragmatic Viterbi approach makes it possible to use the same decoder described in the case of binary modulations to treat all the M-ary modulations with $M = 2^m$, where m is a positive and even integer. The principle consists of coding the binary data with a product code then de-multiplexing the binary flow at the encoder output according to m binary streams which will address the modulator (see Figure 6.22). The labeling of the modulator symbols is generally carried out using a Gray code. The output of the coherent receiver is then given by relation [6.73] where the

e_i symbols are the *M*-ary symbols of the modulator considered and the attenuation ρ_i is constant and equal to 1 (Gaussian channel).

From the observations we calculate the LLR of the *m* binary symbols associated with each *M*-ary symbol; LLRs are then used as the turbo decoder input [PYN 95].

To illustrate the principle we will consider the case of a QAM modulation with 16 states. In this case the constellation may be considered as the superposition of two independent amplitude modulations (4-PAM); one following the phase path and the other following the quadrature path.

The labeling of the symbols of a 4-PAM according to a Gray code is illustrated in Figure 6.23.

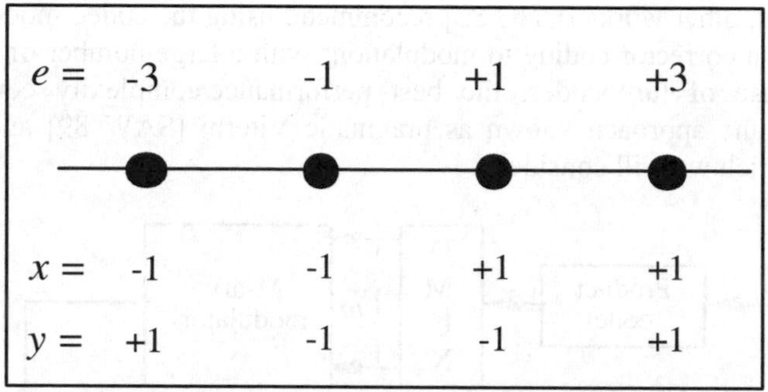

Figure 6.23. *Labeling of a 4-PAM constellation*

We may then calculate the LLR of the two binary symbols of the phase and quadrature paths independently. The two cases are treated identically. Considering Figure 6.23, the LLR of the binary symbols noted *x* and *y* are defined by:

$$\lambda_l = \ln\left(\frac{\Pr\{l=+1/r\}}{\Pr\{l=-1/r\}}\right) \quad [6.74]$$

with *l*=*x* or *y*. Let us take the case of the binary symbol called *x*. Taking into account the notations used we show that:

$$\lambda_x = \ln\left(\frac{\Pr\{e=+1/r\}+\Pr\{e=+3/r\}}{\Pr\{e=-1/r\}+\Pr\{e=-3/r\}}\right) \quad [6.75]$$

Supposing that the data to be transmitted are independent and have mutually equal probabilities we shows using Bayes and preserving only the dominating term in the numerator and the denominator of the sum that:

$$\lambda_x \approx \left(\frac{2}{\sigma^2}\right) \times \begin{cases} r \text{ if } |r| \leq 2 \\ 2(r-1) \text{ if } r > 2 \\ 2(r+1) \text{ if } r < 2 \end{cases} \qquad [6.76]$$

Similarly, for the binary symbol called y we show that:

$$\lambda_y \approx \left(\frac{2}{\sigma^2}\right) \times \begin{cases} +(r-2) \text{ if } r > 0 \\ -(r+2) \text{ if } r < 0 \end{cases} \qquad [6.77]$$

Relations [6.76] and [6.77] make it possible to very easily determine the LLR of the binary symbols x and y of the component in phase and quadrature on the basis of the observation R at the output of the coherent receiver.

These LLR are supplied to the iterative decoder described previously.

The curves of BER according to E_b/N_0 for the product code $(128,113,6)^2$ associated with a 16-QAM modulation are given in Figure 6.24.

We observe that each additional iteration makes it possible to improve the BER for a $E_b/N_0 > 6.0$ dB.

Figure 6.25 compares the performances of the 16-QAM associated with various product codes. The curves of BER according to E_b/N_0 are plotted up to iteration-4.

By way of comparison we also provide the performances of the TCM (trellis coded modulation) with 64 states associated with a 16-QAM for a spectral efficiency of 3 bits/s/Hz. We notice that the BTC yields better performances for low BER.

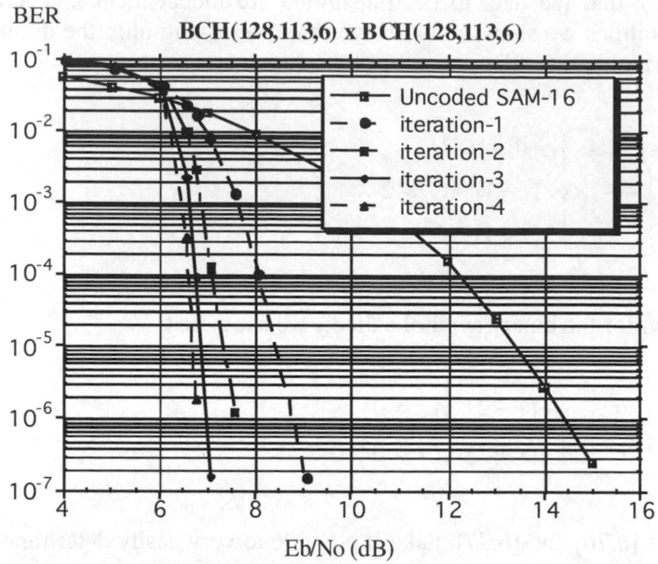

Figure 6.24. *Curves of BER according to E_b/N_0 in the case of the product code $(123,113,6)^2$ associated with a 16-QAM modulation*

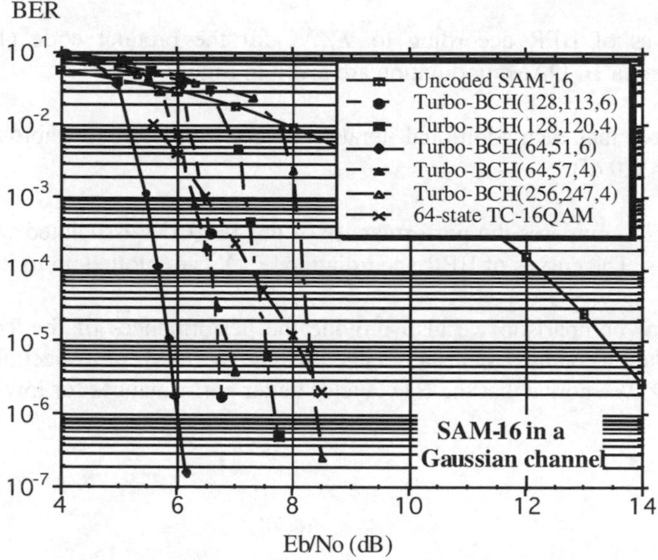

Figure 6.25. *Curves of BER according to E_b/N_0 in the case of a 16-QAM modulation associated with various product codes*

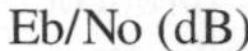

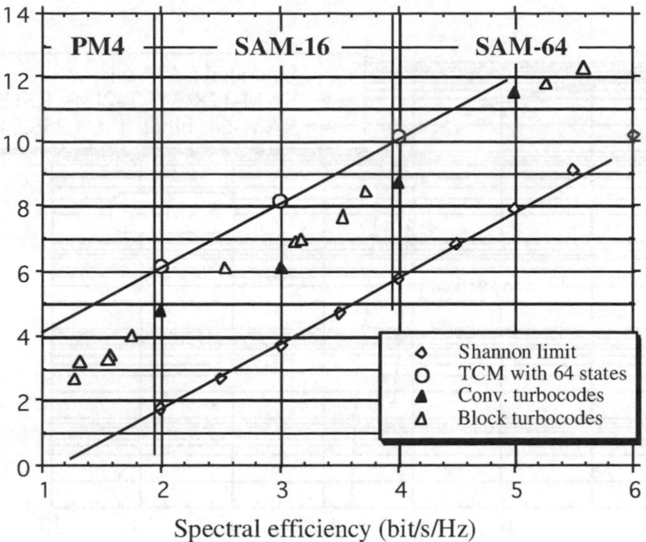

Figure 6.26. *Comparison of the performances of turbocodes associated with QAM modulations with their Shannon limit. The performances of the TCM are also given for comparison*

The method described for 16-QAM applies to other QAM modulations. Figure 6.26 carries out a comparison of the performances of turbocodes associated with QAM modulations with the Shannon limit. The performances are given for a BER of 10^{-5} and four iterations. The TCM performances are also provided by way of comparison. We note that turbocodes make an improvement of more than 1 dB compared to the TCM. The performances of turbocodes are located approximately 2.5 dB from the Shannon limit. We notice that this limit is optimistic because it supposes a Gaussian modulation and a coding of infinite size.

Lastly, Figure 6.27 compares the performances of a traditional concatenated coding scheme with those of a turbocode. The traditional concatenated code comprises an external code of the RS type (204,188) concatenated by a two-dimensional TCM associated with a 128-QAM. This coding scheme is currently normalized for digital television broadcasting. The turbocode solution consists of a product code $(256,247,4)^2$ associated with a 64-QAM using the pragmatic approach. The performances of the turbocode are given for four iterations. The spectral efficiency of the turbocode solution is slightly higher (5.59 bit/s/Hz instead of 5.53 bit/s/Hz) than that of the concatenated code recommended by the standard. The BTC solution makes it possible to improve the connection balance by approximately 1 dB. In addition, as we will see later on, the complexity of the turbocode circuit is even lower, leading to an excellent complexity/performances compromise.

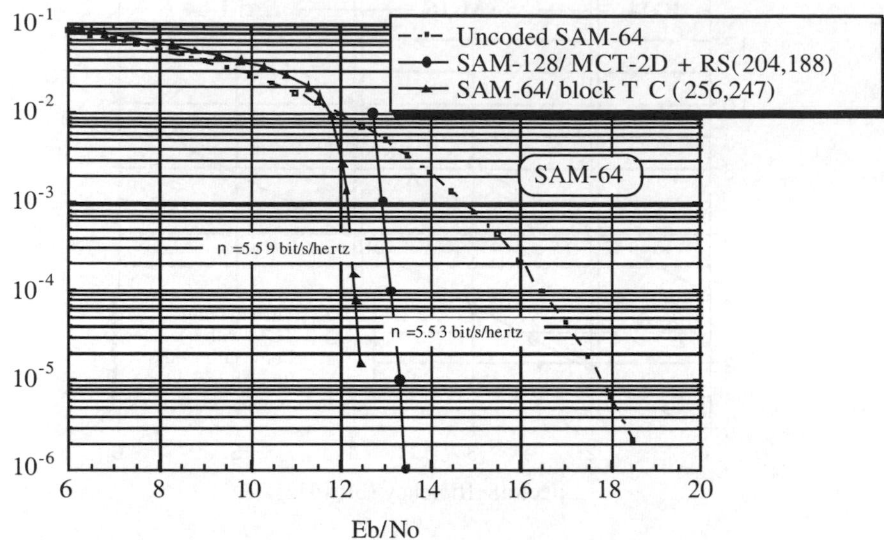

Figure 6.27. *Comparison of the performances of a traditional concatenated coding scheme (RS+TCM) with those of a turbocode*

d) Performance of BTC in a binary symmetric channel (BSC)

Certain digital systems do not have a reliability measurement associated with the binary symbols in the receiver, mainly for technological or economic reasons. By way of an example we can cite optical transmission, data storage or network layer coding. The channel is then called a BSC and is modeled by:

$$r_i = e_i \oplus \varepsilon_i \qquad [6.78]$$

where r_i, e_i and $\varepsilon_i \in \{0,1\}$; e_i is the transmitted binary symbol, r_i is the received binary symbol and ε_i is a variable that takes the value "one" when the received binary symbol is wrong. In this case we do not have the reliability measurement at the decoder input and we will have a degradation of the performances.

For the traditional codes, the optimal decision consists of finding the codeword with the minimum Hamming distance from the observation, which leads to a degradation from 2 to 2.5 dB.

In the case of turbocodes we can use SISO decoding after having applied the transformation $\{0,1\} \to \{-1,+1\}$ to the received binary symbols.

The first decoding is based only on binary data and the performances are identical to those of binary decoding.

For the following decodings the decoder benefits from the contribution of extrinsic information which tends towards a Gaussian distribution and SISO decoding then acquires its full significance.

Figure 6.28 provides the performances of three BTC $(128,120,4)^2$, $(128,113,6)^2$ and $(128,108,8)^2$. We note that degradation in the SBC channel is relatively low (between 1 and 1.5 dB) compared to the optimal case (Gauss).

We can reduce this degradation in the case of the SBC channel with deletion (SBCD) and the loss lies between 0.5 and 1 dB.

Also, let us note that degradation results in a simple shift of the curves of BER according to E_b/N_0 to the right without notable change of slope.

Figure 6.29 presents a comparison of the performances of BTC in a Gaussian channel and SBC with the iteration 8 for a BER of 10^{-5} with their Shannon limit.

We notice that the variation with respect to Shannon is relatively low even in the case of the SBC channel (1 to 1.5 dB).

In addition, the difference between the theoretical performances in a Gaussian and a SBC channel decreases and tends towards "zero" when the output of the code tends towards "one". Thus, the BTC with high output can be used in a channel SBC with a very light loss of performance. This result is very important and paves the way for many BTC applications.

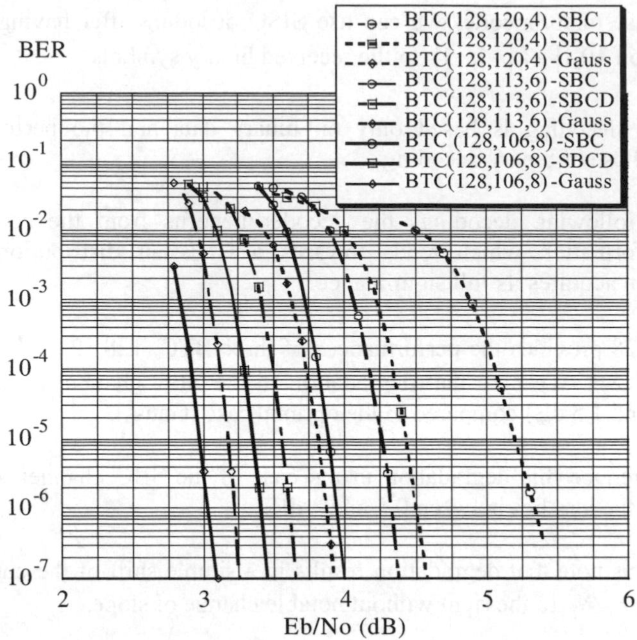

Figure 6.28. *Comparison of the performances of BTC (128, K, δ) ^{2} for different channels (Gaussian, SBC and SBC with deletion) with iteration 8*

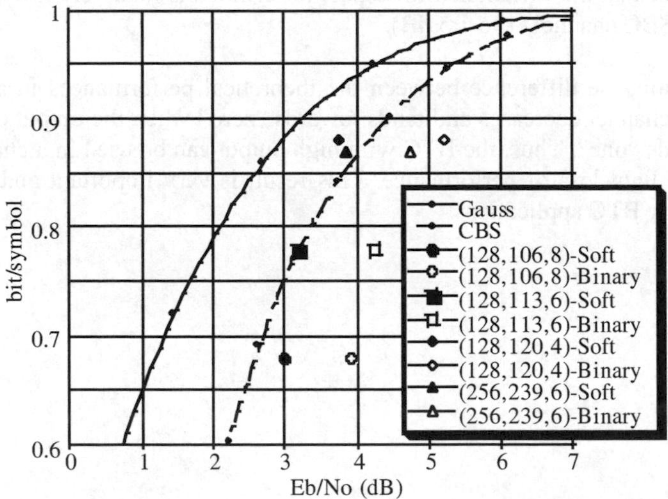

Figure 6.29. *Comparison of the performances of BTC for different channels (Gaussian and SBC) with the Shannon limit*

6.5. Conclusion

In this chapter we have tried to present as simply and clearly as possible the concept of BTC. At first, we have considered the theoretical aspects, such as the concatenation of block codes and the theoretical optimal performances. We then dealt with the soft decoding of block codes considering various algorithms and emphasized that the Chase algorithm [CHA 72] offers the best compromise between complexity and performance. We then introduced the concept of SISO decoding of block codes and described an algorithm of weighting of the decisions provided by the Chase algorithm. Let us note that this algorithm of weighting can be exploited by other algorithms of soft decoding [FOS 98, FAN 00]. Lastly, we have described the iterative decoding of product codes, which constitutes the heart of turbocode and provided several results of simulation considering various product codes, modulations (PSK and QAM) and channels (Gaussian and Rayleigh).

Certain points evoked in this chapter deserve to be underlined. In addition to the exceptional performances of the BTC, we have error probability limits that are very precise for low BER. The comparison between simulated performances and the lower limit of the probability of error shows than the limit is reached for low BER and than the BTC does not present an effect known as flattening. This result shows that the proposed process of iterative decoding of the product codes is quasi-optimal for low BER. Moreover, we can predict the performances of the BTC for very low BER (lower than 10^{-9}) thanks to the limit of thc probability of error.

In conclusion, the BTC is a very promising technology and complementary to the CTC. It is very attractive for certain applications, such as: data transmission by packages of small size, transmission with high yield and large spectral efficiency and transmission with a very broad band. At present many transmission systems consider the use of the BTC, and the IEEE-802.16 standard has already adopted it as an option. This choice is justified by the fact that the BTC not only demonstrates very good performances, but also has a very low implementation complexity [ADD 97, GOA 97] and that it has many other qualities such as flexibility and error detection. These aspects are considered in the following chapter.

6.6. Bibliography

[ADD 96] ADDE P., PYNDIAH R., BERROU C., "Performance of Hybrid Turbocodes", *Electronics Letters*, vol. 32 no. 24, November 1996.

[ADD 97] ADDE P., PYNDIAH R., RAOUL O., INISAN J. R., "Block Turbo Decoder Design", in *Proc. of IEEE Int. Symposium on Turbocodes and Related Topics*, vol. 1/1, p. 166-169, Brest, September 1997.

[ADD 00] ADDE P., PYNDIAH R., "Recent Simplifications and Improvements in Block Turbocodes", 2nd *International Symposium on Turbocodes & Related Topics*, p. 133-136, Brest, September 2000.

[AIT 95] AITSAB O., PYNDIAH R., "Performances des turbocodes en blocs Q-aires Reed Solomon", *GRETSI-95*, Juan les Pins, September 1995.

[AIT 96] AITSAB O., PYNDIAH R., "Performance of Reed-Solomon Block Turbocodes", in *Proc. of IEEE Globecom'96 Conference*, vol. 1/3, November 1996, London.

[AIT 97.1] AITSAB O., PYNDIAH R., SOLAIMAN B., "Optimisation conjointe d'un décodeur source et d'une modulation codée pour la transmission numérique d'images sur un canal bruité", *GRETSI-97 Conference*, Grenoble, September 1997.

[AIT 97.2] AITSAB O., PYNDIAH R., "Performance of Concatenated Reed-Solomon/Convolutional Codes with Iterative Decoding", *IEEE Globecom'97 Conference*, Phoenix, November 1997.

[BAH 74] BAHL L. R., COCKE J., JELINEK F., RATIV J., "Optimal Decoding of Linear Codes for Minimizing Symbol Error Rate", *IEEE Trans. on Inf. Theory*, p. 284-287, March 1974.

[BAR 94] BARBULESCU A. S., PIETROBON S. S., "Interleaver Design for Turbocodes", *Electronics Letters*, vol. 30, p. 2107-2108, December 1994.

[BAR 95] BARBULESCU A. S., PIETROBON S. S., "Rate Compatible Turbocodes", *Electronics Letters*, vol. 31, p. 535-536, March 1995.

[BEE 86] BE'ERY Y., SNYDERS J., "Optimal Soft Decision Block Decoders Based on Fast Hadamard Transform", *IEEE Trans. on Inf. Theory*, vol. IT-32, no. 3, p. 355-364, May 1986.

[BER 68] BERLEKAMP E. R., *Algebraic Coding Theory*, McGraw-Hill, 1968.

[BER 93-1] BERROU C., GLAVIEUX A., THITIMAJSHIMA P., "Near Shannon Limit Error-Correcting Coding and Decoding: Turbocodes (1)", *IEEE Int. Conf. on Comm. ICC'93*, p. 1064-1071, vol. 2/3, May 1993.

[BER 93-2] BERROU C., ADDE P., ANGUI E., FAUDEUIL S., "A Low Complexity Soft-Output Viterbi Decoder Architecture", *Proc. of IEEE ICC'93,* Geneva, May 1993.

[BER 95] BERROU C., GLAVIEUX A., PYNDIAH R., "Turbo-codes: principes et applications" *GRETSI-95*, Juan les Pins, September 1995.

[BER 96] BERROU C., GLAVIEUX A., "Near Optimum Error Correcting Coding and Decoding: Turbocodes", *IEEE Trans. on Comm.,* vol. 44, October 1996.

[BER 97] BERROU C., "Some Clinical Aspects of Turbocodes", *Proc. of IEEE Int. Symp. on Turbocodes and Related Topics*, p. 26-31, Brest, France, September 1997.

[BEN 96] BENEDETTO S., MONTORSI G., "Design of Parallel Concatenated Convolutional Codes", *IEEE Trans. on Communications*, vol. 44, no. 5, p. 591-600, May 1996.

[CAV 00] CAVALEC K., PYNDIAH R., “Block Turbocodes for Transmit and Receive Diversity Systems”, *2nd International Symposium on Turbocodes and Related Topics*, p. 379-382, Brest, September 2000.

[CHA 72] CHASE D., “A Class of Algorithms for Decoding Block Codes with Channel Measurement Information”, *IEEE Trans. Inform. Theory*, vol. IT-18, p. 170-182, January 1972.

[DIV 95] DIVSALAR D., POLLARA F., “Turbocodes for Deep-Space Communications”, *IEEE Communication Theory Workshop*, Santa Cruz, VA, 23-26 April, 1995,

[DOU 95] DOUILLARD C., DIDIER P., JEZEQUEL M., BERROU C., GLAVIEUX A., “Iterative Correction of Intersymbol Interference: Turbo Equalization”, *European Transactions on Telecommunication*, special issue vol. 6, no. 5, 1995.

[ELB 98] ELBAZ A., AITSAB O., PYNDIAH R., SOLAIMAN B., “Channel Decoding of Block Codes with A Priori Information”, *IEEE Globecom'98 Conference*, Sydney, December 1998.

[ELI 54] ELIAS P., “Error-Free Coding”, *IRE Trans. on Inf. Theory*, vol. IT-4, p. 29-37, September 1954.

[FAN 00] FANG J., BUDA F., LEMOIS E., “Turbo Product Code: A Well Suitable Solution to Wireless Packet Transmission for Very Low Error Rates”, *2nd International Symposium on Turbocodes & Related Topics*, p. 101-111, Brest, September 2000.

[FOR 66] FORNEY G. D., *Concatenated Codes*, MIT Press, Cambridge, MA, 1966.

[FOS 98] FOSSORIER M. P. C., SHU LIN, “Soft-Input Soft-Output Decoding of Linear Block Codes based on Ordered Statistics”, in *Proc. of IEEE GLOBECOM'98 Conference*, vol. 5, p. 2828-2833, Sydney, 1998.

[GOA 97] GOALIC A., PYNDIAH R., “Real-Time Turbo-Decoding of Product Codes on a Signal Processor”, *IEEE GLOBECOM'97 Conference*, Phoenix, November 1997.

[HAG 89] HAGENAUER J., HOEHER P., “A Viterbi Algorithm with Soft-Decision Outputs and its Applications,” *Proc. of IEEE Globecom'89 conf.*, p. 1680-1686, Dallas, November 1989.

[HAG 96] HAGENAUER J., OFFER E., PAPKE L., “Iterative Decoding of Binary Block and Convolutional Codes” *IEEE Trans. on Inf. Theory*, vol. 42, p. 429-445, March 1996.

[HAR 76] HARTMANN C. R. P., RUDOLPH L. D., “An Optimum Symbol-by-Symbol Decoding Rule for Linear Codes”, *IEEE Trans. on Inf. Theory*, vol. IT-22, no. 5, p. 514-517, September 1976.

[IMA 77] IMAI H., HIRIKAWA S., “A New Multi-Level Coding Method using Error-Correcting Codes”, *IEEE Trans. on Information Theory*, vol. IT-23, p. 371-377, 1977.

[JAC 95] JACQ S., PYNDIAH R., PICART A., “Algorithme turbo: un nouveau procédé de décodage des codes produits”, *GRETSI-95*, Juan les Pins, September 1995.

[JUN 94] JUNG P., MASSHAN M., BLANZ J., “Application of Turbocodes to a CDMA Mobile Radio System using Joint Detection and Antenna Diversity”, *Proc. of 44th IEEE Vehicular Technology Conference VTC'94*, p. 770-774, Stockholm, 1994.

[KSC 95] KSCHISCHANG F. R., SOROKINE V. “On the Treillis Structure of Block Codes”, *IEEE Trans. on Info. Theory*, vol. 41, no. 6, p. 1924-1937, November 1995.

[LOD 93] LODGE J., YOUNG R., HOEHER P., HAGENAUER J. “Separable MAP Filters for the Decoding of Product and Concatenate Codes”, *IEEE ICC’93*, p. 1740-1745, May 1993.

[MAC 78] MACWILLIAMS F. J., SLOANE N. J. A., *The Theory of Error-Correcting Codes*, p. 567-580, North-Holland Publishing Company, 1978.

[MAS 69] MASSEY J. L., “Shift-Register Synthesis and BCH Decoding”, *IEEE Trans. on Inf. Theory*, vol. IT-15, p.122-127, January 1969.

[MOR 93] MORELLO A., MONTOROSI G., VISINTIN M., “Convolutional and Trellis Coded Modulations Concatenated with Block Codes for Digital HDTV”, *Int. Workshop on Digital Comm.*, p. 237-250, September Tirrenia, Italy, 1993.

[NIC 97] NICKL H., HAGENAUER J., BURKERT F., “Approaching Shannon’s Capacity Limit by 0.27 dB using Simple Hamming Codes”, submitted to *IEEE Comm. Letters*, 1997.

[PIC 96] PICART A., PYNDIAH R., “Performance of Turbo-Decoded Product Codes used in Multilevel Coding”, in *Proc. of IEEE ICC’96 Conference*, Dallas, TX, June 1996.

[PRO 89] PROAKIS J. G., *Digital Communications*, 2nd edition, Chapter 2, McGraw-Hill International Editions, 1989.

[PYN 94] PYNDIAH R., GLAVIEUX A., PICART A., JACQ S., “Near-Optimum Decoding of Products Codes”, in *Proc. of IEEE GLOBECOM’94 Conference*, vol. 1/3, p. 339-343, San Francisco, November-December 1994.

[PYN 95] PYNDIAH R., PICART A., GLAVIEUX A., “Performance of Block Turbocoded 16-QAM and 64-QAM Modulations”, in *Proc. of IEEE GLOBECOM’95 Conference*, vol. 2/3, p. 1039-1044, November 1995.

[PYN 97] PYNDIAH R., “Iterative Decoding of Product Codes: Block Turbocodes”, in *Proc. of IEEE Int. Symp. on Turbocodes and Related Topics*, vol. 1/1, p. 71-79, September 1997.

[PYN 98] PYNDIAH R., “Near Optimum Decoding of Product Codes: Block turbocodes”, *IEEE Trans. on Comm.*, vol. 46, no. 8, p. 1003-1010, 1998.

[RAO 95] RAOUL O., ADDE P., PYNDIAH R., “Architecture et conception d’un circuit turbo-décodeur pour un code produit”, *GRETSI-95*, Juan les Pins, September 1995.

[RED 70] REDDY S. M., “On Decoding Iterated Codes”, *IEEE Trans. Inform. Theory*, vol. IT-16, p. 624-627, September 1970.

[RED 72] REDDY S. M., ROBINSON J. P., “Random Error and Burst Correction by Iterated Codes”, *IEEE Trans. Inform. Theory*, vol. IT-18, p. 182-185, January 1972.

[ROB 94] ROBERTSON P., VILLEBRUN E., HOEHER P., “A Comparison of Optimal and Sub-Optimal MAP Decoding Algorithms Operating in the Log Domain”, *Proc. of ICC’95*, p. 1009-1013, Seattle, WA, June 1995.

[SHA 48] SHANNON C. E., "A Mathematical Theory of Communication", *Bell System Technical Journal*, vol. 27, p. 379-423, July 1948, and p. 623-656, October 1948.

[UNG 82] UNGERBOECK G., "Channel Coding with Multilevel Phase Signals", *IEEE Trans. on Information Theory*, vol. IT-28, p. 55-67, 1982.

[VIT 77] VITERBI A. J., OMURA J. K., *Principles of Digital Communication and Coding*, Mc Graw-Hill, New York, 1977

[VIT 89] VITERBI A. J., ZEHAVI E., PADOVANI R., WOLF J. K., "A Pragmatic Approach to Treillis-Coded Modulation", *IEEE Commun. Mag.*, vol. 27, no. 7, p. 11-19, July 1989.

[WIB 95] WIBERG N., LOELIGER H., KOTTER R., "Codes and Iterative Decoding on General Graphs", *European Trans. on Telecommunications*, vol. 6, no. 5, p. 513-526, October 1995.

[WOL 78] WOLF J. K., "Efficient Maximum Likelihood Decoding of Linear Block Codes Using a Treillis", *IEEE Trans. on Info. Theory*, vol. IT-24, no. 1, p. 76-80, October 1995.

Chapter 7

Block Turbocodes in a Practical Setting

7.1. Introduction

The turbodecoding algorithm of product codes (iterative decoding algorithm with soft input and output of product codes) described previously offers remarkable performances in terms of correction of transmission errors (residual bit error rates (BER) for a given level of noise). Its complexity of integration onto microchips and DSP is the subject of this chapter.

The impact of integration constraints and the quantification of data on theoretical performances obtained previously are evaluated first. The number of quantification bits directly conditions the complexity of the circuit, in particular with respect to the size of the memories. Two types of architecture are then presented before proposing a decomposition of the elementary decoder into blocks and evaluating its complexity.

7.2. Implementation of BTC: structure and complexity

7.2.1. *Influence of integration constraints [RAO 97]*

7.2.1.1. *Quantification of data*

In a digital circuit, data is represented by discrete values coded with a finite number q of binary characters. The higher this number is, the more precise the data representation is and the better its performances are. On the other hand, the

---

Chapter written by Patrick ADDE and Ramesh PYNDIAH.

complexity of the circuit increases with the value of the parameter q (size of the memories and the processing unit). It should be optimized in order to obtain the best performances/complexity compromise.

Using simulation we may study the evolution of the BER for the product code $\text{BCH}(64,57,4) \otimes \text{BCH}(64,57,4)$ according to the signal-to-noise ratio E_b/N_0. For four decoding iterations the degradation of the BER at the decoder output was measured according to the number of quantification levels used. In these simulations the data was coded successively for $q = 3$, 4 then 5 bits (sign bit included). The coefficients α and β, used respectively in the iterative algorithm to modulate the negative feedback and to calculate extrinsic information were optimized according to the value of q and for a BER close to 10^{-6}.

The transmission used is a QPSK in a Gaussian channel. Figure 7.1 depicts the results obtained. The BER curve of an iterative decoding using the binary data ($q = 1$) is also provided. It makes it possible to appreciate the improvement of the performances brought around by balanced decoding (gain higher than 2 dB for a BER of 10^{-5}).

The simulation results show that a quantification of the samples at the demodulator output for $q = 4$ bits is sufficient to ensure quasi-optimal operation of the decoder. Indeed, the degradation of the coding gain remains below 0.1 dB and the BER curve merges quickly with that of the real data. The best performances/complexity compromise is obtained for this value.

We find virtually identical variation with the product codes constructed on the basis of block codes, such as the BCH (32,26,4) corrector of $t = 1$ error or the BCH (32,21,6) corrector of $t = 2$ errors.

This low degradation is explained by the approximations introduced into the calculation of the Logarithm Likelihood Ratio (LLR). They mask the negative effect of the quantification.

For $q \leq 3$ the degradation of performances is no longer negligible. The coding gain for a BER of 10^{-5} is thus decreased by 0.3 dB when q equals 3. We also note that beyond $q = 3$ the slope of the curves practically does not vary any longer when the number of the quantification levels increases.

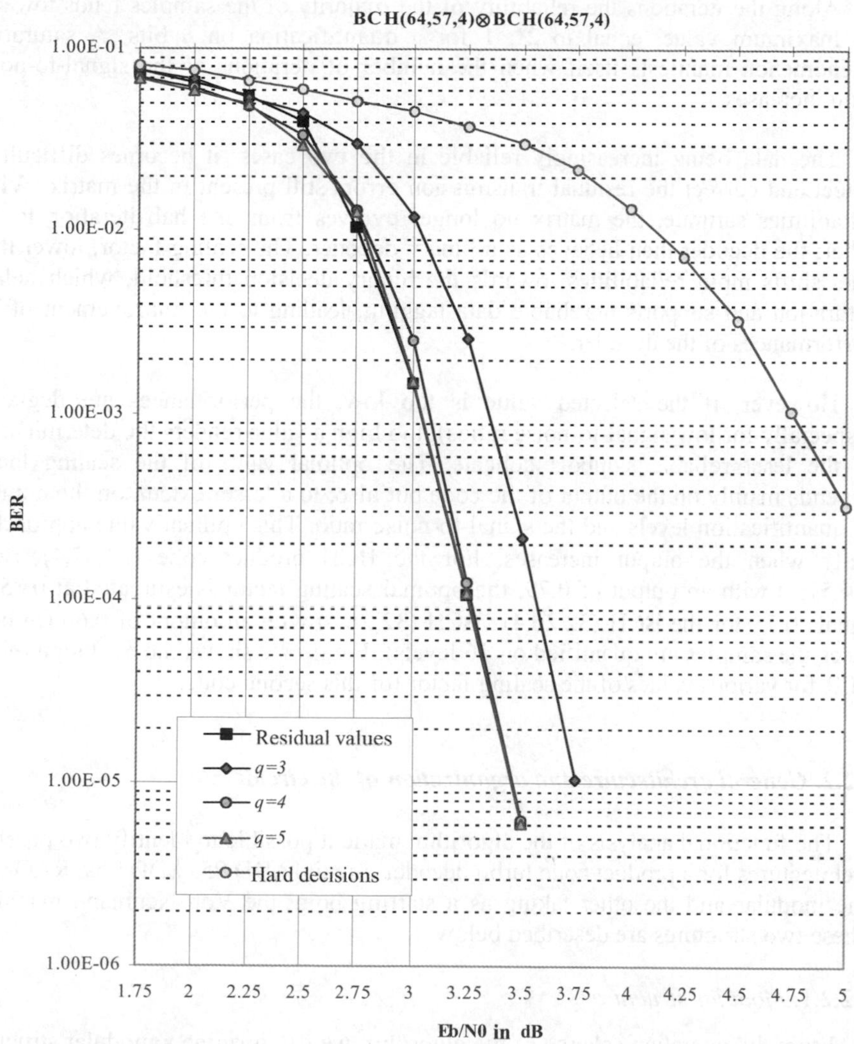

Figure 7.1. *Influence of quantification on performances*

7.2.1.2. *Choice of the scaling factor*

The "scaling factor" makes it possible to balance the quantified data presented at the circuit input. It acts as a multiplicative coefficient applied to the received data. In practice, it is a purely combinative block functioning as a conversion chart placed after the Analog-to-Digital Converter (ADC). Strictly speaking, it does not form part of the turbo decoder.

Along the iterations the reliability of the majority of the samples tends towards its maximum value, equal to 2^{q-1}-1 for a quantification on q bits. A saturation phenomenon manifests itself when the number of iterations or the signal-to-noise ratio increases.

The data being increasingly reliable in the two cases, it becomes difficult to detect and correct the residual transmission errors still present in the matrix. When reliabilities saturate, the matrix no longer evolves from one half-iteration to the other; the decoder then behaves as a binary decoder. The scaling factor, lower than one, shifts input reliabilities towards the binary decision threshold, which delays saturation and supports unreliable data tagging, leading to the improvement of the performances of the decoder.

However, if the selected value is too low, the performances are degraded (especially for low signal-to-noise ratios); too high a value renders the determination of the least reliable symbols delicate. The optimal value of the scaling factor depends mainly on the output of the code but also, to a lesser extent, on the number of quantification levels and the signal-to-noise ratio. This optimal value approaches unity when the output increases. For the BCH product code (64,57,4)⊗BCH (64,57,4) with an output of 0.79, the optimal scaling factor is estimated at 0.85. It equals 0.75 for the BCH (32,26,4)⊗BCH (32,26,4) with an output of 0.66 (in both cases the samples are quantified on 16 levels). Figure 7.2 shows the evolution of the BER for various values of the scaling factor for this second code.

7.2.2. *General architecture and organization of the circuit*

The functional analysis of the algorithm made it possible to identify two possible architectures for a product code turbo decoder circuit [ADD 95, ADD 96a, RAO 95], one modular and the other taking as a starting point the Von Neumann machine. These two structures are described below.

7.2.2.1. *Modular structure*

From the operation scheme of the algorithm we can imagine a modular structure for the turbo decoder, in which each sub-circuit carries out a half-iteration of decoding (i.e. a decoding of the rows or columns of a data matrix, [R] or [R']).

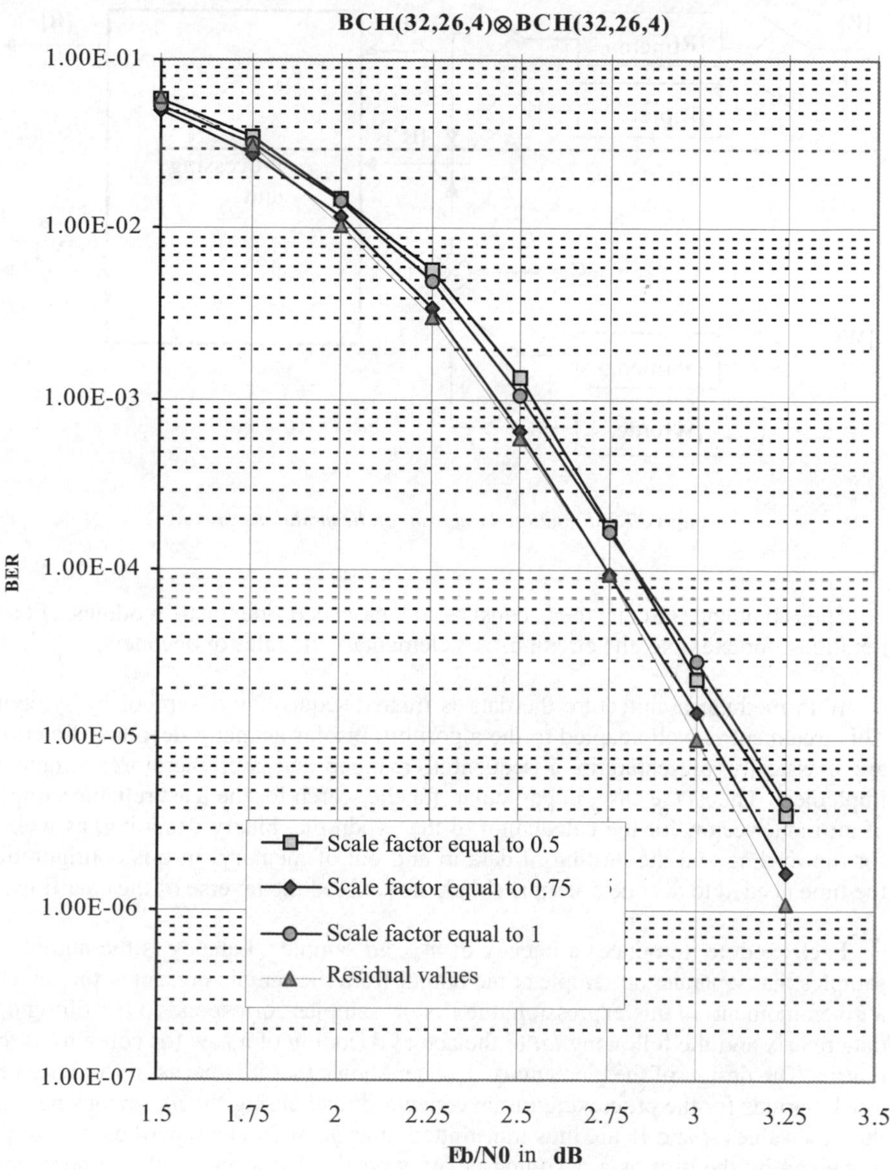

Figure 7.2. *Influence of the scaling factor on the performances of a block turbocode (QPSK modulation, Gaussian channel)*

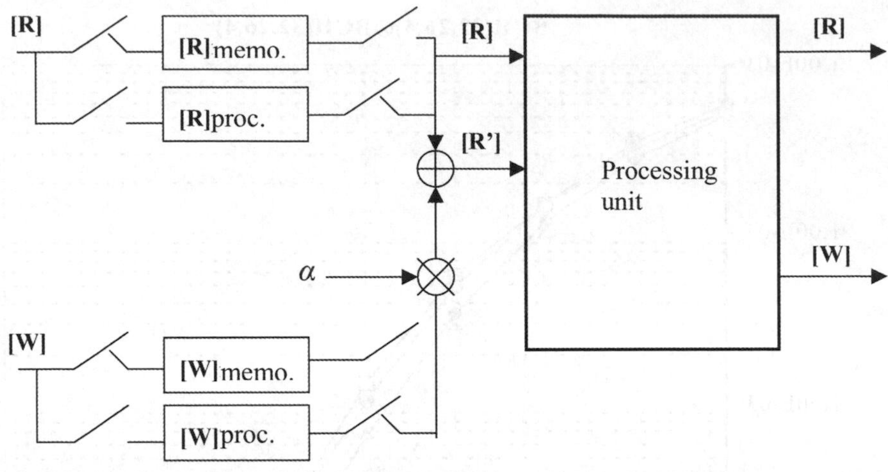

Figure 7.3. *Functional diagram of the modular structure*

The complete circuit then consists of cascaded identical modules. For 4 iterations, for example, the circuit uses 8 elementary modules or decoders.

With modular architecture the data is treated sequentially (symbol by symbol). This treatment is well adapted to the algorithm insofar as many decoding functions are classically presented in a sequential manner and are, therefore, simple to implement. This is the case, in particular, for the search for the least reliable samples of an input vector, for the calculation of the syndrome, binary decoding, as well as for the reading and the writing of data in and out of memory. In this configuration the time needed to decode a sample should not exceed the inverse of the data flow.

Each module introduces a latency of $n^2 + lat$ samples. Latency is the number of samples that separate the sample at the output from the sample present at the input at a given moment. In this expression, the first n^2 samples correspond to the filling of a data matrix and the following *lat* to the actual decoding of a row (or column) of this matrix. The design of the elementary decoder shows that it is necessary to reserve n clock periods for the processing of the codeword read during the n previous periods: the new values R and W are thus transmitted after $lat = 2n$ clock periods. Latency is perceived by the user as a decoding delay, whose value grows with the number of cascaded modules.

7.2.2.2. *Von Neumann architecture*

The second architecture is linked to the Von Neumann sequential machine. It uses the same processing unit to carry out several iterations. Compared to the previous architecture, this solution mainly aims at reducing the obstruction of the turbo decoder. Moreover, it has the advantage of limiting to a maximum of $2n^2$ symbols the total latency introduced by the circuit, independently of the number of iterations carried out (n^2 samples to fill an additional matrix and n^2 for decoding).

This time the data can be processed either sample by sample, i.e. sequentially, or on the contrary, vector by vector (row or column of a matrix) described hereafter as "block processing". In both cases, each sample must be decoded in a time not exceeding the inverse of the product of the data flow by the number of half-iterations to be performed. Thus, for four iterations, the data flow can only exist at a rhythm at least eight times lower than that of their treatment. In the case of sequential treatment this implies that between the modular and the Von Neumann architectures the maximum data flow is divided by a factor equal to the number of half-iterations used. Latency is less for the Von Neumann architecture ($2n^2$ samples at most for the first case against $8n^2 + 16n$ for the other) but the flow is lower for the same speed of data processing.

Generally, for both types of treatment the maximum number of iterations that can be integrated in the circuit is limited by the flow that we wish to achieve on the one hand, and by the maximum frequency of operation that the technology used allows on the other hand. This frequency conditions the maximum speed, at which the processing unit can function. To reduce the processing time of a matrix we could possibly decode several rows or columns in parallel. This multiplication of the resources (in a number of processing units) certainly makes it possible to achieve higher flows, but increases the obstruction of the circuit by an equal amount (see section 7.2.5). If we consider the BCH product code $(64,57,4)^2$, the complexity of the elementary decoder is approximately 5,000 gates and the memory size is 64 Kbits (for samples quantified for 4 bits including the sign bit), which gives a gate equivalent of 100,000. The sequential structure for 4 iterations is of 840,000 gates, whereas the Von Neumann architecture requires only 205,000 gates. It should be noted that more than 95% of total surface is occupied by the memories.

The second possibility, called "block processing", is specific to block codes. More complex to implement than the previous possibility, it combines a high data flow with the limitation of latency inherent for the Von Neumann architecture. This time we work directly with the vectors of n coded samples and no longer sequentially with each of them, which makes it possible to reduce the average time needed for the decoding of data.

In the case of the Von Neumann architecture, the maximum value of the data flow in block processing is higher for the same iteration count than that which can be achieved with sequential treatment. Reciprocally, if the flow is fixed, we can carry out more iterations using block processing than using sequential treatment. However, block processing requires, as we will see, a particular architecture of the memories in order to preserve a high user frequency. It also induces increased complexity and obstruction of the processing unit.

The solution consisting of associating block processing with modular architecture is not of practical interest except for very high flows.

7.2.3. *Memorizing of data and results*

The obstruction of the circuit stems primarily from the size and the number of the memories used. Regardless of the general architecture selected it is, indeed, essential to memorize the matrices [R] and [R'] (or [W]) during half-iteration in progress (a half-iteration corresponds to a decoding of the rows or columns of a data matrix). Data processing in rows then in columns forces the use of a first memory unit to receive the data and a second unit to treat it. These two memories work alternatively in writing and reading mode with an automaton managing the sequencing. Each memory is organized as a matrix and composed, for a code of length n and a quantification of the data for q bits, of q memory planes of n^2 bits each.

7.2.3.1. *Modular structure*

In the case of the modular structure, the general organization of the circuit for a half-iteration is that of Figure 7.4 below.

The data coded for q bits that arrive at the decoding module is arranged by rows of a first memory A functioning in writing mode. In parallel, the data of the matrix received previously is taken by columns from the second memory B functioning in reading mode. Once memory A fills and the contents of memory B are read, the operating modes are inverted. Memory A passes to reading – we then start to decode it – while memory B passes to the writing mode in order to store the data corresponding to the following codeword. Cascading two modules, one for the decoding of columns and the other for the decoding of rows of a coded matrix, we carry out a complete iteration.

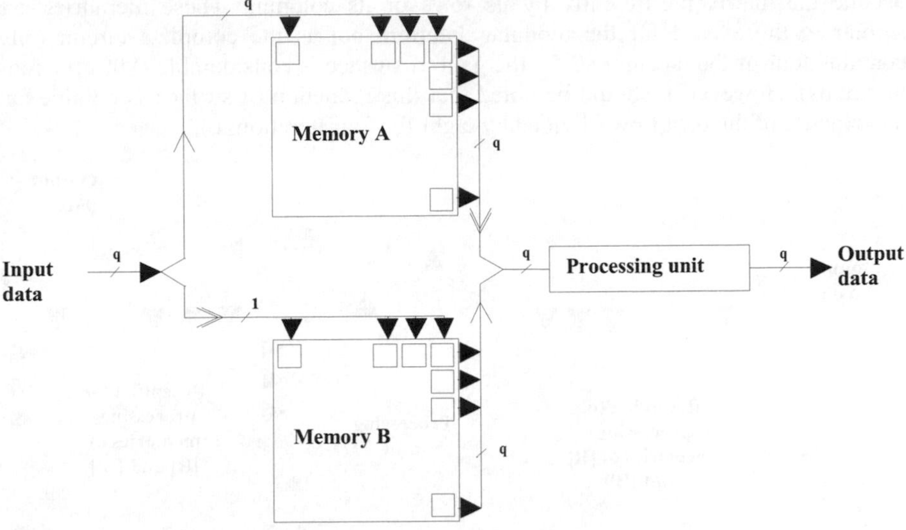

Figure 7.4. *Diagram of a decoding module*

The memories used can be easily conceived on the basis of classical RAM (Random Access Memory), that can be addressed by row and column. Other solutions can be considered (shift registers, etc.) but they are more cumbersome.

From a practical point of view the modular solution has the advantages of allowing a high operational frequency and having great usage flexibility. On the other hand, setting several modules in cascade involves an increase in latency and obstruction of the circuit. These parameters quickly become crippling as the iteration count and/or the length of the code increase.

7.2.3.2. *Von Neumann architecture*

This time the circuit performs several iterations using only one memory unit and only one processing unit for the whole of the iterative process.

We short a decoding module in on itself. With this architecture the complete circuit includes only four memories regardless of the number of iterations performed. These memories must, however, be able to be read and written by rows as well as by columns.

7.2.3.2.1. Sequential processing

The memory used is classical RAM organized in a matrix where we can read or write data located by its address. As we reach each symbol directly, it is possible to decode the matrix indifferently by its rows or its columns. These memories are similar to those used for the modular solution, but as the complete circuit only contains four of them (Figure 7.5), the gain in surface is considerable (80% for four iterations). However, it should be noted that this reduction of surface is obtained at the expense of the data flow (divided by eight for four iterations).

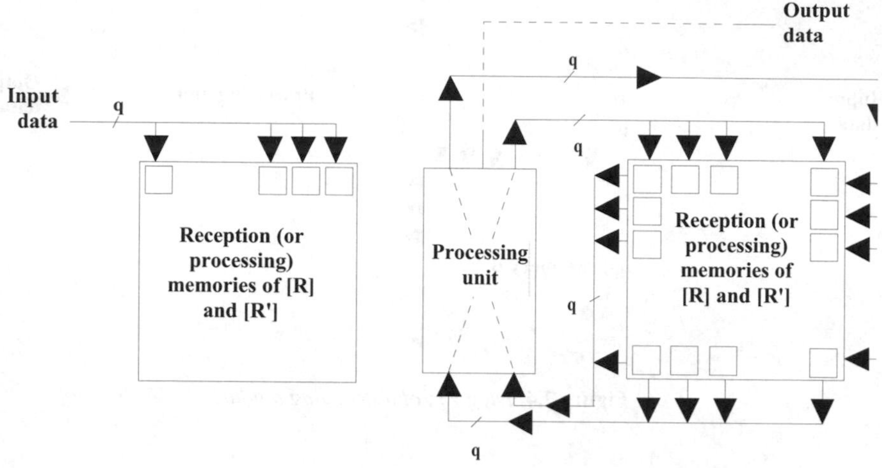

Figure 7.5. *Von Neumann architecture with sequential treatment*

7.2.3.2.2. Block processing

This time the data is decoded directly by blocks of n samples rather than sequentially. The principal advantage of this solution is to cumulate an increased flow with the reduction of latency inherent to Von Neumann's structure. The memories, for reading and writing by row or column, must give access a complete data vector in only one clock period instead of n. Such memories are *a priori* more cumbersome and more complex to implement that those used in sequential treatment. However, we still need only four of them to memorize [R] and [R'] (Figure 7.6).

In the case of the product code BCH(64,57,4) $\otimes$ BCH(64,57,4) integrated over four iterations we have 4,096 symbol-times (the time of filling the following matrix) to decode 4 times the 64 rows and the 64 columns of a data matrix, which leaves a maximum of eight symbol-times per codeword.

Since the standard libraries of the founders do not have such memories accessible by block, they should be especially conceived for this application. Complex and delicate to manufacture, their production cost is necessarily higher than that of conventional memories. The various solutions envisaged to carry them out are divided into two families. They impose constraints on data processing, as well as on topology of the circuits used.

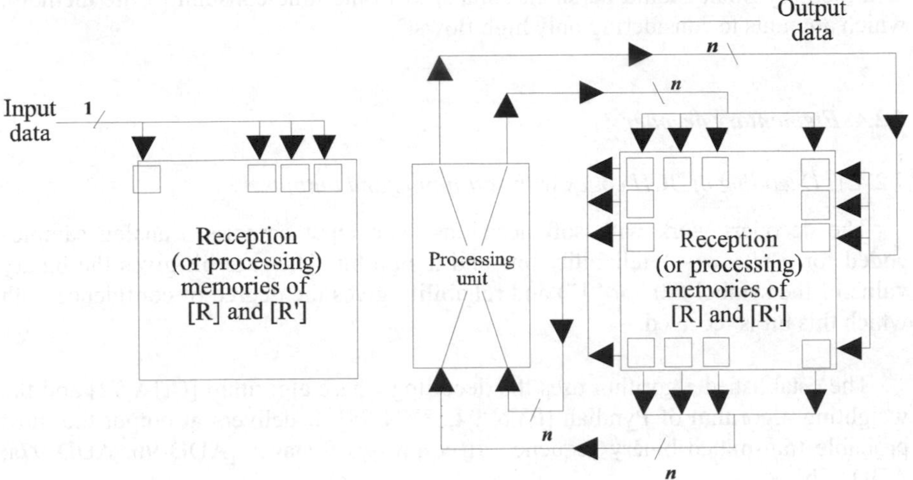

Figure 7.6. *Von Neumann architecture with block processing*

The first possibility again uses the principle of data shift. The structure obtained is comparable to a set of shift registers by row and column. The second solution uses the technique of address shift. The memory point then consists of oscillators and has a particular device to address the rows and columns. We can also use a memory point comparable with that found in traditional RAM, but whose more complex addressing system here uses the row and column address amplifiers.

In all the cases, the elementary memory point can be of static or dynamic type [PUC 88]. With a dynamic cell the information stored in the form of an electric charge in a stray capacity tends to disappear because of the escape resistances of this capacity. The dynamic memory point does not have a durable state and the information which it contains must be restored (or "refreshed") periodically in order not to be lost. If it is compact (3 transistors, or even just one), the dynamic cell consumes more electric power than its static equivalent. In a static cell, the output signal is looped on to the input making it possible to preserve the data without having to regenerate it (at least as long as the memory is fed). Because of this defect, the static memory point is more cumbersome than its dynamic counterpart (approximately 10 transistors). It also has a slightly more elaborate design, but, on the other hand, remains simpler to use. In order to be freed of the cooling problem, it is preferable to use only the memories of the static type. These memories, moreover, make it possible to reach higher flows than dynamic memories (they have faster access [WES 93]). Let us note that it would nevertheless be fully possible to use dynamic cells while avoiding having to regenerate their contents. For that it would be necessary that the interval of time separating the reading and writing operations of a memory point should be small compared to the time-constant of the memory, which amounts to considering only high flows.

7.2.4. *Elementary decoder*

7.2.4.1. *Decoding of BCH codes with soft inputs and outputs*

The decoders work with soft decisions. The input comprises analog samples coded for q bits: $q - 1$ reliability bits and a sign bit. The sign bit gives the binary value of the symbol ("0" or "1") and reliability gives the degree of confidence with which this bit is received.

The established algorithm uses the decoding Chase algorithm [CHA 72] and the weighting algorithm of Pyndiah [PYN 94, PYN 98]. It delivers at output the most probable transmitted binary sequence. It comprises 9 stages [ADD 96, ADD 97a, ADD 97b]:

– reception of the codeword of input R'_K;

– search for the least reliable samples;

– construction of the test vectors;

– calculation of the syndrome S_1 (and S_3 for a code with correction power $t = 2$) of the first test vector;

– calculation of the syndromes S_1' (and possibly S_3') of the other test vectors;

– calculation of the metrics of each test vector and selection of the "decided" and "competitor" vectors;

– calculation of the new reliability for each sample of the decided codeword;

– calculation of extrinsic information;

– addition to the codeword received from the channel and correction of this codeword.

Example of the choice of test vectors, calculation of the corresponding syndromes and calculation of the metrics:

Let us consider 8 test vectors comprising 32 samples in the case of extended BCH code (32,21,6). The first C_0 vector simply comprises the sign bits of the received codeword. The 7 other vectors (C_1 to C_7) are constructed, for example, on the basis of this one by reversing the positions of the 4 least reliable samples (LRS_i, i = 1.., 4):

– C1 to 4 where the LRS1 to 4 sample is reversed;

– C5 to 7 where the 2 samples LRS1 and LRSi (i=2 to 4) are reversed.

The syndromes of each of these vectors are then calculated in the following manner:

– $S_1' = S_1 \oplus (1^{st}$ reversed position$) \oplus (1^{st}$ reversed position$)$

– $S_3' = S_3 \oplus (1^{st}$ reversed position$)^3 \oplus (1^{st}$ reversed position$)^3$ where S_1 and S_3 are the syndromes of C_0 and the reversed positions are located in the Galois field GF (2^5).

S_1' and S_3' allow the algebraic decoding of each test vector and give positions PC_1 and PC_2 that have to be corrected: the calculation of the syndromes after the corrections verifies that the word obtained is a codeword.

The calculation of the Euclidean distances between each test vector and the input vector take into account only the positions where the symbols are different:

$$M_L = \sum_K \left\| R_K' - C_L^K \right\|^2 = \sum_{K=0}^{N-1} \left| R_K' C_L^K \right|$$

for the bits K such that sign $\left(C_L^K\right) \neq$ sign $\left(C_0^K\right)$.

7.2.4.2. *Functional structure and sequencing*

After the analysis of the various functions of the decoding algorithm, the functional diagram of the decoder can be defined [ADD 99a, ADD 99b]. It consists of several parts:

– the sequential part at input, which gathers all the blocks whose calculation progresses at the same rate as the arrival of each sample of the received codeword (S1 and S3 syndromes, parity and positions of the least reliable samples);

– the non-sequential part where calculations are carried out after complete reception of the codeword, and thus do not require any synchronization with respect to the arrival of the samples (binary decoding, calculation of metrics and choice of the decided and competitor vectors). This block is connected on the outside to a combinatory function (ROM) which gives the positions of the symbols to be corrected according to the syndromes. This function is useless for Hamming codes (the S1 syndrome indicates the position to be corrected in the Galois field);

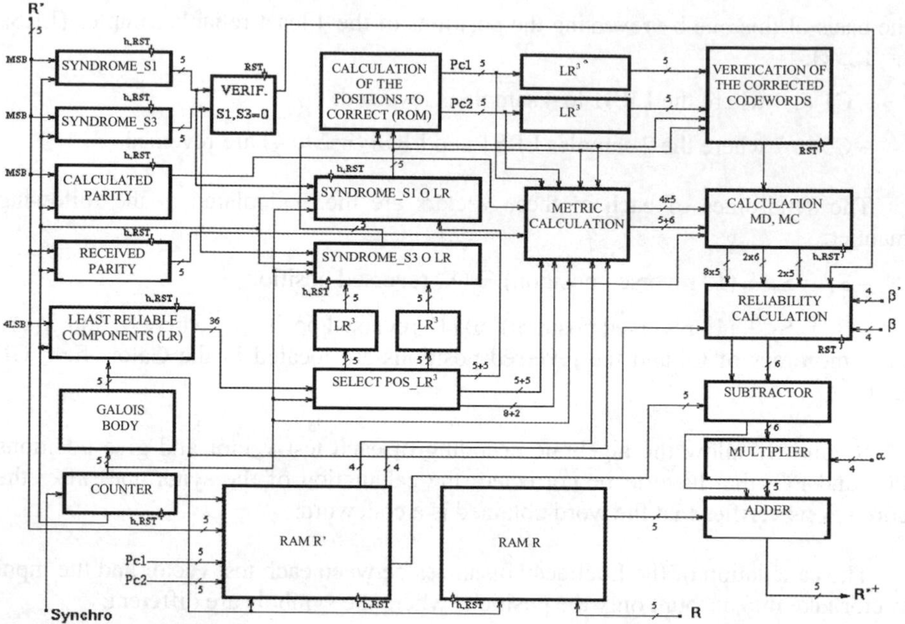

Figure 7.7. *Block diagram of the decoder*

– the sequential part at output, which groups the blocks working at the same rate as the transmission of new samples (calculation of the reliability of each sample and its extrinsic information, construction of the following codeword);

– the memory part, consisting of two RAM inserted between each half-iteration propagating the samples [R] and [R'] (or [W]) and compensating for the latency due to decoding.

The processing of each codeword is thus divided overall into three parts: a sequential part at the input, a non-sequential codeword processing part and a sequential part at the output.

The "word processing" part consists of a processing for each test vector: calculation of the syndromes and metrics of each vector and calculation of the positions to be corrected for each vector.

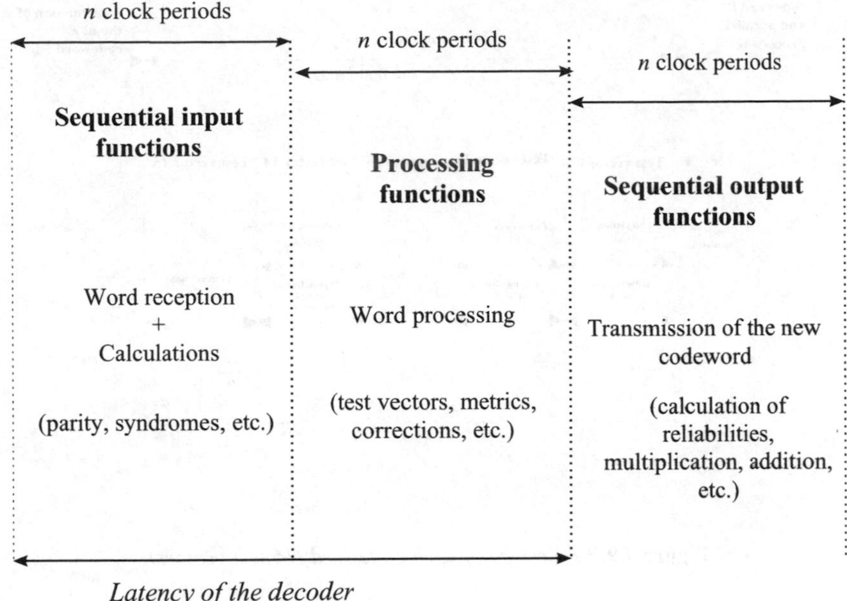

Figure 7.8. *Temporal cutting*

The time necessary for the processing is n clock periods.

Thus, for a decoder working with L vectors, n/L clock periods are allocated to each vector.

At the end of this processing, we can select the vectors giving the "Decided" and "Competitor" codewords that make it possible to calculate the reliability of the data at the output of the half-iteration.

The latency of the decoder, i.e. the time that passes between the reception of the first symbol of the codeword R_k' at the input and the transmission of the first symbol of the codeword $R_k'^{+}$ at the output, is 2^n clock periods.

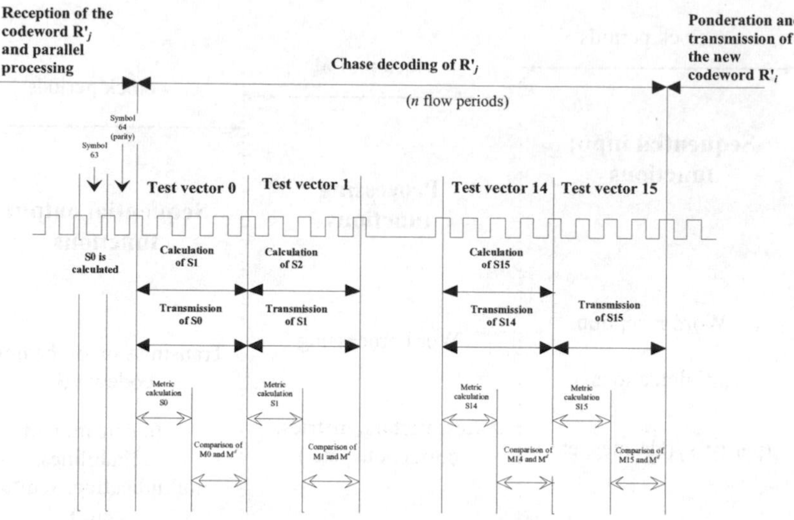

Figure 7.9. *Processing of each codeword (16 test vectors)*

7.2.4.3. *Installation of a decoder on a silicon microchip*

Among the blocks requiring a lot of resources we may cite those seeking the least reliable samples, those calculating the metrics and those selecting the decided and the competitor codewords.

For the latter, in particular, it is possible to reduce the number of competitors necessary for the calculation of reliability while minimizing the degradation of decoding performances [ADD 97a, ADD 97b].

Figure 7.10 below represents the BER of the product code BCH(128,120,4)⊗ BCH(128,120,4) for various values of the number of competitors [KER 99].

We may realize that degradation is minimum (0.1 to 0.2 dB) when there is one competitor whereas complexity can be reduced by a ratio of 10. The following table shows the increase in surface and the performance gain when the number of competitors varies (the reference is the decoder with soft input and outputs BCH(128,120,4) with one competitor and 16 test vectors).

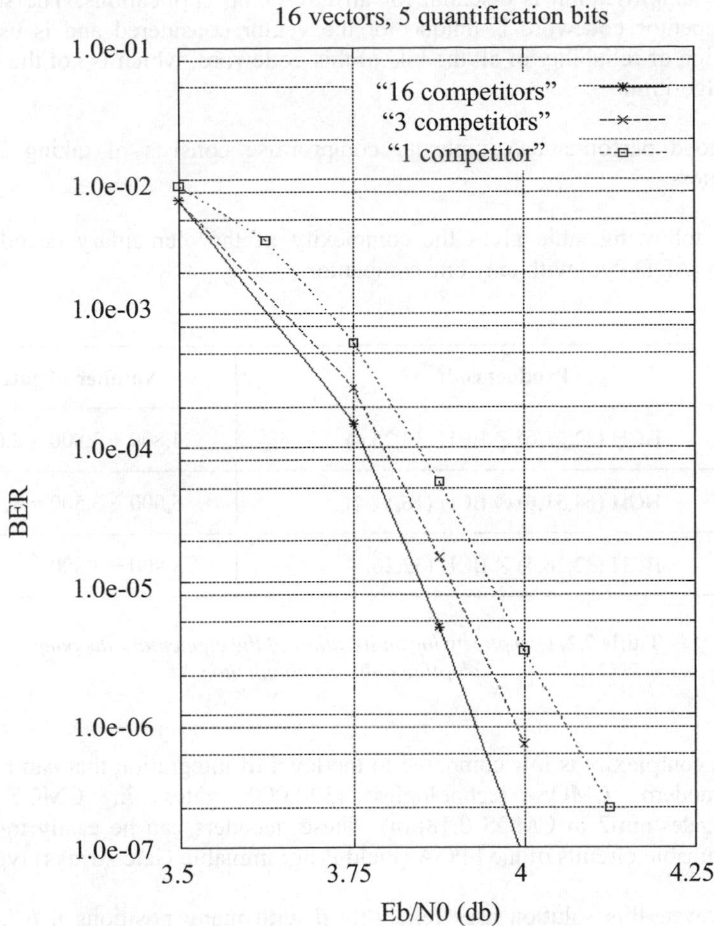

Figure 7.10. *Number of competitors and performances*

	1 competitor	3 competitors	16 competitors
Performance gain (at 10^{-6})	Ref.	0.06 dB	0.13 dB
Increase in surface	Ref.	13.5%	90%

Table 7.1. *Performance gain and increase in surface according to the number of competitors for turbocode BCH(128,120,4)^2*

This simplification is essential for all microchip applications. The selection of the competitor codeword is unique for the vector considered and is used for the calculation of reliability of all the bits of this codeword, which is not the case in the initial algorithm.

A good performance/complexity compromise consists of taking 2, even 3, competitors.

The following table gives the complexity of the elementary decoders for an iteration [ADD 99c] with only one competitor.

Product code	**Number of gates**
BCH (32,21,6) ⊗ BCH (32,26,4)	4,800 + 3,800 = 8,600
BCH (64,51,6) ⊗ BCH (16,11,4)	5,000 + 3,500 = 8,500
BCH (32,26,4) ⊗ BCH (32,26,4)	3,800 + 3,800 = 7,600

Table 7.2. *Complexity for an iteration of the elementary decoders adopting only one competitor*

This complexity is low compared to the level of integration that can be obtained with modern CMOS technologies (300,000 gates in CMOS 0.09µm, 50,000 gates/mm2 in CMOS 0.18µm). These decoders can be easily integrated in programmable circuits of the FPGA (Field Programmable Gate Arrays) type.

However, this solution uses reliability β with many positions j. β is taken by default in the absence of a competitor; it is constant, optimized by simulation and depends on the current half-iteration. A better estimate of this reliability $\Lambda'(d_j)$ is given by the following equation [ADD 00a]:

$$\Lambda'(d_j) = |\, rj \,| + \sum_{k=a}^{b} |\, rk \,|$$

where a and b are selected among the least reliable positions of the vector considered $[R'_m]$. $\Lambda'(d_j)$ then depends on this vector.

The following curves compare the results when β is fixed and variable for the same half-iteration.

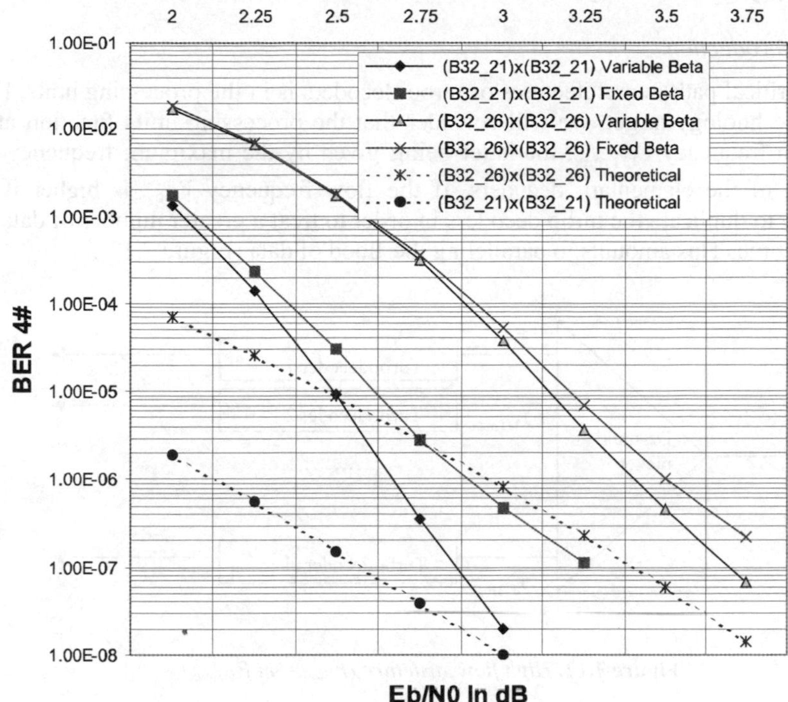

Figure 7.11. *BER function of the Eb/N0 ratio when β is fixed and variable for two product codes*

When β is variable, the results are close to the theoretical curve obtained with the union bound [ADD 00a] (variation of 0.1dB to 0.2dB). Its implementation does not influence the low complexity of the elementary decoder [KER 00].

Another estimate of $\Lambda'(d_j)$ is given by the following expression:

$\Lambda'(d_j) = |r_j| + (\sum_{k=0}^{n} |r_k|)/A$ where A is a constant (A is optimized by simulation).

7.2.5. *High flow structure [ADD 00b, ADD 01, CUE 02, CUE 03a, CUE 03b]*

7.2.5.1. *Introduction*

The critical path in architectures of turbo decoders is in the processing units. For a given technology target, we will consider that the processing units function at a maximum frequency F_{UTmax}, the latter being given by the maximum frequency of operation of the elementary decoders. If the flow frequency F_{Flow} is higher it is necessary to duplicate the turbo decoders in order to treat a greater number of data at the same time. This amounts to paralleling the flood of data (Figure 7.12).

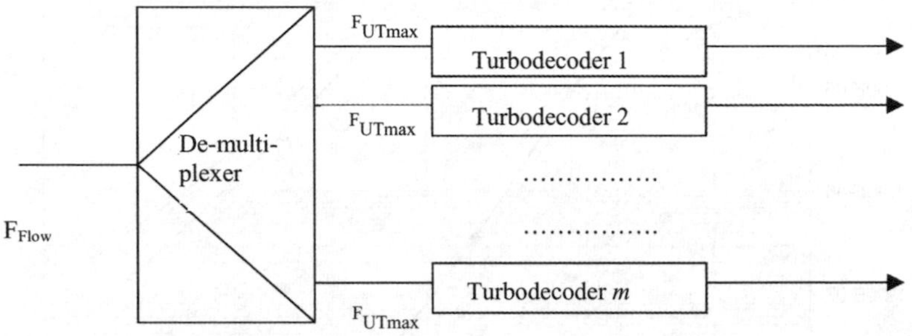

Figure 7.12. *High flow structure ($F_{Flow} = m.F_{UTmax}$)*

We have seen that product codes had the property of having codewords in all the rows (or columns) of the initial matrix C. We may thus, in the same manner as has just been seen (Figure 7.12), parallel the decoding of the matrix samples by duplicating the number of elementary decoders of the C_1 code (or C_2). We can thus process a maximum number n_1 (or n_2) of codewords under the condition, however, that the memory access during reading or writing takes place at different moments (several memory points of a matrix cannot be read or written at the same time, unless using "multiport" RAM). If this constraint is respected, it is possible to increase a factor n_2 (or n_1) in the F_{Flow}/F_{UTmax} ratio, since there can be n_2 (or n_1) samples processed at a given moment.

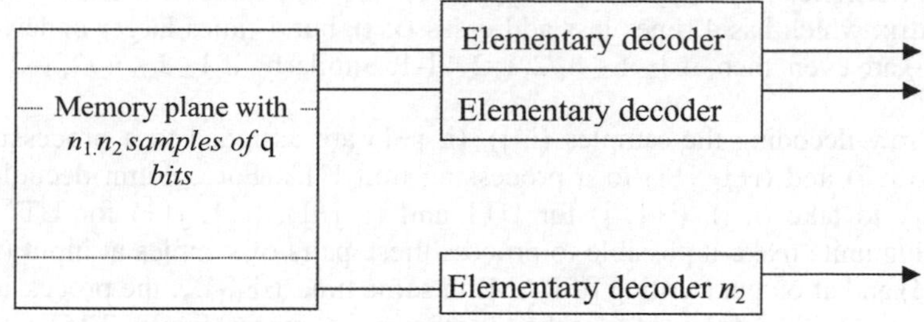

Figure 7.13. *High flow structure decoding several columns at the same moment*

The major disadvantage of this architecture is that the memory must function at a frequency $n_2.F_{UTmax}$, if there are n_2 elementary parallel decoders.

If we wish to keep the same speed of operation for the memory and increase the flow, we can memorize several data units at the same address. However, it is necessary to use these data by row as well as by column. Hence the following organization: at this address will be found the data adjacent in reading (or writing), by row as well as by column.

Let us consider two adjacent rows and two adjacent columns of the initial matrix.

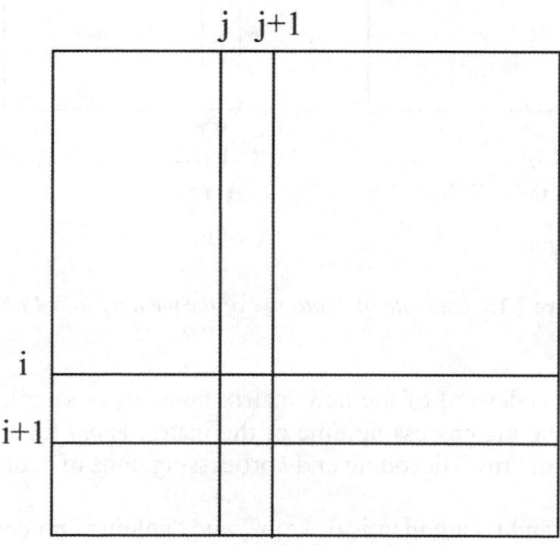

Figure 7.14. *Obtaining new memory*

The 4 samples (i, j), (i, j+1), (i+1, j) and (i+1, j+1) constitute a codeword of the new matrix which has 4 times less addresses (i, j), but 4 times larger codewords. If n_1 and n_2 are even, then, if $1 \leq I \leq n_1/2$, i=2 * I-1. Similarly, if $1 \leq J \leq n_2/2$, j=2 * J-1.

For row decoding the samples (i, j), (i, j+1) are assigned to a processing unit UT1, (i+1, j) and (i+1, j+1) to a processing unit UT2. For column decoding it is necessary to take (i, j), (i+1, j) for UT1 and (i, j+1), (i+1, j+1) for UT2. If the processing units make it possible to process these pairs of samples at input (reading of RAM) and at output (writing of RAM) in same time $1/F_{UTmax}$, the processing time of the matrix will be 4 times faster than for the initial matrix (Figure 7.15).

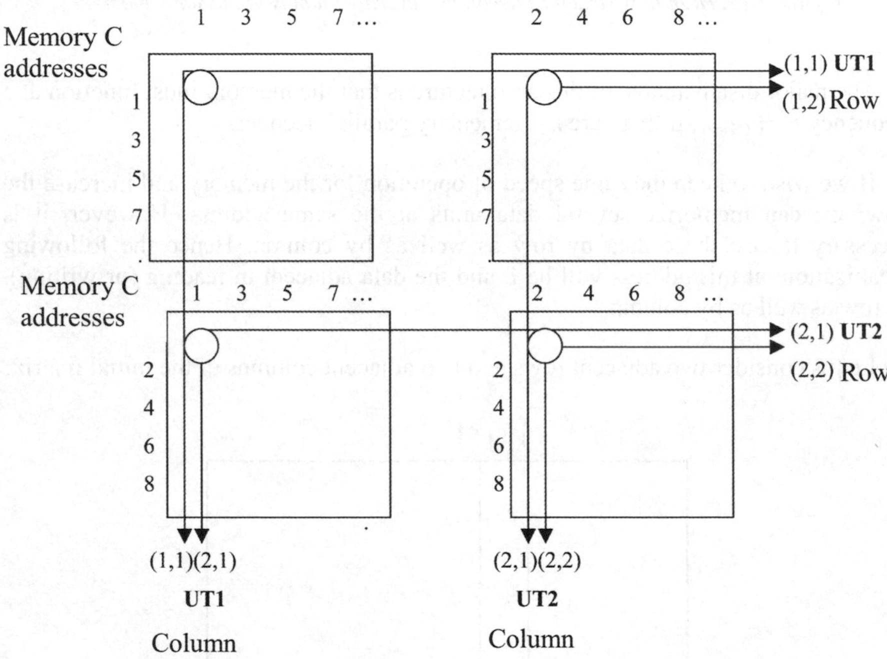

Figure 7.15. *Example of "cutting" of the memory into 4 parts*

In general, if a codeword of the new matrix contains *m* samples of a row and *l* samples of a column, the processing time of the matrix is *m.l* times faster with only *m* processing units of "row" decoding and *l* processing units of "column" decoding.

If the codes C_1 and C_2 are identical, "row" and "column" processing units are as well: then *m*=*l* and *m* processing units are necessary (Figure 7.16). This organization of data matrices requires neither particular memory architectures nor greater speed. In addition, if the complexity of the processing unit remains lower than *m* times that of the preceding processing unit, the total complexity is less for a m^2 times greater speed (the latter result could have been obtained using m^2 processing units, as indicated in Figure 7.12).

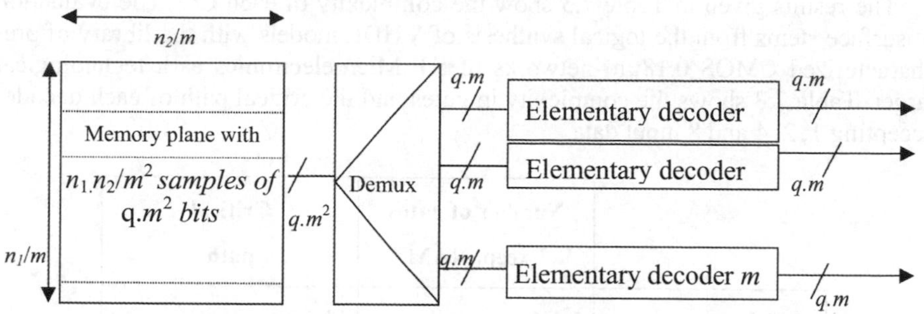

Figure 7.16. *Block diagram when C_1 and C_2 are identical. The speed of decoding is m^2 times faster using m elementary decoders*

The memory comprises m^2 times fewer codewords than the initial matrix. Since the technology is identical, its access time will therefore be less.

7.2.5.2. *High flow turbo decoder in a practical setting*

Using the method described previously, it is possible to increase the turbodecoding flow of product codes by a factor of m^2 if m decoders are used. In the results given hereafter we will consider an elementary decoder studied in [KER 00]. The selected product code was code BCH(32,26,4) $\otimes$ BCH(32,26,4). The characteristics of the practical setting were as follows:

– 5 levels of quantification;

– 16 test vectors for the Chase algorithm [CHA 72];

– 3 "competitor" vectors for the implementation of the Pyndiah weighting algorithm [PYN 94].

With 7.5 iterations this turbo decoder was close to the theoretical bound described in [ADD 00a]. The decoding of a codeword (a row or a column of the matrix C) breaks up into 3 parts: reception, decoding and transmission. Each one of them requires 32 clock periods and the latency of each elementary decoder is 64 clock periods.

In order to estimate this new architecture, it was necessary to model elementary decoders which simultaneously process m data units (2, 4 or 8) using $32/m$ clock periods (16, 8 or 4 respectively) for the decoding of a codeword. The design of these

3 elementary decoders (CU) is an extension of the initial model, written in VHDL language. The synthesis was made using the Synopsis Design Compiler tool for each decoder.

The results given in Table 7.3 show the complexity of each CU. The evaluation of surface stems from the logical synthesis of VHDL models with the library of pre-characterized CMOS 0.18µm networks of ST Microelectronics as a technological target. Table 7.3 shows the complexity in gates and the critical path of each decoder accepting 1, 2, 4 and 8 input data.

	Number of gates Except RAM	**Critical path**
CU-1data	5,400	11.5 ns
CU-2data	6,750	10.5 ns
CU-4data	11,000	12.5 ns
CU-8data	18,800	11.5 ns

Table 7.3. *Surface and critical path of various decoders*

These results show that critical path is virtually identical for each of the 4 CU. New syntheses were made with F_{Tumax} = 100Mhz (clock period =10ns). The results are provided in Table 7.4. The decoder CU-1data noted CU1 is the reference, CU2, CU4 and CU8 are respectively the decoders CU-2data, CU-4data and CU-8data.

CU	***m***	**Flow**	**Latency**	**Surface (gates)**	**Surface *m* CU (gates)**
CU1	1	D1	L	5,500=S	S
CU 2	2	4 D1	L/2	7,000	1.25x2xS
CU 4	4	16 D1	L/4	11,200	2.0x4xS
CU 8	8	64 D1	L/8	21,800	3.95x8xS

Table 7.4. *Principal characteristics of various decoders*

The latency of CUm is divided by m. The flow of CUm is multiplied by m^2 and the surface necessary to set up m CUm is multiplied roughly by $m.m/2$. The memory size (RAM) used to store $[R_k]$ and $[W_k]$ is the same in all the cases.

Thus, with a CU8 decoder the flow is 6.4 Gbits/s and the surface necessary is multiplied by 32. To obtain the same flow with a decoder CU1 it would be necessary to use 64 decoders and 64 RAMs. In Table 7.5 a turbo decoder using CUm is compared with that obtained on the basis of a CU1 decoder, both functioning with a flow m^2 D_1.

	CUm	CU1
Flow	$m^2.D_1$	$m^2.D_1$
CU numbers	m	m^2
Surface	$\approx S.\ m^2/2$	$S.m^2$
Latency	L/m	L
Cut RAM	S_{RAM}	$m^2.\ S_{RAM}$

Table 7.5. *Comparison of the characteristics of a turbo decoder using CUm with that using CU1, and functioning at* m^2 D_1

This architecture of decoding of the product codes allows a high flow operation. It can be applied to convolutional codes or linear block codes. It consists of modifying the initial organization of the memory C in order to accelerate the decoding time. During a time $1/F_{UTmax}$ m samples are treated by each of the m elementary decoders, which enables an increase of m^2 in flow, latency then being divided by m. If the processing of these m samples does not increase the surface of the elementary decoder considerably, the surface increase is close to m when we compare this solution with the one requiring m^2 decoders (Figure 7.12). In addition, the number of memory samples being fewer, the access time to this memory is reduced. The flows achieved provide hope for applications in optics and data storage.

7.3. Flexibility of turbo block codes

It has been seen how product codes are constructed using elementary extended BCH codes. Let C^1 and C^2 be two systematic block codes with respective parameters (n_1, k_1, d_1) and (n_2, k_2, d_2), where n_i indicates the length of the codewords C^i, k_i is the length of its messages and d_i is its minimum Hamming distance. The product code $C=C^1 \otimes C^2$ is composed of all the matrices [C] with n_1 rows and n_2 columns, in which:

the binary information symbols form a sub-matrix [M] with k_1 rows and k_2 columns;

- each of the k_1 rows of [M] is coded by the C^2 code;
- each one of n_2 columns of [C] is coded by the C^1 code.

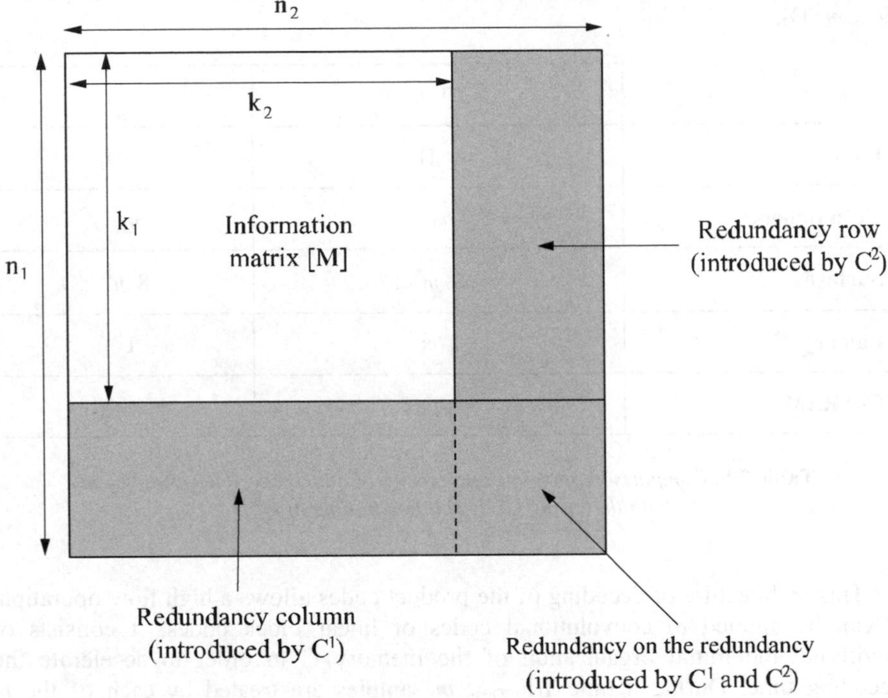

Figure 7.17. *Coding principle of the product code C*

[BUR 65] has demonstrated that the parameters (n, k, d) of the product code C result from those of C^1 and C^2 through: $n = n_1 \times n_2$, $k = k_1 \times k_2$, $d = d_1 \times d_2$ and a remarkable property of these codes, obtained on the basis of linear block codes, is that if n_2 columns of a [C] codeword are, by construction, C^1 codewords, the n_1 rows of [C] are in turn C^2 codewords.

The coding output of C is also given by $R = R_1 \times R_2 = (k_1 \times k_2)/(n_1 \times n_2)$, where $R_i = K_i/n_i$ indicates the rate of code C^i. The parameters of a product code are thus expressed overall as the product of the respective parameters of the elementary codes used.

In practice, it is desirable to have a programmable BTC circuit that makes it possible to process storage blocks of different sizes with a variable number of redundancy bits (in order to adjust the output, for example). We may thus use the same BTC circuit for various applications.

Two techniques can be used to meet this need: shortening and puncturing [PYN 96a, PYN 96b, PYN 97]. The latter technique is very often used to adapt the rate of convolutional codes [YAS 83, YAS 84].

The shortening of a block code is a well-known technique to modify the parameter k of a block code. The programming of the encoder consists of providing the number X representing the difference between the number of bits k of the matrix [M], to which the product code is applied, and the number of bits k-X to be coded per block.

Using this number X the encoder determines X positions in the matrix [M] containing bits of given values (for example, 0), which are found in corresponding positions of the matrix [C] during the processing of each block and are excluded from the transmitted bits. Several techniques can be used: uniform distribution in the matrix [M] (Figure 7.18) or removal of rows and/or columns in the matrix [C] (Figure 7.19).

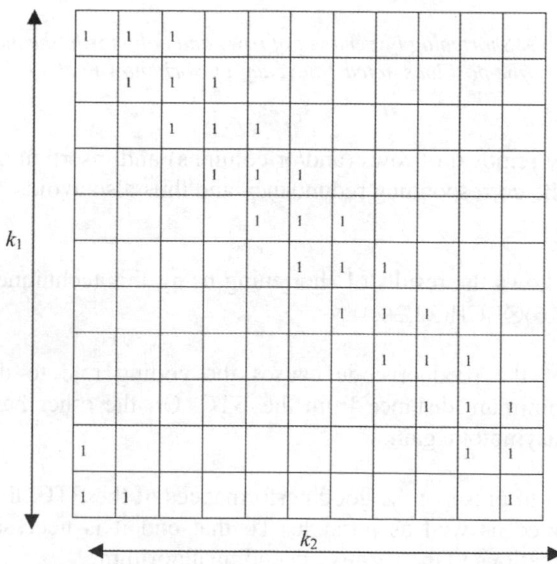

Figure 7.18. *Shortening by uniform distribution in the matrix [M] (the positions noted 1 have an a priori known value)*

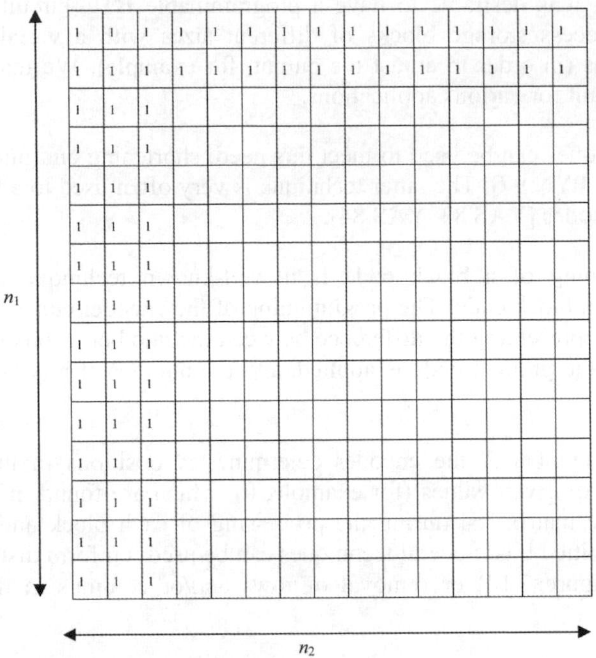

Figure 7.19. *Shortening by removal of rows and columns in the matrix [C] (the positions noted 1 have an a priori known value)*

Shortening by removal of rows (and/or columns) and insertion of "0" avoids the transmission of the corresponding redundancy and thus also avoids "decreasing" this redundancy.

Figure 7.20 shows the results of shortening using this technique for the product code BCH(32,21,6)⊗BCH(32,26,4).

Shortening of the product code causes the coding rate to decrease without modifying the minimum distance from the BTC. On the other hand, it leads to a reduction in the asymptotic gain.

Thus, in order to preserve the good performances of the BTC, the predefined bits should be exploited as well as possible. To that end it is necessary to make the following modifications to the iterative decoding algorithm:

– we assign an increased reliability to the predefined bits that have not been transmitted;

– we grant an increased value to the extrinsic information associated to predefined bits;

– in the case of shortening by row (or column), these rows are not decoded.

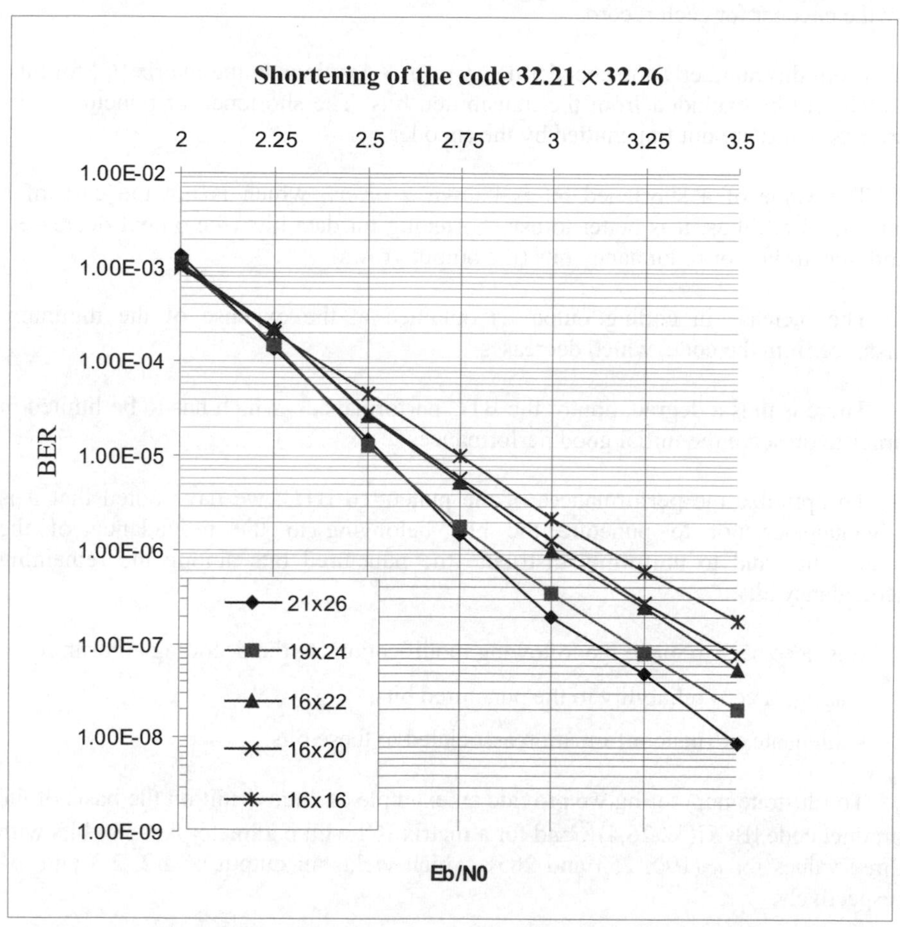

Figure 7.20. *Performances of various shortening codes obtained on the basis the product code BCH(32,21,6)⊗BCH(32,26,4) by the technique of removal of rows (and columns). The channel is Gaussian, the number of iterations is 7 and the data is quantified for 5 bits. 21*26 data. output = 0.53 --- 19*24 data. output = 0.51 --- 16*22 data. output = 0.47 --- 16*20 data. output = 0.46 --- 16*16 data. output = 0.43*

The difference in slope between these curves is explained by the reductions in asymptotic gain and the size of the code.

With respect to puncturing, programming consists of providing the encoder and the decoder with a number *Y* that represents the difference between the *n-X* number of unknown bits of the matrix [C] resulting from the application of the product code (*X=0*, if code shortening is not carried out) and the number of *n-X-Y* bits transmitted by the encoder for each record.

From this number *Y* the encoder determines *Y* positions in the matrix [C] for bits, which will be excluded from the transmitted bits. The shortened or punctured bits are bits which are not transmitted by the encoder.

The value of a shortened bit is known *a priori*, which is not the case of a punctured bit: thus, it is better to use shortening for data bits (the output decreases) and puncturing for redundancy bits (the output grows).

The increase in coding output is obtained at the expense of the minimum distance from the code, which decreases.

There is thus a degradation of the BTC performances, which has to be limited in order to preserve the initial good performances.

To optimize the performances of the punctured BTC, we have noted that it is advantageous not to puncture the bits belonging to the redundancy of the redundancy and to uniformly distribute the punctured bits among the remaining redundancy bits.

It is necessary to make the following modifications to the decoding algorithm:

– assign a zero reliability to the punctured bits;

– attenuate extrinsic information associated to these bits.

To illustrate puncturing we provide an example of codes built on the basis of the product code $[BCH(32,26,4)]^2$ and for a matrix [C] with parameter $n = 380$ bits with three values for k (190, 253 and 285), which yields an output of 1/2, 2/3 and 3/4 respectively.

The parameters of shortening and puncturing are indexed in the table below.

n	k	R	No. of rows with zero weight	No. of columns with zero weight	No. of bits with 0 on the diagonal	No. of punctured bits
380	190	1/2	12	12	6	14
380	253	2/3	10	10	3	101
380	285	3/4	9	9	4	145

Table 7.6. *Input parameters of the BTC considered*

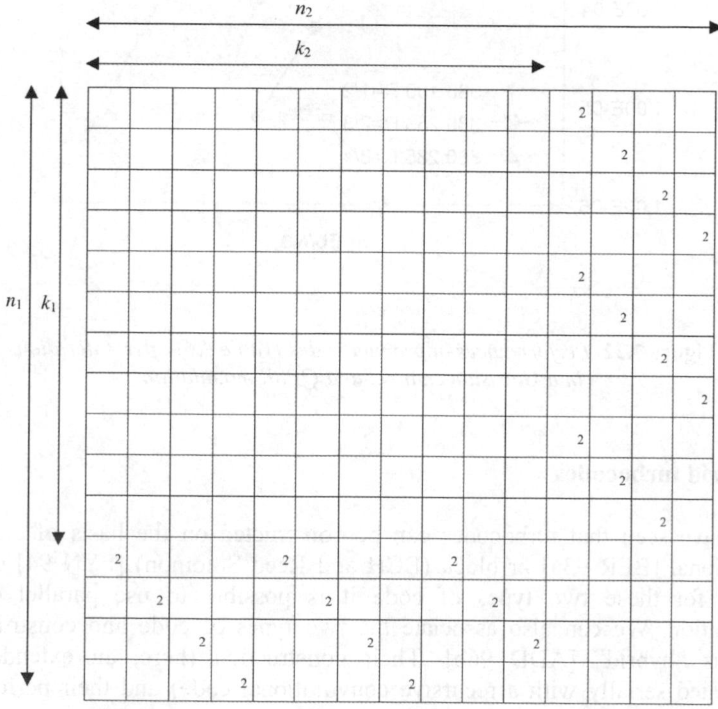

Figure 7.21. *Example of distribution of the punctured bits in the matrix* ***[C]*** *(located by "2")*

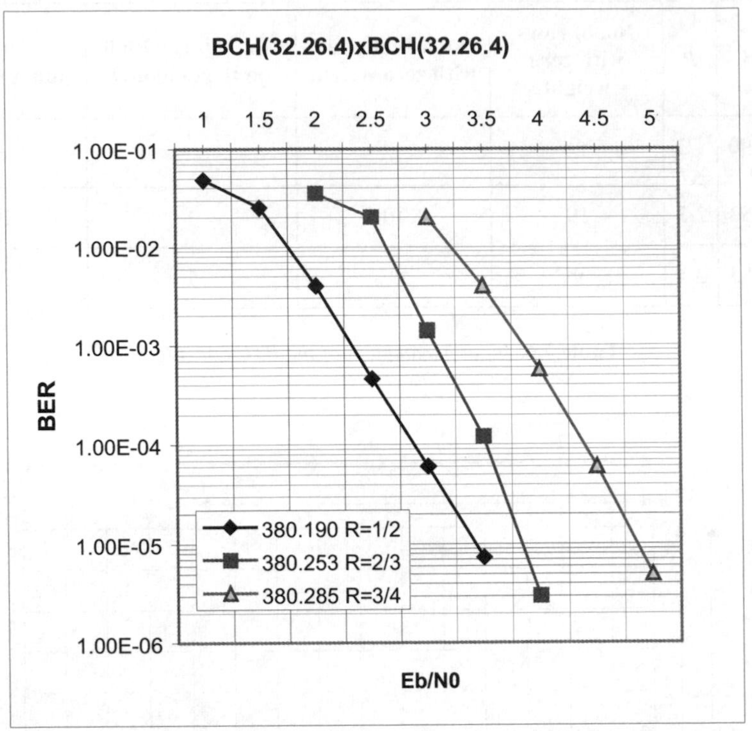

Figure 7.22. *Performances of product codes (Table 7.6) after 4 iterations in a Gaussian channel and QPSK modulation*

7.4. Hybrid turbocodes

We have seen that turbocodes can be constructed on the basis of elementary convolutional [BER 93a] or block (BCH and Reed Solomon) [PYN 94] codes. In addition, for these two types of code it is possible to use parallel or serial concatenation. We can also associate the two types of code and construct codes known as "hybrid" [ADD 96b]. Their construction (here, an extended BCH concatenated serially with a recursive convolutional code) and their performances are presented in this chapter.

7.4.1. *Construction of the code*

Let us consider the extended BCH code (n, k, δ). The hybrid turbocode is obtained placing $N_c.k$ information bits in a matrix with k rows and N_c columns. The

N_c columns are coded using an extended BCH code. The new matrix with N_c columns of n symbols is used as a uniform interleaving matrix.

Each row is coded with a recursive systematic convolutional (RSC) code with a memory depth ν as indicated in Figure 7.23. The redundancy Y of this code can be punctured: it is organized in N_y columns with N symbols

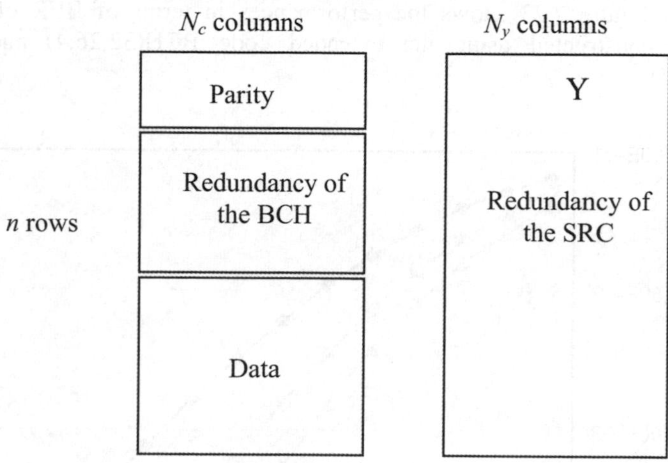

Figure 7.23. *Construction of the hybrid turbocode*

The output of coding R of the hybrid turbocode is:

$$R = \frac{K.N_c}{N.\left(N_c + N_y\right)}$$

where $\frac{K}{N} = R_{BCH}$ is the output of the BCH code

and $\frac{N_c}{N_c + N_y} = R_{RSC}$ is the output of the RSC code

7.4.2. *Binary error rates (BER) function of the signal-to-noise ratio in a Gaussian channel*

Iterative decoding requires reliability information at the output of each elementary decoder. For the RSC code, the Viterbi algorithm described in [BER 93b] may be used or, more generally, the MAP (maximum *a posteriori*) algorithm [BAH 74] or its simplified version SUB-MAP [ROB 97]. For an extended BCH code the measure of reliability is obtained on the basis of the estimate of the LLR [PYN 98]. Figure 7.23 shows the performances in terms of BER of the hybrid turbocode constructed using the extended code BCH(32,26,4) and the code RSC(23,35).

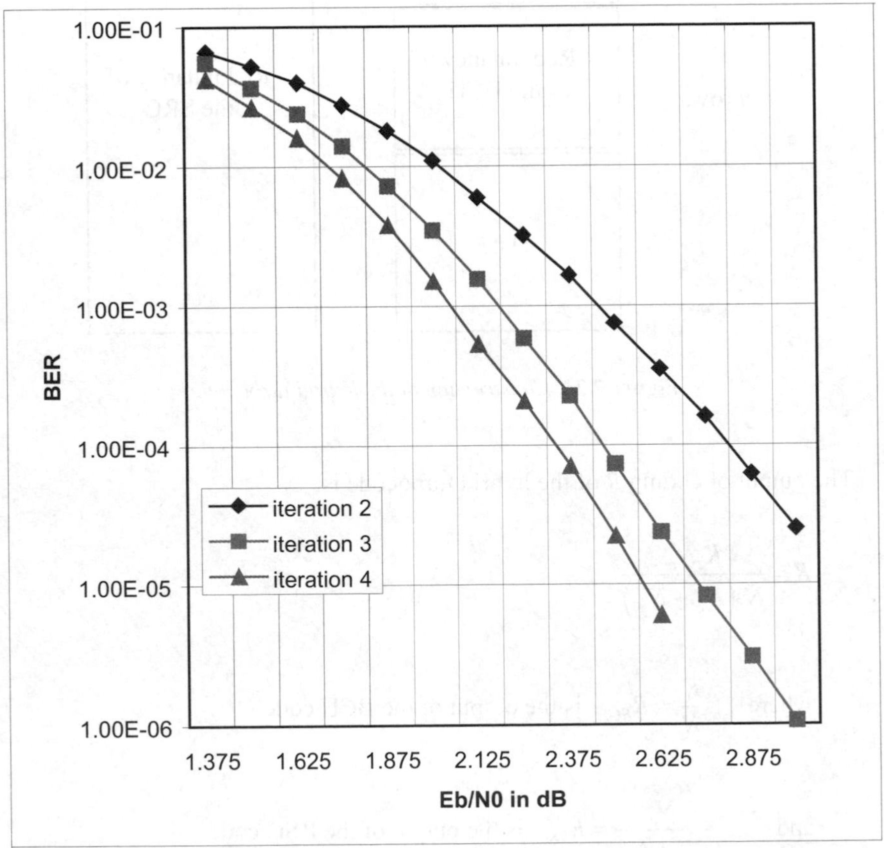

Figure 7.23. *Bit error rates (BER) for the product code BCH(32,26,4)⊗RSC(23,35) with q = 4 quantification bits, R = 1/2 and after 2, 3 and 4 iterations*

The symbols are coded with 4 bits (a sign bit and 4 reliability) and the total coding output is 0.5. The results are shown in Figure 7.24.

Taking the same total output we can associate different extended BCH codes to a RSC code $\nu = 4$ (23,35) or $\nu = 6$ (133,171). The results are shown in Figure 7.24.

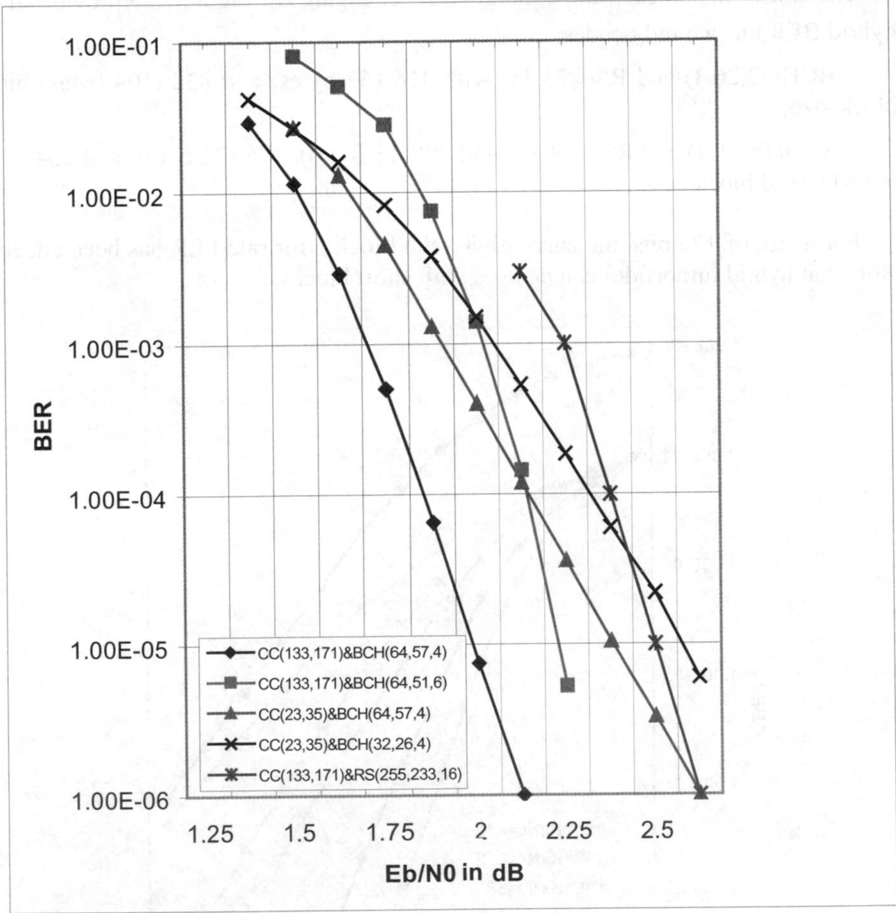

Figure 7.24. *Performances of different hybrid turbocodes ($q = 4$, $R = 1/2$ and after 4 iterations)*

The signal-to-noise ratio necessary for a BER of 10^{-5} to the 4th iteration has a 1.7dB theoretical Shannon bound [SHA 48] for the concatenation of code BCH(64,57,4) and the code RSC(133,171). We have compared this association with the serial concatenation of a Reed-Solomon code (255,223, t = 16) and a convolutional code (133,171) [POL 89] (in this case the rate is lower: 0.43). The coding gain is approximately 0.5dB for the hybrid turbocode.

7.4.3. *Variation of the size of the blocks*

The size of the blocks can be modified by changing N_c. Figure 7.25 provides the hybrid BER for two turbocodes:

– BCH(32,26,4) and RSC(23,35) with 416 (52 bytes) and 832 (104 bytes) bits block size;

– BCH(16,11,4) and RSC(23,35) with 88 (11 bytes), 176 (22 bytes) and 264 (33 bytes) bits of block size.

For a size of 176 bits, the curve giving the block error rate FER has been added. Note that hybrid turbocodes can be used with short blocks.

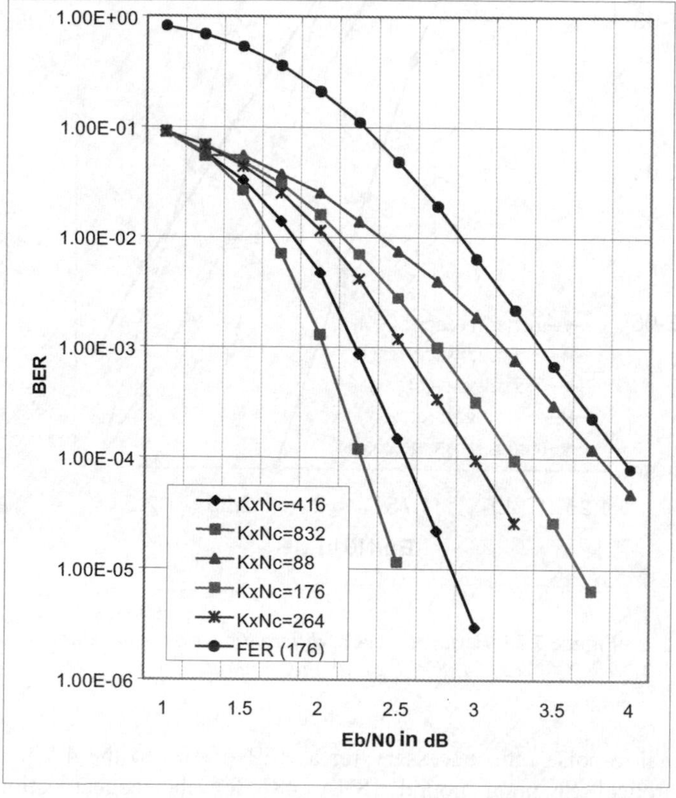

Figure 7.25. *BER function of the block size and the signal-to-noise ratio*

($q = 4$, $R = 1/2$ and after 4 iterations).

7.4.4. *Variation of the total rate*

By puncturing the redundancy Y of the RSC code it is possible to change R_{RSC} (and R). Figure 7.26 provides the BER of the concatenation of BCH(32,26,4) and RSC(23,35) according to various outputs. The results of this turbocode hybrid are compared with the serial concatenation of a convolutional code ($\nu = 6$, $R = 3/4$) and RS code (255, 239, $t = 8$), used in high definition digital television [MOR 93].

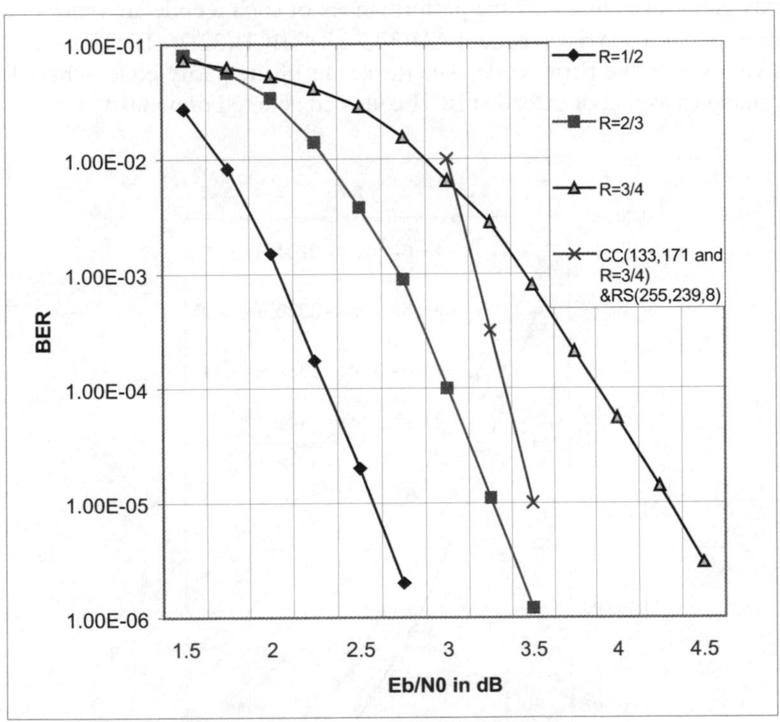

Figure 7.26. *BER function of the output and signal-to-noise ratio obtained by concatenating the code BCH(32,26,4) and the code RSC(23,35) with $q = 4$ and after 4 iterations ($R = 1/2$, $2/3$ and $3/4$)*

7.5. Multidimensional turbocodes

The product turbocodes are constructed on the basis of two elementary block codes. We have seen that in terms of performances, their minimum distance, which fixes the asymptotic gain, is equal to the product of the minimum distances from the elementary codes. This result can, of course, be extended to product codes constructed with a number of elementary codes higher than two. The minimum distance can become very large (43 = 64 or 63 = 216, for example, for a three-dimensional structure). Their principal disadvantages are low output and considerable block sizes. It is possible to remedy that using the techniques of shortening and puncturing demonstrated previously. Another solution can be envisaged in order to deteriorate the output to the smallest possible extent: it consists of using a parity code in the third dimension. This code has an output close to 1 (4/5, 6/7, 7/8, or more). The total output changes little compared to that of the two-dimensional code, but the total minimum distance is multiplied by 2, distance from the parity code. In Figure 7.27 the performances of such a code are represented: the two-dimensional code is a code BCH(32,26,4)⊗BCH(32,26,4). The size of the blocks varies with the parity code. The decoding of the parity code is based on the same principles as that of extended BCH codes considered previously.

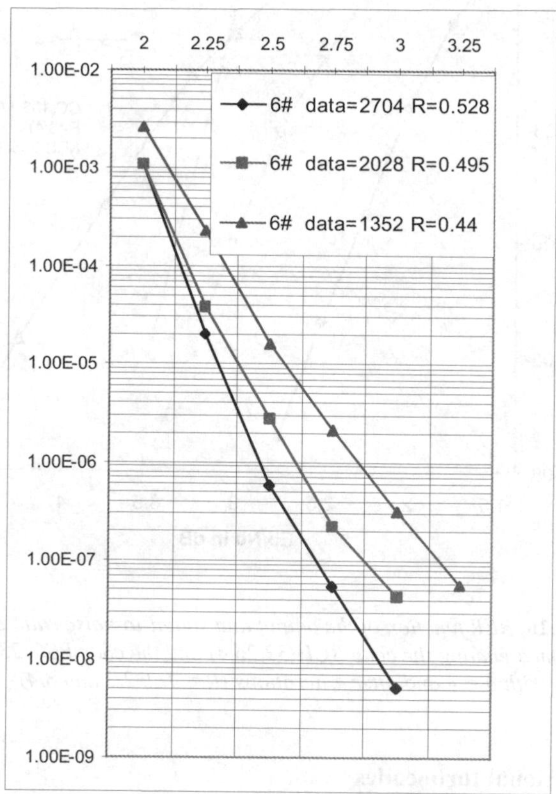

Figure 7.27. *Performance of a three-dimensional turbocode in a Gaussian channel using a parity code*

It is possible to construct two-dimensional product codes using a parity code in its second dimension. These codes will have high outputs. An example is provided in Figure 7.28 (the output is 0.87, close to 6/7 and 7/8).

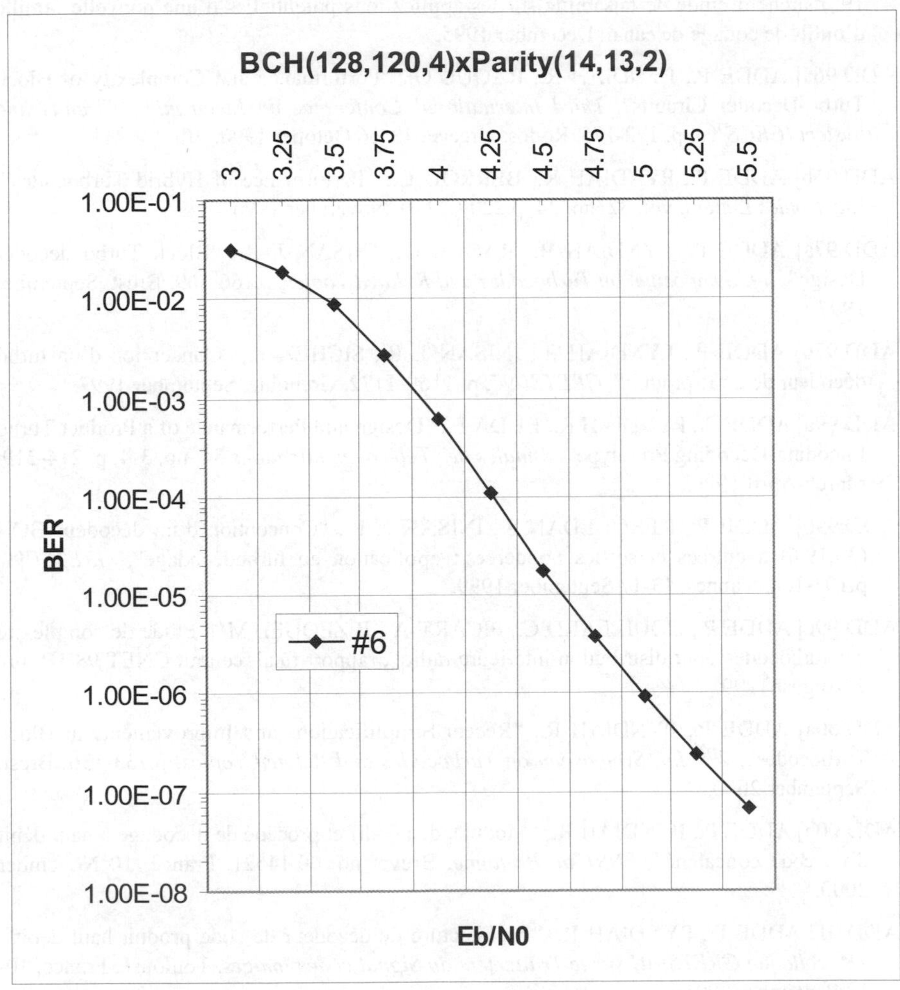

Figure 7.28. *Performance of a product code BCH(128,120,4)⊗P(14,13,2) in a Gaussian channel*

output = 0.87 and bloc size = 188 bytes.

7.6. Bibliography

[ADD 95] ADDE P., PYNDIAH R., RAOUL O., Etude de la mise en œuvre des turbocodes en blocs: Evaluation de complexité matérielle, rapport final, contrat CCETT no. 94 ME 19, marché d'étude de faisabilité sur les applications potentielles d'une nouvelle famille d'outils de codage de canal, December 1995.

[ADD 96a] ADDE P., PYNDIAH R., RAOUL O., "Performance and Complexity of Block Turbo-Decoder Circuits", *Third International Conference on Electronics, Circuits and System ICECS'96*, p. 172-175, Rodos, Greece, 13-16 October 1996.

[ADD 96b] ADDE P., PYNDIAH R., BERROU C., "Performance of Hybrid Turbocodes", *Electronics Letters*, vol. 32, no. 24, p.2209-2210, November 1996.

[ADD 97a] ADDE P., PYNDIAH R., RAOUL O., INISAN J. R., "Block Turbo decoder Design", *Int. Symposium on Turbocodes and Related Topics*, p.166-169, Brest, September 1997.

[ADD 97b] ADDE P., PYNDIAH R., INISAN J. R., SICHEZ Y., "Conception d'un turbo décodeur de code produit", *GRETSI'97*, p. 1169-1172, Grenoble, September 1997.

[ADD 99a] ADDE P., PYNDIAH R., BUDA F., "Design and Performance of a Product Turbo Encoding-Decoding Prototype", *Annales des Télécommunications* 54, no. 3-4, p. 214-219, March-April 1999.

[ADD 99b] ADDE P., KEROUEDAN S., INISAN J. R., "Conception d'un décodeur BCH (30,19,6) à entrées et sorties pondérées : application au turbodécodage", *GRETSI'99*, p.103-106, Vannes, 13-17 September 1999.

[ADD 99c] ADDE P., DOUILLARD C., PICART A., JEZEQUEL M., "Etude de complexité des turbocodes pour distribution intérieure radio", rapport final, contrat CNET 98 1B, no. 2, August 1999.

[ADD 00a] ADDE P., PYNDIAH R., "Recent Simplifications and Improvements in Block Turbocodes", *2nd Int. Symposium on Turbocodes and Related Topics*, p.133-136, Brest, September 2000.

[ADD 00b] ADDE P., PYNDIAH R., "Module, dispositif et procédé de décodage à haut débit d'un code concaténé", *ENST de Bretagne*, Brevet no. 00/14521, France, 10 November 2000.

[ADD 01] ADDE P., PYNDIAH R., "Architecture de décodeur de code produit haut débit", *18e colloque GRETSI'01 sur le Traitement du Signal et des Images*, Toulouse, France, 10-13 September 2001.

[BAH 74] BAHL L. R., COCKE J., JELINEK F., RAVIV J., "Optimal Decoding of Linear Codes for Minimizing Symbol Error Rate", *IEEE Trans. on Info. Theory*, vol. IT-20, p. 284-287, March 1974.

[BER 93a] BERROU C., GLAVIEUX A., THITIMAJSHIMA P., "Near Shannon Limit Error-Correcting Coding and Decoding: Turbocodes", *IEEE Int. Conf. on Comm. ICC'93*, vol. 2/3, p. 1064-1071, May 1993.

[BER 93b] BERROU C., ADDE P., ANGUI E., FAUDEIL S., "A Low Complexity Soft-Output Viterbi Decoder Architecture", *Proc. of IEEE ICC'93*, Geneva, p. 737-740, May 1993.

[BUR 65] BURTON H. O., WELDON E. J. JR., "Cyclic Product Code", *IEEE Trans. Inform. Theory*, vol. IT-11, p.433-439, January 1965.

[CHA 72] CHASE D., "A Class of Algorithms for Decoding Block Codes with Channel Measurement Information", *IEEE Trans. Inform. Theory*, vol. IT-18, p. 170-182, January 1972.

[CUE 02] CUEVAS ORDAZ J., ADDE P., KEROUÉDAN S., KEROUÉDAN R., "New Architecture for High Data Rate Turbo Decoding of Product Codes", *GLOBECOM 2002 IEEE Global Telecommunications Conference*, vol. 2, p. 1363-1367, Taipei, Taiwan, 17-21 November 2002.

[CUE 03a] CUEVAS ORDAZ J., ADDE P., KEROUÉDAN S., "Very Powerful Block Turbocodes for High Data Rate Applications", *3rd International Symposium on Turbocodes and Related Topics*, p. 251-254, Brest, France, 1-5 September 2003.

[CUE 03b] CUEVAS ORDAZ J., ADDE P., KEROUEDAN S., ARZEL M., LE MASSON J., "Turbodécodage de code produit haut débit utilisant un code BCH étendu", *GRETSI'03: 19e colloque sur le traitement du signal et des images*, vol. 1, p. 349-352, Paris, 8-11 September 2003.

[ELI 54] ELIAS P., "Error-Free Coding", *IRE Trans. on Inf. Theory*, vol. IT-4, p. 29-37, September 1954.

[KER 99] KEROUEDAN S., ADDE P., FERRY P., "Comparaison performances/complexité de décodeurs de codes BCH utilisés en turbodécodage", *GRETSI'99*, p. 67-70, Vannes, 13-17 September 1999.

[KER 00] KEROUÉDAN S., ADDE P., "Implementation of a Block Turbo decoder on a Single Chip", *2nd Int. Symposium on Turbocodes and Related Topics*, p.243-246, Brest, September 2000.

[MOR 93] MORELLO A., MORELLO G., VISINTIN M., "Convolutional and Trellis Coded Modulations Concatenated with Block Codes for Digital HDTV", *Proceedings of 6th Tirrenia International Workshop on Digital Communications*, Audio and video digital radio broadcasting systems and techniques, p. 237-250, Tirrenia, 5-9 September 1993.

[POL 89] POLLARA F., CHEUNG K.-M., "Performance of Concatenated Codes using 8-bit and 10-bit Reed-Solomon Codes", *TDA progress report 42-97*, p. 194-201, January-March 1989.

[PUC 88] PUCKNELL D. A., ESHRAGHIAN K., *Basic VLSI Design System and Circuits*, Prentice Hall, New York, 1988.

[PYN 94] PYNDIAH R., GLAVIEUX A., PICART A., JACQ S., "Near Optimum Decoding of Product Codes", *IEEE GLOBECOM'94 Conference*, vol. 1/3, p. 339-343, San Francisco, November-December 1994.

[PYN 96a] PYNDIAH R., ADDE P., Turbocode raccourci, brevet France no. 96 10501, France Télécom, August 1996.

[PYN 96b] PYNDIAH R., ADDE P., Turbocode poinçonné, brevet France no. 96 10521, France Télécom, August 1996.

[PYN 97] PYNDIAH R., "Iterative Decoding of Product Codes: Block Turbocode", *Int. Symposium on Turbocodes and Related Topics*, p. 71-79, Brest, September 1997.

[PYN 98] PYNDIAH R., "Near Optimum Decoding of Product Codes: Block Turbocodes" *IEEE Trans. on Comm.*, vol. 46, no. 8, p. 1003-1010, August 1998.

[RAO 95] RAOUL O., ADDE P., PYNDIAH R., "Architecture et conception d'un circuit turbodécodeur de codes produits", *GRETSI'95*, Juan Les Pins, p. 981-984, September 1995.

[RAO 97] RAOUL O., "Conception et performances d'un circuit intégré turbodécodeur de codes produits", PhD Thesis, University of West Brittany, Brest, 21 November 1997.

[ROB 97] ROBERTSON P., HOEHER P., VILLEBRUN E., "Optimal and Suboptimal Maximun *A Posteriori* Algorithms Suitable for Turbodecoding", *European Trans. Télécommunications*, vol. 8, p. 119-125, March-April 1997.

[SHA 48] SHANNON C. E., "A Mathematical Theory of Communication", *Bell System Technical Journal*, vol. 27, p. 379-423, July 1948 and p. 623-656, October 1948.

[WES 93] WESTE N. H. E., ESHRAGHIAN K., *Principles of CMOS VLSI Design: A Systems Perspective*, second edition, Addison-Wesley, Reading, MA,1993.

[YAS 83] YASUDA Y., KASHIKI K., HIRATA Y., "Development of Variable-Rate High-Rate Viterbi Decoder and its Performances Characteristics", 6th *Int. Conf. Digital Satellite Commun.*, Phoenix, AZ, p. XII-24 – XII-31, September1983.

[YAS 84] YASUDA Y., KASHIKI K., HIRATA Y., "High-Rate Punctured Convolutional Codes for Soft Decision Viterbi Decoding", *IEEE Trans. Commun.*, vol. COM-32, no. 3, p. 315-319, March 1984.

List of Authors

Patrick Adde
ENST Bretagne
Technopôle de Brest-Iroise
Brest, France

Gérard Battail
Formerly ENST
Paris, France

Claude Berrou
ENST Bretagne
Technopôle de Brest-Iroise
Brest, France

Ezio Biglieri
Department of Electronics
Politecnico di Torino
Turino, Italy

Catherine Douillard
ENST Bretagne
Technopôle de Brest-Iroise
Brest, France

Alain Glavieux
Formerly ENST Bretagne
Technopôle de Brest-Iroise
Brest, France

Michel Jézéquel
ENST Bretagne
Technopôle de Brest-Iroise
Brest, France

Annie Picart
ENST Bretagne
Technopôle de Brest-Iroise
Brest, France

Alain Poli
AAECC/IRIT
Paul Sabatier University
Toulouse, France

Ramesh Pyndiah
ENST Bretagne
Technopôle de Brest-Iroise
Brest, France

Sandrine Vaton
ENST Bretagne
Technopôle de Brest-Iroise
Brest, France

Index

K

L

M

P

R

S

T

V